Isle of Fire

University of Chicago Geography Research Paper, number 246

Titles published in the Geography Research Papers series prior to 1992 and still in print are now distributed by the University of Chicago Press. For a list of available titles, please see the end of the book. The University of Chicago Press commenced publication of the Geography Research Papers series in 1992 with number 233.

ISLE OF FIRE

The Political Ecology of Landscape Burning in Madagascar

Christian A. Kull

THE UNIVERSITY OF CHICAGO PRESS
CHICAGO AND LONDON

CHRISTIAN A. KULL is a senior lecturer in the School of Geography and Environmental Science at Monash University in Australia.

The University of Chicago Press, Chicago 60637
The University of Chicago Press, Ltd., London

Printed in the United States of America

13 12 11 10 09 08 07 06 05 04   1 2 3 4 5

ISBN: 0-226-46140-8 (cloth)
ISBN: 0-226-46141-6 (paper)

Library of Congress Cataloging-in-Publication Data

Kull, Christian A.
Isle of fire : the political ecology of landscape burning in Madagascar / Christian A. Kull.
p. cm.
Includes bibliographical references (p.).
ISBN 0-226-46140-8 (cloth : alk. paper) — ISBN 0-226-46141-6 (pbk. : alk. paper)
1. Burning of land—Madagascar. 2. Fire management—Madagascar I. Title
S608.85 2004
333.76′152′09691—dc22

2003022755

♾ The paper used in this publication meets the minimum requirements of the American National Standard for Information Sciences—Permanence of Paper for Printed Library Materials, ANSI Z39.48-1992.

Contents

Acknowledgments

THIS RESEARCH was made possible by the generous funding of an Environmental Protection Agency STAR Graduate Fellowship, a National Science Foundation Doctoral Dissertation Improvement Grant (SBR 9811046), a Rocca Scholarship for Advanced African Studies, a University of California at Berkeley University Fellowship, and by McGill University. Monica Lepe, Simon Roger, and Gary Swinton expertly drafted the maps. Portions of this book have previously appeared elsewhere. Earlier versions of the chapter on the *tapia* woodlands appeared in the *Journal of Cultural Geography*, the chapter on resistance and criminalization in *Political Geography*, and the chapter on community-based fire management in *Development and Change*. Bits and pieces of a paper originally published in *Environment and History* appear in chapters 2 and 5; and some of the book's general themes appeared in *Environment* magazine. I thank all of the above.

I AM INDEBTED to a wide array of individuals for this study of fire in Madagascar. They sparked the ideas, enflamed the discussions, and kept the embers burning throughout the process.

To begin with, I cannot adequately express my gratitude to the people of Afotsara and the other field sites for their patience, humor, and camaraderie. Their willingness to share their time and love, to invite me into their humble homes—whether a cramped, smoky mud-floored *tranobonga* up in the hills or a drafty, wood-floored house in the village—and to accept my intrusions with good humor have forever endeared them to me. I look forward to meeting again, soon.

My intellectual debts are first and foremost to Nancy Peluso. I went to her seeking to challenge my thin understanding of the political and social side of "political ecology," and found a person of great insight, strong support, and wonderful laughs. I also thank Lynn Huntsinger, Louise Fortmann, and Michael Watts for their enthusiasm, advice, and inspiration. The original research proposal benefited from numerous rounds of criticism from Jennifer Sowerwine, Kurt Spreyer, Janet Sturgeon, Lea Borkenhagen, Suraya Affif, Jake Kosek, Jen Sokolove, Anand Pandian, and Lis Grinspoon; additional intellectual debts have piled up over the years to Gary Gaile, Jim Wescoat, Terry McCabe, Rod Neumann, Ray Bryant, four University of Chicago reviewers, and Jim Scott and Yale's *agraristas.*

This research would not have been possible without the support of Joelisoa Ratsirarson, now *directeur adjoint* of the School of Agricultural Science (ESSA) at the University of Antananarivo. At ESSA, I also thank Gidehona Randriamboavonjy, Gérard Rambeloarisoa, Daniel Razakanirina, Gabrielle Rajoelson, "Jo" Ratsirarson, the late Mamy Razafindrabe, and "RaFred" Rakotoarivelo. Also at the University of Antananarivo, I also thank in particular geographers Joselyne Ramamonjisoa, Céline Ratovoson, Gabriel Rabearimanana, and Arsène Rabarison.

I have benefited from contact with a host of individuals working on or in Madagascar. First of all, I am indebted to Daniel Gade for his unfailing support, excellent advice, and challenging criticisms. Alison Richard and Bob Dewar have selflessly contributed time from very busy lives. Erik Smith was a constant source of inspiration, ideas, and fellowship. Richard Marcus fearlessly slogged through an early version of chapter 8 and generously kept me informed on many details of Malagasy law and politics. Jean-Pierre Raison has provided advice, support, and useful data since we first corresponded about Leimavo; he even visited Afotsara in 1998. I should also thank, for various kinds of academic help, inspiration, or camaraderie at different stages along the way: David Burney, Alain Bertrand, Bill McConnell, Nabiha Megateli, Ned Horning, Susan Kus and Victor "Kim" Raharijaona, Paul Laris, Pier Larson, Henry Wright, Gunilla Olson, Chantal Blanc-Pamard, Christophe Maldidier, Simon Peers, Jim Allman, Lalaina Rakotoson, Mark and Karen Freudenberger, Jürg Brand, Jean-Laurent Pfund, Joanna Durbin and Frank Hawkins, Jon Randby, Dan Turk, Joe Peters, Gerald Berg, Bob Sussman, Chris Barrett, Rheyna Laney, Ueli Müller, Steve Goodman, Mimi Gaudreau, Bruno Locatelli, Diane Henkels, Oliver Pierson, and Hervé Rakoto Ramiarantsoa.

Numerous people aided travel, language, and research logistics—or just gave of their hospitality and friendship. I thank, in particular, Liliane Koziol, Lalao Ramarosanjy, Edouard Andriamendrikaja, Zanoa Rasanjison and her family, Jimmy and Eliane Ramiandrison, Shannon McClellan, David

McAfee, Sharon Duckett, many Peace Corps volunteers, the CARE and CRS folks, Franz and Rita Stadelmann, Edline Rakotovao, Renée Raza, Fanja Harinirina, Harizo Nambinina Ralambominosoa, Jean-Louis Razafimahefa, Allan Mackay, Hasina Razafimampanana, and the staff and children of Sekoly Norveziana in Antsirabe. All translations of materials and quotes are my own, with the help of Frédérique Lehoux (French), Erik Gabrielsen (Norwegian), and Liliane Koziol (Malagasy).

Ever since I hand delivered the mail from Gland to Antananarivo in 1992, WWF has held its door wide open. I thank Sheila O'Connor in Tana and Dominique Halleux, Alexandris Lesira, Jean Claude Tsaramila, and intern Kristin Sakaël in Andapa. I am particularly grateful to Peter Schachenmann, former technical counselor to WWF's Andringitra project, who invited me along on several expeditions into the distant corners of the reserve, and whose energy and camaraderie inspired and facilitated my work there, along with Hanta Rabetaliana, Joseph Ralaiarivony, Michel Randriambololona, Bernardin Rasolonandrasana, Ramosa, Rajaona, Kazimena, Martin, and many others.

I thank the people and institutions who helped me access documents and information: the National Archives (Mme. Razoharinoro), FOFIFA (Bruno Ramarorazana and Mme. Mimose), Multi-Donor Secretariat (Andy Keck and Bienvenu Rajaonson), INSTAT, FTM, PACT (Jean-Michel Dufils), ONE (Rivo Ratsimbarison), CIDST, the Musée d'Art et Archeologie (Jean-Aimé Rakotoarisoa), the Académie Malgache, the Ministère de la Justice, the Archives d'Outre Mer in Aix-en-Provence, the CIRAD library in Montpelier, the archives of the Norwegian Missionary Society in Stavanger (Nils Kristian Höimyr), and last but not least the foresters of the Ministère des Eaux et Forêts.

McGill University's Department of Geography was the ideal venue for the writing of this work on Madagascar, following in the footsteps of Sherry Olson. I thank all my colleagues for their camaraderie and support, and in particular Oliver Coomes and Mark Brown for their intellectual stimulation. Heidi LaJoie and Eby Heller, my able research assistants, contributed more than they realize.

Christie Henry and Michael Koplow at the University of Chicago Press have ably seen this project through to completion, and I cannot thank them enough.

IN MADAGASCAR I met Frédérique Lehoux, with whom I was able to share my enthusiasm about the Malagasy landscape and its fires (and much more).

May our own fires keep on burning. I also thank my family for their love and for giving me the drive to learn and the freedom to explore.

Finally, as a geographer, I cannot resist recognizing the contributions of a number of places to the last decade of work and learning. They provided the quiet reflection, the natural inspiration, or the physical and spiritual home that kept this project going: Moosilauke, Adelboden, Nordmarka, the Flatirons, East Rock, the Berkeley Hills, the open ridges of Afotsara, the sunset balcony at Antsahatsiroa, the bougainvillea- and mango-blessed villa on Morne Brun in Pétionville, Parc Lafontaine, and Mont Royal.

Misaotra indrindra ianareo rehetra.

Glossary

ANM: Archives Nationales de Madagascar (Arisivim-Pirenena Malagasy), Antananarivo.

AOM: Archives d'Outre Mer, the French overseas archives, Aix-en-Provence, France.

arrêté: In French and Malagasy law, an executive regulation enacted by the president or prime minister is called a *décret,* while an executive regulation from lower-level officials is an ***arrêté.*** Acts of parliament (the National Assembly and the Senate) are *lois* (laws). *Ordonnances* are used in Madagascar for laws enacted by the president under conditions of urgency or with special power delegated by the National Assembly.

circulaire: A communication from the French colonial governor to colonial officers explaining policies and regulations.

commune rurale: Currently the primary state-sanctioned rural political unit. Run by an elected mayor and councilors, with appointed assistants and secretaries. Receives a budget from the central state, and administers voting, schools, projects, markets, and health centers in conjunction with the central state. Consists of multiple ***fokontany*** and forms part of a larger ***fivondronana.*** Known during the Second Republic (1975–1993) as *firaisana;* sometimes rendered in Malagasy as *kaominina.*

dahalo: General term for bandits. Originally restricted to intertribal cattle rustling, especially in the West, now associated with all forms of banditry. Also called *malaso* or *fahavalo.*

décisions locales: Regulations promulgated by provincial or district-level authorities during the colonial period.

décret: See ***arrêté.***

DEF: The colonial or postcolonial Forest Service in Madagascar, called, depending upon the administrative structure of the day, the Service des Eaux et Forêts, the Direction des Eaux et Forêts, or the Ministère des Eaux et Forêts. Headquartered in the Nanisana neighborhood of Antananarivo, with regional offices. In this book, Forest Service refers to the organization itself, while **DEF** refers to documentary sources from the organization (unless otherwise noted, sources consulted at Nanisana; when a year is given, e.g., "DEF 1996," the citation refers to a publication listed in the bibliography).

dina: Decisions and rules at the level of local governance (***fokonolona, fokontany, commune rurale***). Traditionally, *dina* are local socially guaranteed agreements; since 1960, *dina* have been increasingly given formal legal status.

Direction des Eaux et Forêts: See **DEF.**

doro tanety: Hill fires, bushfires, or, in French, *feux de brousse.* Also *hain-tany,* or land fire.

famadihana: Reburial ceremony (see introduction to part 1, note 2).

feux de brousse: See ***doro tanety.***

fivondronana: The administrative level between the ***commune rurale*** and the six provinces (*faritany*). Sometimes called *sous-préfectures;* I often refer to them simply as "districts."

fokonolona: The community group. A concept with multiple overlapping meanings, including a village assembly, the collected citizens of a hamlet, ***fokontany,*** or ***commune rurale,*** or a sort of collective will of the people.

fokontany: Unit of governance below ***commune rurale;*** the lowest official political unit. Serves, on average, 400–1,200 citizens in one or several villages or hamlets. Led by a **PCLS.**

Forest Service: See **DEF.**

GELOSE: *Gestion locale sécurisée.* Denotes a recent policy effort to devolve control of some renewable natural resources to local communities; more specifically, refers to the law that makes this possible (Loi 96-025).

gendarmes: Malagasy police force responsible for rural zones.

KASTI: Komiten'ny Ala sy ny Tontolo Iainana (Forest and Environment Committee). Village environmental committees set up by the **Forest Service.**

JOM: The weekly governmental record; titles vary over the years from *Journal Officiel de Madagascar et Dépendences,* to *Journal Officiel de la République de Madagascar,* to *Gazetim-Panjakan'ny Repoblika Demokratika Malagasy.* References to JOM are limited to the issue date.

lambamena: A burial shroud woven from silk.

landibe: Wild, endemic silkworms (*Borocera madagascariensis*); domesticated fine silkworms (*Bombyx mori*) are *landikely.*

lavaka: Gullies formed by erosion.

loi: See ***arrêté.***

m.a.s.l.: Meters above sea level.

mimosa: An exotic Australian tree that grows spontaneously in many unburned pastures in zones over about 1,200 **m.a.s.l.,** and is frequently harvested for wood fuel as young saplings. The name applies to both *Acacia mearnsii* (black or tan wattle) and *A. dealbata* (silver wattle).

Ministère des Eaux et Forêts: See **DEF.**

NMS: Archives of the Norwegian Missionary Society (det Norske Misjonsselskap), Stavanger, Norway.

ordonnance: See ***arrêté.***

PCLS: The president of a ***fokontany.*** The PCLS, appointed by the state, signs certificates of birth, death, sale, etc. From "Président du Comité Local de Sécurité."

procès-verbal: A report or citation issued by a policeman, ***gendarme,*** or **Forest Service** officer for an infraction of forest or fire laws.

Quartier Mobile: A community member appointed by the ***commune rurale*** to part-time police duty.

rambiazina: A wild shrub (*Helichrysum bracteiferum*) burned to provide fertilizer for sweet potatoes.

ray-aman-dreny: The elders of a village or hamlet. Literally means "fathers and mothers"; in practice, refers to influential and respected male senior community members, who must approve important local decisions and who are often called upon to arbitrate conflicts.

savoka: Secondary vegetation in areas of ***tavy*** cultivation. Can include secondary forest, woody brush, stands of bamboo, or grassy-bushy fallow.

Service des Eaux et Forêts: See **DEF.**

tanety: Hillsides, all lands not cultivable with irrigated rice.

tapia: The tree *Uapaca bojeri,* principal food of the ***landibe.*** Forms nearly monospecific stands in the highlands, known as *tapia* woodlands.

tavy: Forest-based slash-and-burn cultivation, also known as swidden or shifting cultivation.

WWF: The World Wide Fund for Nature, known in the U.S. as the World Wildlife Fund, a large international conservation organization.

Part One

The Fire Problem

MADAGASCAR IS AFLAME. Every year, farmers and cattle herders burn from around a quarter to a half of the island's grasslands and pastures, and some 2,000 to 7,000 km^2 of rain forest and secondary bush. The effect is both eerie and dramatic. Standing on a hill among the vast grassy rangelands of the Middle West (the lower western part of the highlands), a Malagasy herder watching over his cattle will almost always be able to point out a fire in the distance, at least in the dry months from June through November. During the afternoon, several plumes of smoke will billow up from hillsides near and far; in the evening, the orange flicker of flames will silently light the night sky all around the horizon. Further east, on the central plateau, urban dwellers know to expect smoke and haze in the months before the rainy season, as slash-and-burn fires in the upwind eastern rain forest prepare fields for cultivation.

Burning is ubiquitous across much of Madagascar. From my window in a small village in a deep, intensively cultivated valley of the central highlands, I watched each evening for an entire fire season as red yellow garlands of flame decorated hillsides like necklaces, or as the nighttime clouds quietly reflected the orange glow of conflagrations on the other sides of the nearby mountains. Twice I watched ashes from unseen distant fires slowly float to the ground, gently settling out of the sky. The fires of Madagascar are often lonely actors, working by themselves with an independent spirit; away from the slash-and-burn fires in the eastern forests there is hardly ever a person in sight. In some cases, arcs of flame slowly work their way across a slope, quietly crackling as the fire consumes the dry grass and leaves behind a black surface of dusty ashes. In others, the flames rush and leap, popping and exploding under a good wind or in thick, dry brush.

MADAGASCAR IS NOT just aflame; it has a fire problem, a century-long conflict over natural resource use and protection. On the one side, a broad group of actors and institutions believe that the island sees too much fire, that fires should be stopped. They blame the fires for a variety of ills, including deforestation, desertification, rangeland impoverishment, soil degradation, and accelerated erosion. Being powerful—the past colonial rulers, the Forest Service, the post-Independence leaders, the international environmental agencies—they have used their power, including laws and propaganda, to enforce their views. On the other side, the Malagasy farmers and herders rely on fire as a tool of resource management, as a tool to meet their own livelihood needs, and they are trapped by the antifire laws. Being resourceful people who know and understand the intricacies of fire and the weaknesses of the state, they keep on burning, taking care to be out of sight, and letting fire do its work for them. Fire itself is oblivious to this conflict and to the different ideas of the peasants and the state; it will burn anything, given the right ingredients of ignition, fuel, temperature, and wind. Sometimes it burns more than what the people (or some of the people) intended, such as the grass roof of a house or a plantation of pines.

Taken together, this has created a problem, a fire problem. Today, as for the last century, landscape burning in Madagascar is a conflictual, tense subject, fraught with misunderstandings and worries, but ever present in both the night sky and the national imagination. The touchy and political character of the fire problem is perhaps best illustrated by some anecdotes from several years of research on the island. Enchanted by the island as a tourist in 1992, I have returned at least every second year for research, from fieldwork for my master's thesis on land-use change in 1994 through my most recent trip in 2003.

IN JULY 1994 I climbed into a WWF[1] land rover in the misty northeastern town of Andapa, where raindrops seemed to continuously drip from corrugated roofs and lush, green leaves. We drove west on the puddled road into the rain forest of Anjanaharibe-Sud Special Reserve, rich with the island's exotic wildlife, including *babakoto,* the indri, the largest of the lemurs. At various overlooks on the incredibly steep, windy dirt road, we gazed at the rugged expanses of forest, as well as at numerous slash-and-burn cultivation clearings, called *tavy.* Black stumps of charred trees rose in some of the fallow

1. WWF facilitated a number of field visits (to Andapa and especially Andringitra) by allowing me to accompany their teams at work.

plots. Due to a couple of decades of waning enforcement, and to the recent construction of this road connecting Andapa to the west, an influx of farmers has begun cutting and burning forest plots within the nature reserve.

We parked at a gravel pit and headed straight into the forest, following a faint path that one of my companions—all of whom were Forest Service agents and interns funded under a WWF program—seemed to know. We marched around a hill, down a ridge, across a river, through dense secondary forest, all the time brushing leeches off our skin. After wading for hundreds of yards in a knee-deep stream, we popped out into an open *tavy* clearing, well hidden from the authorities, deep in the forest. A family had created its farm here, a small bamboo hut surrounded by coffee bushes, vanilla vines, sugar cane stalks, banana trees, *tavy* clearings, and fallow bush.

For a long time, the foresters talked with a young woman, Lele, who sat on a burned log nervously petting a peeping chick. She looked strong and beautiful, in a simple earth-stained skirt and shirt, with braided hair and her *coupe-coupe* machete resting nearby. The foresters reminded Lele that at an April meeting in Befingotra, the nearest village, local authorities had been told to spread the message that illegal cultivation and burning in the reserve would no longer be tolerated. She and her family (her husband was probably hiding in the forest) would have to leave after the next harvest or face arrest. The foresters reasserted that her slash-and-burn fires were destroying precious native forest. After the complicated hike back through the stream and forest to the road, we visited two other hidden forest settlements. By the next year, one of the foresters later reported, Lele had moved back into the Andapa basin.

DURING MY NEXT stay on the island, in 1996, dedicated to five months of intensive study of the Malagasy language, I decided to focus my doctoral research on fire. So, from March through December 1998, I returned and moved into a small village in the mountainous grassland-dominated central highlands. My purpose was to observe firsthand a full fire season in its social context, including all of the associated management practices. The village, Afotsara, lies in the deep valley of the Manandona and Sahatsiho rivers about 200 km south of the capital, Antananarivo. (Afotsara is a pseudonym, as are the names of other villages and people encountered during my 1998–1999 research. As the state considers most fires illegal, the use of pseudonyms is intended to protect these farmers and their communities).

In mid-April 1998, towards the end of the rainy season, I joined four boys from Afotsara for a day to weed a taro field in a hollow on a distant, high mountaintop pasture. Three of the boys belonged to a local rich family; the eldest were in the village on Easter holidays from school in the city. The fourth was being raised by the family in exchange for working as cattle herder. We climbed for nearly two hours along a narrow footpath through the yellowing green pastures, accompanied by their father's two steers. After finishing the weeding, we returned along the same path. We rested on a steep, rocky descent with grand views of the valley, sitting on a boulder, swatting at the *mokafohy* (black flies) that had been bothering us all day, biting and itching. One of the boys pulled out a lighter and lit the grass between a big rock and the trail, and we all jumped into the resulting smoke to escape the black flies. As the grass was still green, it burned poorly, and soon all three boys were busy lighting little fires, laughing and chatting. One of the schoolboys chastised his brother—"hey, don't burn too much, the mayor might see us"—pointing at the village in the distance. He continued to me, "You, a foreigner, shouldn't be seen near fire," and pointed at the fire: "look, it's burning, we are destroying the environment!" We did not stay long, for a thunderstorm was brewing. All the same, one boy took a smoldering branch with him as insect repellent until the downpour drenched it (and all of us).

A FEW DAYS after the Malagasy national holiday (26 June), I climbed to the increasingly dry pastures high above Afotsara. A large blaze smoked all day long from behind the hills to the southwest, and in the breezy afternoon another fire burned on the high ridge to the west (where I had been in April with the four boys). On the hilltop, the ground was black from a twenty-hectare fire that had burned during the holiday, climbing from the edge of an April fire scar up to the top of the ridge, bounded on one side by a stream and on the other presumably by the wind. At a lonely new house built alongside some new rice terraces in the highest reaches of a small watershed, I found a young woman carrying her little baby. I asked about the purpose of the fire that caused the scar; she said it was for pasture renewal, in order to provoke the green grass resprouts that feed the cattle at the end of the dry season. I continued on my way, noting smoke from additional fires on the high plateau to the east and on the mountainous ridge of Ivato to the northwest.

A month later I climbed to the area again. At a hamlet below the holiday fire scar I was told that that fire had not been for pasture renewal; this family claimed not know what the fire had been for, as it had come from the other side of the mountain. A boy along the trail, however, reconfirmed the pasture renewal explanation. Returning to the lone house, I found the young woman's husband. He claimed the fire had perhaps been an escaped blaze lit to control locusts, which had passed by the area recently. He claimed it was too early in the year for pasture renewal fires, for there was no rainfall and the grass would not resprout yet. I returned down the hill confused.

ON 21 JULY, while bathing in a natural pool across the road from Afotsara, I noticed thick smoke rising to the south. Drying off, my companion Radesy and I climbed out of the stream gully onto the trail. High flames, starkly orange, marched downhill through the scrubby *tapia* woodlands behind the village. We walked back to the village, and Radesy stated that if the government caught the person responsible for that fire, they would be in trouble since it was in a forest. At the first village building, Sahondra's small adobe shack, from where she sells coffee and snacks, we found a small crowd. They were looking at Nambinina's photographs from a recent *famadihana* (reburial ceremony).[2] Inside, a gendarme, in fatigues with his automatic rifle slung over his shoulder, was eating lunch; another gendarme sat outside. By this point, the fire had caught the attention of several people, and a discussion ensued about what to do and who should take charge. The gendarmes suggested calling the local part-time police officer, the *Quartier Mobile*, who was not present; the villagers suggested that normally, someone would blow on a whistle to summon people to fight the fire.

The gendarmes sprang to their feet, one delivered two short blasts on his whistle, and they began walking towards the fire. Ratsiky, a schoolteacher, came running by, and Radesy blew on his own whistle, usually used for soccer games. All of the able-bodied men present that day—those not off at the weekly market or in distant fields—climbed the hill behind the village church, with children tagging along to the edge of the village. Ahead, the huge fire advanced downhill in a 100 m–wide front with flames consistently at least 3 m high, tickling the sky at 10 m occasionally, orange and infernally hot. Along an old roadbed that formed a natural barrier, the residents of a nearby house were frantically cutting grass to make a firebreak. We arrived there when the fire was still some ten meters uphill, yet advancing steadily. A woman from the nearby house repeatedly shouted that the fire came from "up there" to whomever would listen.[3]

2. In July and August, highland families stage large multiday festivals, including music and drink, to honor the ancestors assembled in a tomb. They exhume the corpses and wrap them in new silk cloths. A single tomb may be exhumed every decade or so; these ceremonies serve ritual, social, and prestige purposes (Dahl 1998).

3. In retrospect, this unprovoked announcement was too much information for a standard Malagasy interaction; I believe she was trying to protect herself or her family from accusations of arson.

We broke off large leafy tree branches—pine, eucalyptus, *tapia,* whatever was available—and attempted to contain the fire by swatting the flames. The villagers had done this before: they automatically chose good-sized branches, and easily succeeded in snuffing out the flames as opposed to fanning them. As a neophyte fire fighter, I chose branches that were too small; a friend repeatedly tried to instruct me in the subtle difference between snuffing and fanning the flames. Meanwhile, the heat of the fire was sometimes unbearable, and we had to retreat momentarily. The crackling, the bright orange flame front, and the heat were overwhelming. Before the fire approached the old roadbed, Radesy used his cigarette lighter to ignite a backfire, to burn the tall grass uphill from the roadbed towards the wildfire.[4] Ratsiky did the same, lighting a wad of dry grass in the wildfire and carrying it over to an unburned section. Another man, Ramavolahy, told me not to worry about the spot where the roadbed was patchy vegetation or where there was cassava growing—"the fire will not be able to cross there, and over there the water content of the grass is too high to burn," he said. We succeeded in stopping the fire along the roadbed and moved east. Just above several large tombs, a group of twenty had already contained the fire; finally, we entered the woodland and put out smoldering sections. Soon, the fire was controlled. The total area that burned was approximately four hectares.

After a short discussion, the gendarmes disappeared down the road. People asked each other worriedly if the gendarmes would write a citation, a *procès-verbal.* Ramavolahy invited me to his house, on the side of the charred hill, for a plate of boiled cassava. His family had worried about the fire, he and his mother told me, but it did not crest the hill. According to Ramavolahy, the fire escaped from a poorly executed crop field–clearing fire performed by some children. I walked home, noting smoke from fires in at least four different spots around the mountainous horizon. A few days later, Ratsiky told me that the gendarmes had left, satisfied by the explanation that the fire had escaped from a fire lit by a child playing at cooking outdoors. Other people claimed ignorance of the fire's cause. Three months later, villagers spent many days working in the burned area, employed by a contractor from a nearby city, smashing the granitic rocks littering the forest floor into gravel for a road project. Most likely, as several people hinted to me, the fire had been lit to clean up the tall brush in this area for the gravel project,

4. A backfire is a well-known fire-fighting technique where one lights a controlled fire from a defensible firebreak (in this case, the roadbed) to burn towards the approaching fire, in order to head off the wildfire and not allow it to jump the firebreak.

and other explanations involving children were fabricated to satisfy the gendarmes.

A MONTH LATER, on 22 August, another fire swept across the steep slopes east of Afotsara. Fires had been burning all day up and down the valley, and in the evening, an arc of flame stretched across the mountainside to the east. I was eating dinner in my room when I heard whistles outside, calling the men to fight the fire. I grabbed my headlamp, sweater, and backpack, and joined a group of men on the dark road walking south. We walked in no particular hurry down the road, then turned off onto a trail that leads up the mountain alongside a dramatic cascade. There was no moonlight, but the stars were out. The orange ring of fire stretched for about a kilometer across the side of the mountain ahead of us, descending with monstrous flames.

We climbed in a line of eight people, by the light of my headlamp and another flashlight. At the last house, where the trail steepened, an old man behind us asked how many people were up ahead. It was Rapaul, and he was surprised to see me there. He explained to the assembled men that he wanted to make a list of those present, so in case the gendarmes came to make an investigation, he could document that the village had rallied to fight the fire, thus avoiding being fined for not fighting a fire. A little later his son Tiana explained to me that Rapaul was the one who had blown the whistle.

I climbed the slope with Tiana as the others surged ahead. There was occasional hooting and hollering, whistles and shouts, from up and down the mountain. Tiana said that the fire started up near the top of the river cascades, which I confirmed (I had seen the first smoke that morning). He had been up in that area building a house for Ravalala; he said the fire escaped from someone's careless attempt to burn the brush off the edge of an irrigation canal. The fire crossed several pastures and now burned into the mountainside woodlands. Tiana told me how these woodlands, composed chiefly of the short, endemic *tapia* tree, are a good thing, how many birds and animals they house, how they help the rain, and how it is nice to be inside of them. He also spoke of how houses frequently burned down; two years ago there were three or four houses that burned in this region, due to both escaped kitchen fires and wildfires.

We climbed and climbed the steep dirt footpath through the dark woodland. High up the mountain, where woodland meets pasture, we arrived at a brushy zone where several people were already fighting the fire. Orange flames spread in arcs throughout the hillsides below us to the south; in fact we could even see that there was another fire in the far distance, on the northeastern slopes of Ivato mountain. The fire flared up pine trees, crackling filled the air. We grabbed nearby branches, *tapia,* eucalyptus, pine, or brush, and swatted at the flame front when it was possible. A big owl flew by, and people spoke of birds waking up and escaping. We contained the fire along a brushy, grassy zone where it was slowly backing down the hill, but further over it had already descended far down the hill below us and we had a tougher time fighting the vertical fire line. The fire caught piles of pine branches and flared up, or entered areas of thicker grass, brush, and ferns and crackled like wild. The area was full of man-deep holes, practically invisible in the dark; I once fell in up to my waist. The villagers were fighting the fire barefoot, so they could not walk far through the still-hot ashes. We often just stood and watched; depending on the vegetation and gusts of wind the fire was hopeless to fight, even dangerous.

A nighttime fire is quite a show, better than fireworks. The sparks flowed to the sky, twirling and then floating, the flames curving with the wind and dancing or pounding or roaring. The firelight made my flashlight irrelevant. After a particularly dazzling and dangerous inferno, we rejoined another group of men—we were at least fifteen by then—and continued to contain the side of the fire. But most of the time was spent standing around and discussing, laughing, joking, and occasionally an opportunity was seized to put out the fire's advance. Between the wind, our efforts, and the steep topography, the fire eventually looked confinable, a 200 m–long front slowly moving downhill. At this point the group, with long roundabout discussion, decided to go home; the main danger was past, people were tired, and they were confident that the fire would go out by itself. We left the fire around 9:45 P.M., descending the steep trails by flashlight and creating great clouds of dust. We walked in the starlight down the road and people peeled off one by one to their houses.

FIRES CONTINUED throughout the rest of the dry season in Afotsara that year, from large pasture fires to small crop field fires. Most large fires burned quietly, uncontrolled, in the grasslands—the village whistle was sounded only once more, though households and hamlets occasionally fought fires that approached their houses. Some property was inevitably damaged—a lean-to used for lunch cooking and to wait out rains burned accidentally on 31 August, a large mountainside of pines burned after a crop-fertilizing fire escaped control on 6 November, and four hectares of pine saplings roasted on 13 November, possibly a case of arson or conflict. The owner of half of the pines told me that his trees, which he grew for construction wood, were very damaged. He suggested that perhaps a neighbor—pointing to a house downhill to the north—was angry or jealous and lit the trees out of spite over an irrigation water feud. However, he and another neighbor stated that they would not bring this matter to the authorities, for there were no witnesses and they did not want to risk a resulting villagewide fine.

By mid-November, over half of the pastures had been burned, the rains began their tentative approach, and crop field fires were more common. On 23 November, I sat on the balcony of a mountainside home chatting with an elderly couple. As we watched, a lone man lit a fire in a flat area on the next ridge to the north. The flames reached high, up to three meters, and the man walked around the fire's edge. It was actually the first time I had seen anyone light and monitor a fire (other than the small insect-repelling fires set by the kids, or the backfire). The elderly couple stated that the man would cultivate cassava in the area he was burning, and that the fire was bounded by crop fields so it could not escape. Before the fire died down, the man grabbed his sack and descended down the hill.

THESE STORIES document the omnipresence of fire, in both useful and destructive forms. They show the close relation the villagers have to fire, their familiarity and comfort in using fire and their experience in controlling fire. They also hint at the relation of the state—the foresters, the mayor, the gendarmes—with fire, including antifire propaganda and rules, but little enforcement. Finally, they show how people cope with outside concerns about fire, by evading explanation, creating alternative stories, and rarely being present at a fire. These anecdotes encapsulate many of the central issues and concerns about fire, showing its conflictual and complex nature.

THE "FIRE PROBLEM" of Madagascar is not that there are too many fires (though this is certainly true in certain places and moments); the real problem is a century-long conflict over appropriate resource use. This view counters a well-established conviction that fires are ruining the island and that something must be done to stop the burning. This conviction, which I will call the "antifire received wisdom," permeates almost all discussions of Madagascar as part of a larger discourse of environmental degradation. As a result of this received wisdom, in my first visits to the island I approached everything through the lens of environmental degradation. I took pictures of cut trees, of erosion gullies, and of fires on the horizon. Yet as I began to study land-use change, I focused on the farmers and their interactions with the land, and I started to see fire as not just a cause of rapid degradation, but also as one of the farmers' key land management tools.

It is true that there are sometimes too many fires, and some should be better controlled (as fire is not always the best-behaved servant). Yet the more I thought about the historical and political context—together with the rural livelihood needs of Malagasy peasants in the context of a tropical landscape with a long dry season—the more I was convinced that the "fire problem" is not the fires. They would burn anyway, much more catastrophically, if humans allowed fuel to build up (as experience in the western United States shows). The problem is the political struggle over appropriate land use—shaped by the antifire received wisdom on one side and livelihood necessities on the other, not to mention by the lively actor fire itself.

THE CENTRAL argument of this book is that contrary to the antifire received wisdom, fire is an integral part of many Malagasy rural production systems. Fire, however, is a complex tool. It is self-propagating, unpredictable, and easily used anonymously. Due to this complexity, the visible damages of fire, and the antifire received wisdom, an elite coalition (including colonial bureaucrats, government foresters, and international environmental agencies) has criminalized burning through rhetoric and repression. However, 100 years of antifire repression have only succeeded in worsening the problem, polarizing the peasantry against the state, and giving meaning to fire itself as a tool of resistance. Farmers and herders resent state interference with a practice they consider crucial for environmental management and for their livelihoods. They take advantage of fire's complex character and of internal inconsistencies within the state in order to continue burning, leading to today's stalemate.

This book comprises three major parts that support this overall argument. The first sets the stage. Chapter 1 lays out the fire problem, documents the theoretical and empirical bases of this work, and introduces the reader to Madagascar. Chapter 2 investigates ideas about the nature of fire. It traces the development and persistence of the antifire received wisdom in Madagascar and elsewhere. I argue that fire is neither simply bad nor good, but a complex tool of resource management and a natural ecological process. I investigate in detail the factors that make fire a complex process—its self-ignition and self-propagation, its easy anonymity, its unpredictability—as well as the environmental debates over fire.

The second part looks at how fire fits into land users' livelihood strategies around the island. I demonstrate that from a farmer's or herder's point of view, fire can be an appropriate, efficient, and inexpensive tool for land management. Chapter 3 focuses on grassland and cropland fires, particularly in the highlands. I document fire's various uses, and, based on five case studies, show that demographic and economic pressures sometimes serve to squeeze fire out of the land use strategies of intensively utilized regions. Chapter 4 focuses on the special case of the *tapia* woodlands, a fire-dependent vegetation type that provides resources such as wild silk, fruit, and firewood to surrounding communities. I argue that villagers actively shape these woodlands by manipulating woodland fire ecology. Chapter 5 addresses the crucial question of *tavy* fires, especially in the humid forest zones of the east. I review the reasons, strategies, and methods of *tavy,* as well as its environmental consequences. Due to its key role in deforestation, burning for *tavy* has been the most actively controlled type of landscape burning, by both the colonial and independent states (though not always successfully).

This discussion of the state repression of *tavy* leads nicely into the third and final part, which focuses on the historical and current politics of fire. First, chapter 6 analyzes the tools used by the state and the peasants in their struggle over fire, framing the analysis with the concepts "criminalization" and "resistance." I show how the state criminalized fire and how peasants resisted, lighting fires on the sly and evading enforcement. Peasant resistance is based around the taking advantage of opportunities such as the physical character of fire (e.g., its easy anonymity and self-propagation) and the weaknesses and hesitations within a multivalent state. Because of these opportunities, resistance has been relatively successful. Then, chapter 7 traces the history of this struggle through the last century. I build a detailed history of fire politics in Madagascar, beginning with the first tentative attempts to control fires at the turn of the century, through the strict legislation of the 1930s, to the more enlightened views of the 1950s, to the incendiary crisis of the early 1970s, and finally to the appearance of the international conservation movement in the 1980s and 1990s.

Chapter 8 brings this discussion up to the current day, focusing on recent attempts to give villagers control over fire management. I contend that a new policy called GELOSE will struggle to meet its goal of decentralizing fire management, due to the weight of the antifire received wisdom (which limits the discursive space within which policy can occur, thus blocking fire-tolerant policies) and due to the realities of village social relations. To contradict this bleak picture, I look to several examples of functioning local-level fire management in Madagascar, suggesting that GELOSE follow their example. Key recommendations are for the state to accept fire as useful tool, to accept real local control, and to simplify the bureaucratic process. The conclusion (chapter 9) unifies the messages of the research at a theoretical level, imagines what unencumbered local fire management would look like, and suggests that the state focus its attention on place-specific control of the environmental consequences of fire, on controlling criminal and accidental fires, and on solving the land use conflict that is at the root of the problem of *tavy* fires.

1 [THE ISLE OF FIRE

Problem, Theory, and Setting

> On the evening of our stay here we had an instance of the great extent to which the grass-burnings are carried on among the wild hills and moorland of Madagascar. The night being clear, and the moon in its first quarter, we counted nearly twenty large fires in different directions.
>
> MOSS (1876, 132)

> *Tany mena tsy mba mirehitra.*
> The red earth isn't on fire, it just looks like it
>
> MALAGASY PROVERB

PROBLEM STATEMENT

Immense areas of Madagascar blaze with flames each year (figure 1.1), charring millions of hectares. Such anthropogenic fires have been ubiquitous since human settlement some 1,500 years ago. These fires clear the land, especially for agriculture and cattle raising. Agricultural fires, especially *tavy* in and around the rain forest paralleling the east coast, are lit just before the rains each October and November, clouding the skies with smoke. On the grasslands that cover almost two-thirds of the island, *doro tanety* (pasture fires) blaze during much of the dry season, the dates varying from region to region. In the more arid western zones, pasture burning commences in May, peaks in August, and continues into November. In cooler highland areas, due to the chill and occasional drizzle of the austral winter, most fires burn in the spring, from August to October. In the humid grasslands of the deforested

FIGURE 1.1. **A fire advances across a pasture adjacent to a eucalyptus woodlot.** In the highlands near Ambositra.

east coast, farmers and herders burn during the somewhat drier months of September to November (Randriambelo et al. 1998).

The actual area that burns each year remains poorly known. Official government statistics for fire (figure 1.2; see appendix 1) report annual burned areas varying between 3,000 and 30,000 km^2, or 0.5 to 5 percent of the island's surface. At the same time, official statistics for fires in Antananarivo Province, which is squarely located in a highland zone well known for vast grassland fires, show annual burned areas ranging from 300 to 25,000 km^2, representing 0.5 to 43 percent of the provincial surface area. As I explain in appendix 1, these

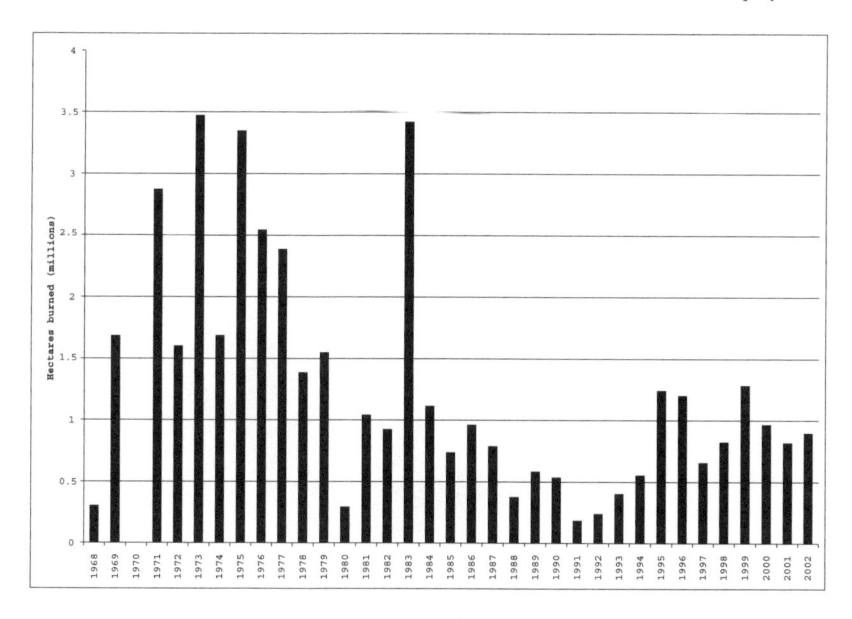

FIGURE 1.2. **Official government statistics of surface area burned in Madagascar, 1968–2002.** For a detailed look at the data, an important disclaimer about its quality, and sources, see appendix 1.

statistics are unreliable and probably underestimate the actual extent of fires, but they might give an indication of annual variations in fire extent.

Recent advances in remote sensing approaches to fire monitoring allow one to compare fire occurrences or intensity through time and across different regions and to estimate actual surface area torched.[1] Nighttime images are used to pinpoint active fire location and are extremely useful in showing the relative extent of fires across the island in different seasons, though for nighttime only (figure 1.3).[2] Daytime images are used to measure burned surface area. Using SPOT images, the Global Burned Areas 2000 project estimated that 11,382 km^2 (about 2 percent) burned in Madagascar in 2000,[3] while Matzke (2003) estimates that 6 to 7 percent of the island burns each year (August to December only) based on Landsat ETM images.

Pasture and grassland fires are responsible for the vast majority—be-

1. See Cahoon et al. (1992), Grégoire (1993), Dwyer et al. (1999), Mbow et al. (2000), Nielsen and Rasmussen (2001), Matzke (2003).

2. See, e.g., the ATSR World Fire Atlas (shark1.esrin.esa.it/ionia/FIRE/AF/ATSR). For Madagascar, see Matzke (2003) and www.ngdc.noaa.gov/dmsp/fires.

3. See www.grid.unep.ch/activities/earlywarning/preview/ims/gba.

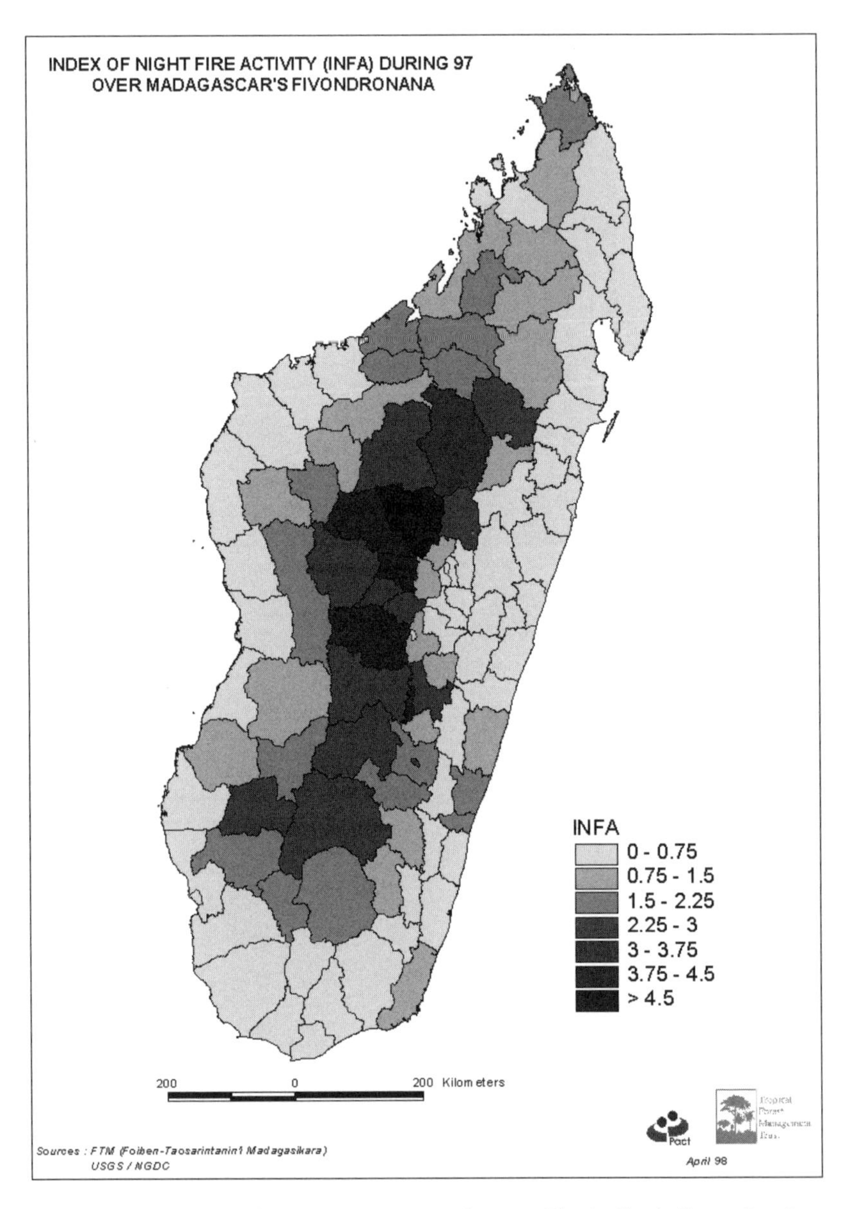

FIGURE 1.3. Nighttime fire intensity across Madagascar. The shading indicates the relative intensity of nighttime fire from August through December 1997. Based on visual and near-infrared data from DMSP (Defense Meteorological Satellite Program) satellites. Source: PACT, Antananarivo.

tween 95 and 99 percent—of surface area burned. Based on a crude interpolation between different sources, I estimate that *in grasslands, roughly one-quarter to one-half the surface area burns annually,* equaling about 15 to 30 percent of the island.[4] The actual amount will of course vary significantly from year to year and region to region.

The fires of *tavy* slash-and-burn agriculture in forest areas have received more attention, due to their impact on deforestation. Between 200,000 and 700,000 ha of brush and forest are cleared and burned each year for agriculture in the island's humid zones (Ramamonjisoa 2001).[5] Only about 5 percent of these fires are in primary forest. Due to repeated short fallow *tavy* cultivation (which impacts nutrient cycles and vegetation succession), the expansion of *tavy* into new forest areas, as well as logging exploitation, over 100,000 ha of forest cover are lost annually (Green and Sussman 1990; see also chapter 5 of this volume).

The immediate physical damages of Madagascar's blazes can be spectac-

4. This figure of "one-quarter to one-half" gives a sense of the magnitude of the fires, but also of lack of certainty in determining fire area. The figure is based upon an interpolation between four categories of sources: local studies and educated guesses (both of which support an estimate of "about one-half") and government statistics and satellite data (which support an estimate of "about one-quarter").

Local studies give widely varying estimates. My own monitoring of fires in Afotsara (see Chapter 3) documented that 54 percent of grasslands burned in 1998. Schnyder and Serretti (1997) found that two-thirds of the prairie was burned in their Ambohitantely study region. Data from Rambeloarisoa (1995) indicate that 2.5 to 12.3 percent of lands burned annually in 1991–1993 in his study region south of Antananarivo. Finally, Andriamampionona (1992) reports that 80 to 90 percent burned near Itasy in 1991, a year of political strife. "About half" seems a fair average between these studies.

Educated guesses also point to an estimate of one-half. Jenkins (1987) estimates that over one-third of the island is burned annually, while Jolly (1990) puts the same figure at one-quarter to one-third. If we assume that 95 percent of fires are in grasslands, and that grasslands cover 64 percent of the island (see table 1.2 later this chapter), then the figure of one-third *of the island* would imply 50 percent *of the grasslands.*

Government statistics give lower estimates. The statistics from Antananarivo Province, which are the most reliable yet still notoriously underreported, show that fires on average burn about 25 percent annually of all provincial lands (which include not just grassland, but also croplands, built-up areas, and forests).

Remote sensing analysis of satellite data also gives a lower estimate. Matzke (2003) estimates that 15–20 percent of Antananarivo Province burns between August and December each year. If one increases this estimate by circa 25 percent (for fires occurring before August, see e.g., figure 3.1) and again by 10 percent (discounting the part of the province covered by cropfields, built-up areas, and forests), then one arrives at a figure of 20–27 percent of grasslands burned annually.

5. These figures make intuitive sense: if the rural population of eastern Madagascar is (very roughly) put at five million, this equals one million households, and if each clears 0.5 ha, the result is 500,000 ha.

ular. To take one year as an example, fires in 1968 destroyed, according to the Forest Service, 16.6 km^2 of reforestations, 38 km^2 of natural forests, 236 houses, 52 granaries, 2 churches, 2 government farms, 6,000 tons of cotton, 1 ha of vineyards, 300 orange trees, and 3,000 pineapple plants; the fires also burned 283 people and caused one death.[6] Escaped fires inflict an annual toll on standing crops, fruit trees, pine and eucalyptus plantations, and natural forest edges. Houses with vulnerable grass roofs frequently go up in flames.[7] However, despite such alarming statistics, the vast majority of fires burn uneventfully. Grassland fires stop at the edge of crop fields for lack of fuel; *tavy* fires burn little more than their allotted plots.

At a broader scale, the island's fires have environmental consequences such as land-cover change and altered nutrient cycles. Deliberate human-lit fires have greatly expanded grassland areas as people use fire to modify the landscape to their needs. Some of these fires continue to nibble at forest edges. Most critically on this island famed for its biodiversity, *tavy* fire is the proximate cause behind much deforestation and resulting loss of habitat. In addition, repeated *doro tanety* keep grassland species diversity low and increase levels of soil erosion. Important nutrients, especially carbon and nitrogen, are volatized by burning, part of a slow process of soil degradation. These and other consequences will be discussed in further detail in chapters 2 and 5.

Due to these consequences, foresters, environmentalists, and the state elite have long seen fire as the chief problem for the Malagasy environment, linking it tightly to deforestation, biodiversity loss, soil degradation, and desertification (Perrier 1921; Humbert 1927; Bosser 1954; IUCN 1972; RDM 1980; Randrianarijaona 1983; Mottet 1988; Rakatonindrina 1989; Rajaonson et al. 1995; Gade 1996). As I show in later chapters, some of these concerns are exaggerated; others are a matter of perspective. However, these concerns play a critical role in the politics of fire. They allow it to be labeled the key environmental problem (Andrianavosoa 1996), a major "handicap to development" (Rakatonindrina 1989, 122), and a *fléau national,* a national "scourge" or "curse" (RDM 1980, 1; RDM 1990, 15; Rambeloarisoa 1995, 4), and they push the issue to national media prominence.[8]

6. Discours, Direction des Eaux et Forêts, au sujet de la Cérémonie Officielle pour ouverture de la Campagne de Reboisement (ANM Vice Présidence 850, subfolder "Conférence internationale . . .").

7. See, e.g., Einrem (1912); *L'Express de Madagascar* (18 Sep. 1996). My interviewees also testified to this.

8. For example the newspaper *L'Express de Madagascar* stated prominently on the front page of its 22 Aug. 2001 edition, "*Doro tanety manomboka sahady*"—hill fires already beginning. (*L'Express* publishes two-thirds in French and one-third in Malagasy—the latter often reserved for bad news like murders and other news they don't want foreigners to see.) See also figure 2.2.

Contrary to this common view, this book argues that the "fire problem" of Madagascar is not the ubiquitous fires, their destructive potential, and the stubborn peasants who continue to ignite them. The more pertinent problem is that after 100 years of strident government antifire efforts, people still light fires as before. Colonial policies beginning 100 years ago, followed by post-Independence legislation, attempted to ban or at least control pasture fires in the name of "rational" land use, safety, and environmental protection. However, fires still burn wild each year, as they did throughout the past century. This leads us to ask *why*—why the rural Malagasy keep igniting the land, why the leaders see fire as an all-around evil that is permanently getting worse (see table 1.1), why the state keeps trying to stop the fires, and why it doesn't succeed. By asking these questions and probing into their depths, we will see that the real "fire problem" in Madagascar is the struggle over the access to and character of natural resources, a struggle fought primarily by the peasants and the state.

TABLE 1.1. The ever-increasing fires of Madagascar.

Year of fires	Phrase used	Source
1903	*recrudescence récente des feux*	report by the director of the Forest Service (Bertrand and Sourdat 1998, 26)
1910	*une extraordinaire recrudescence des tavy*	Coulaud (1973, 325)
1941	*recrudescence notable de ces méthodes néfastes*	circular from Governor Annet (JOM 6 Dec. 1941)
1968	*recrudescence infernale des feux de brousse*	speech by the director of the Forest Service, 1968 or 1969 (ANM Vice-Pres 850/subfolder "Conférence internationale")
1969	*recrudescence spectaculaire*	Desjeux (1979, 25)
1972	*recrudescence inouïe*	annual report of the DEF, 1972
1975	*recrudescence des feux de brousse*	Ordonnance 75-028 (JOM 1 Nov 1975)
1976	*recrudescence actuelle des feux sauvages*	Ordonnance 76-030 (JOM 27 Aug 1976)
1993	*recrudescence des feux de brousse*	Rakotoarivelo (1993, 1)
1995	*recrudescence excessive des feux sauvages*	annual report of the DEF, 1995
mid-1990s	*intensification des feux de brousse*	Aquaterre (1997, 9)
1998	*recrudescence des feux de végétation*	foreword by L. H. Rakotovao to Bertrand and Sourdat (1998, 5)

By looking at the fire problem in Madagascar through the frame of analysis of a peasant-state struggle over resource access and character, we are forced to understand both the motives and methods of the different actors. Why do the peasants continue to burn, against a century of prohibitions and state repression? To understand this, we must investigate the place of fire in peasant agropastoral livelihood strategies and question the relevance of certain ideas linking fire to environmental degradation. Why does the state continue to fight fire, despite the fact that a hundred years of repression hardly worked? For this, we must analyze the interests and ideology of those in power. Who are the actors in this conflict and how do they differ? In looking closer, we will discover that the state and peasants are both internally diverse and intertwined in complex ways that have favored continued burning. Finally, how does the nature of fire itself influence the struggle? We will see that fire's special nature—the way it conflates intention and agency, the way humans do not monopolize ignition and control, and its easy anonymity—makes the conflict even more complex. By framing the fire problem in this way, we will come to a more detailed, nuanced understanding of the complexity of the situation, one that goes far beyond simple declarations that "fire is destroying Madagascar and it must be stopped."

THEORETICAL GROUNDING

This book makes three central assertions that flow from one to the other. Crudely stated, they run as follows. *First,* the colonial and postcolonial states criminalized burning for economic and ideological reasons. *Second,* as a result of the importance of fire in environmental management and in peoples' livelihoods, many farmers and herders resisted this criminalization. *Third,* their resistance was able to succeed, leading to today's stalemate, due largely to the nature of fire and to internal inconsistencies within the state.

These conclusions build upon a number of literatures. Tying them all together is *political ecology,* an approach to environment-society research that examines how people negotiate, cooperate, or fight over the access, control, use, and character of the Earth's environmental resources, with a strong eye to the historical, regional, and ecological contexts. The political ecology approach appeared in academia as a response to concerns that the social, political, and economic context was being ignored in studies of environmental issues (Watts 1985; Blaikie and Brookfield 1987). The relevance of this critique, combined with its call for interdisciplinary and postparadigmatic cross-fertilization, inspired a generation of researchers to take a fresh look at environmental issues previously considered to be simple technical questions (Peluso 1992; Carney 1996; Leach and Mearns 1996; Bryant and Bailey 1997; Neumann 1998; Blaikie 1999; Schroeder 1999; Zerner 2000; Bebbington and Batterbury 2001; Peluso and Watts 2001).

Madagascar's fire problem was also ripe for political ecological analysis. A number of different strands of the literature in political ecology and allied fields have proven most useful in analyzing this fire problem and in developing the three central assertions detailed above. The overall framework of peasants struggling against the state over fire management builds on political ecological work on *resource struggles,* in particular on mechanisms of criminalization and resistance. Lest this struggle be painted too roughly in black and white, I rely on a closely related literature on *differentiated states and peasants* to open up the discussion, leading in part to the third assertion. The first assertion is informed by recent scholarship on the roots and persistence of dominant environmental narratives or *received wisdoms;* the second by a long tradition of inquiry in cultural ecology, economic anthropology and allied fields into *rural livelihood strategies.* Finally, new developments in *ecology,* to which political ecologists have ascribed great importance, also contribute to the second assertion. I review these ideas below.

Struggles over Resources

The history of fire politics in Madagascar is a story of state versus peasant, colonizer versus colonized, urban versus rural, Forest Service versus district administrators, cattle raisers versus tree growers. Each of these groups is interested in controlling the access to or character of certain natural resources, namely the land, the vegetation it carries, and the soil beneath it.

In analyzing this kind of struggle, political ecologists address the conflicting interest groups—split, e.g., by wealth, gender, ethnicity, power—and their mechanisms of struggle, contributing to our understanding of how and why resource struggles unfold. A lynchpin concept in these analyses is *resource access,* which is what people fight over (Berry 1989; Blaikie 1989; Jarosz 1991; Peluso 1992, 1996; Fortmann 1995; Ribot 1998; Batterbury and Bebbington 1999). This concept denotes the possibility, at a certain time and place, of using or harvesting a resource, and as such is broader than strict ideas of "property." Resource access can be legislated, negotiated, purchased, or just established through use. For example, villagers outside Madagascar's Isalo National Park have the legal right to harvest silk cocoons within the park. Similarly, villagers in the Malagasy highlands, by tradition, may enter privately owned woodlots to collect downed wood or lower tree branches for use as fuel. Resource access can be shaped by both everyday negotiation and more dramatic struggle. The result, as demonstrated by Peluso (1992, 1996), is a combination of formal legal strictures, informal "structures of access" (such as patron-client relations or business networks), and "ethics of access" (such as culturally rooted ideas about who deserves access in what situations).

While this study of Madagascar's fire problem is consistent with most political ecology studies of resource conflict, I found the concept of resource access incomplete. Malagasy peasants and the state do not simply struggle over resource *access*, they also struggle over resource *character*, over competing notions of what constitutes good forest or good rangeland. What is grassland worth to a cattle herder—even with unrestricted access—if ferns and shrubs invade it? Are conservationists fighting over access to a forest or over its character—to keep it full of native tree species as opposed to exotic pines or, worse yet, replaced by barren soil? As a result, studies of resource struggle should expand to consider not just resource access, but also *resource character*—the state or condition of the part of the physical environment upon which people make claims.

Resource character is, of course, not unrelated to resource access. By manipulating, shaping, or controlling resource character, people can alter the possibilities of resource access. For instance, if the state causes bush encroachment on a pasture through a fire ban, it has reduced herders' access to suitable rangelands. In turn, when access to a resource is altered, this may alter its character. For example, a slash-and-burn farmer denied access to swidden fields by conservation legislation will be forced to allow fallow fields to mature into forest.

A second lynchpin concept in this book, *criminalization and resistance,* shifts our focus from *what* is being struggled over to *how* the struggle unfolds. In essence, the state has acted to redefine burning from a useful livelihood activity to a criminal act. As this directly affronts many peasants' livelihoods and sense of justice, they resist the restrictions in a variety of ways. I explore these concepts in much further detail in chapter 6.

Differentiated State, Differentiated Peasants

In this book, I describe the fire problem as a struggle over resource access and character *between peasants and the state.* Yet this is only in the first instance. Internal differences within the state and the peasant community, and the way the state and civil society are intertwined, are also crucial to understanding the nature and persistence of the fire problem. The state and the peasants are, in essence, mutually constitutive parts of broader society (Tsing 1993; Gupta 1995; Sivaramakrishnan 1995; Li 1996; Sundar 2000).

For its part, the state is a complex, ever-changing set of institutions and practices, full of competing personal, institutional, and political agendas representing different parts of civil society (Ferguson 1994; Moore 1996; Sundar et al. 1996; Bryant 1997; Espeland 1998). The state may even simultaneously entertain opposing ideologies (Neumann 2000; Head 2000). Take, for instance, the common rift between state foresters and other branches of the administration. Bryant (1998) describes how British foresters in Burma collided with agricultural officers whose goal was to clear forests for cash crop cultivation. The same occurred in Madagascar, where peasants occasionally found themselves fined by a state forester for cutting and burning a field in land designated for that purpose by an agent of the Service d'Agriculture (Ratovoson 1979). In fact, Malagasy foresters were often frustrated by other parts of the state, for example accusing the courts of not upholding forest or fire laws.[9]

States are not only horizontally diverse; they are also vertically diverse. The "state" that a rural villager sees—a poorly paid forester, some young gendarmes, the locally elected mayor—is quite different from the "state" represented by a government minister. Field agents of the state must answer both to villagers as neighbors as well as to their superiors, and they often have their own personal interests—like job security or petty corruption—in mind (Gupta 1995; Neumann 1998; Robbins 2000; Sundar 2000). Dispersed peasants cannot argue against a ministry, but they can make life difficult for field agents. As a result, centralized edicts often look different on the ground.

If it is useful to remember that the state is diverse, it also goes without saying that peasant communities are not without their own internal differences and contests. Rural societies are differentiated by wealth, gender, ethnicity, caste, age, and other characteristics. Some people simply have more money or power than others, and different people within a community may have different interests or goals. For example, near the Manongarivo Special Reserve in the north of the island, longer-term resident farmers compete with recent Tsimihety immigrants for access to forest lands; both cultivate into the forest. However, when the Forest Service comes to punish illegal *tavy* burners, the original residents invariably point their fingers at the newcomers (Spack 2001).

While on the surface the story of fire politics in Madagascar may be one of state versus peasant—these are broad categories, which are, after all, obvious and important—the entangled and internally diverse nature of these

9. Service des Eaux et Forêts annual reports, 1969, 1971, and 1993; interviews, Forest Service agents, 20 May and 10 July 1999; see also Dez (1968) and Andriamampionona (1992).

categories helps shapes the character of the struggle. Specifically, as I argue in chapter 6, successful resistance to state fire criminalization is made possible in part by weaknesses and peasant sympathies within the state.

Received Wisdoms

Let us now return to my first central assertion, that dominant parts of the state criminalized burning for economic and ideological reasons. What role do ideologies play in resource management? By focusing, as suggested above, on resource character as opposed to just resource access, we can highlight the ideological aspects of conflict. This is because resource character is not just measured in physical indicators but also subject to the cultural relativities and biases of observation. It is in this way that the grasslands of Madagascar may be seen simultaneously as "desert" (Sibree 1870; Parrot 1924, 192) and as "fine rich pasture" (Ellis 1838, 91). In Madagascar, the struggle over fire can be viewed as an ideological battle over ideal landscapes. Crudely summarized, it is the forester's love of forests versus the pastoralist's love of pastures (see, e.g., Sauer 1956, 56; Pyne 1997).

In order to address these ideologies, I build the case for an *antifire received wisdom* that has guided policy making in the Malagasy state, both before and after Independence. An environmental received wisdom is an idea that is held as correct by the politically powerful establishment, despite potential flaws or weaknesses, and which shapes the discursive field within which discussions of environmental change and resource management occur (Leach and Mearns 1996). Received wisdoms have also been termed "dominant narratives" (Roe 1991) or "environmental orthodoxies" (Batterbury et al. 1997). Examples include the myth of the pristine pre-Columbian Americas (Denevan 1992), the ignorant swidden farmer myth (Dove 1983), the desertification myth (Swift 1996), the Himalayan or Sahelian degradation crisis narratives (Ives and Messerli 1989; Leach and Mearns 1996), and the exaggerated erosion myth (Stocking 1996). These received wisdoms often persist despite inconclusive evidence (Beinart 1996; Swift 1996; Head 2000). They persist because they are persuasive, simple stories (Roe 1991) that serve the purposes of powerful groups (Bergeret 1993; Leach and Mearns 1996; Batterbury et al. 1997; Blaikie 1999), and because they are institutionalized into state bureaucracies (Espeland 1998). They serve to frame problems such that the answers justify the actions of the interested parties (Ferguson 1994; Escobar 1995) and as such have become a key element of neoliberal development discourse (Bassett and Koli Bi 2000).

Perhaps the most well known work in this direction is that of James Fairhead and Melissa Leach (1996, 1998). They demonstrate how an ideology of deforestation shaped by colonial science and conservation policies led analysts in West Africa to overestimate original forest cover, misinterpret historical data, and misread human-created landscapes. In their widely cited Kissidougou case study (1996), they show that certain forest patches in the Guinean savanna are not relicts of deforested woodlands as previously assumed, but instead human additions to a savanna landscape. In a similar vein, Bassett and Koli Bi (2000) show that the Ivorian savanna is increasingly wooded and bushy, despite a dominant ideology that it is becoming desertified. This book, in its reliance on the received wisdom concept and in its attempts to test and challenge the dominant antifire ideology in Madagascar, broadly corroborates the conclusions of such studies.

Land Users' Livelihood Strategies

If the state's repression of fire is rooted in the antifire received wisdom, the farmers' and herders' resistance is anchored in the means by which they seek their livelihoods (and by extension the meanings embedded in these livelihoods).[10] In this book, I seek to explain the land users' fire-based livelihood strategies based on the ecology of available natural resources and the ways in which locals manage these resources in a changing social and economic context.

Political ecology research, and especially its antecedent cultural ecology, has a long history of detailed attention to the place-based, environmentally rooted practices of rural land users (Conklin 1954; Nietschmann 1973; Turner and Brush 1987; Butzer 1989, 1990; Netting 1993). Studies focus on the specific techniques and tools of peasant agropastoral systems and seek to relate these characteristics to resource ecology (Moran 1982; Zimmerer 1996), demographic patterns and changes (Boserup 1965; Netting 1993; Turner et al. 1993; Tiffen et al. 1994), and broader political and economic forces (Hecht 1985; Coomes and Barham 1997; Muldavin 1997; Grossman 1998; Lambin et al.

10. I do not address the more symbolic or cultural aspects of these livelihood strategies—such as the frequently voiced assertion that *tavy* is the way of the ancestors (Althabe 1969; Ratovoson 1979; Jarosz 1996) or that the Malagasy are culturally predisposed to pyromania (Perrier 1921; Gade 1996; Andriamampianina 1998). Clearly, central livelihood tasks such as burning, planting, or harvesting are incorporated into peoples' symbolic and ritual worlds (Raharijaona and Kus 1989; Beaujard 1995; Razafiarivony 1995). However, such cultural-symbolic aspects do not explain the *persistence* of burning. As part 2 of this book shows, from a farmer's perspective, burning strategies are rational resource management tools in certain agropastoral, economic, and demographic contexts, and they are avoided when they no longer fit (e.g., due to intensification).

2001). In my analysis I have taken inspiration both from cultural ecology's emphasis on understanding rural land use in its local context and from political ecology's focus on "agrarian societies in the throes of complex forms of capitalist transitions" (Peet and Watts 1996, 5). Consistent with the abovementioned research, not only will we try to understand the agroecological logic of burning, but we will also see that growing urban wood fuel markets lead some farmers to forgo oft-burned pastures for wattle-covered hillsides, that market demand for fruit from pyrophytic *tapia* trees contributes to the protection of *tapia* woodlands, and that variations in the export markets for vanilla and coffee influence the amount of *tavy* cultivation in rain forests.

New Ecology

The final theoretical underpinning of the book's three-pronged argument comes from new developments in the ecological sciences. As described in detail in chapter 2, ideas about vegetation community change have evolved from deterministic, equilibrium-based models that saw fire as an outside disturbance to nonlinear, nonequilibrium models that accept fire as simply another factor influencing the development of vegetation communities. These ideas have several implications for my argument. First, field-based evidence of the importance of nonequilibrium pathways in ecology has been crucial to challenging numerous received wisdoms about environmental change. Second, resource management strategies that emerge from this new ecology necessarily have to be adaptive and opportunistic, not unlike many indigenous approaches. Finally, the collapse of a clear, legitimate goal (the climax vegetation) highlights the fact that landscape management is political, to be negotiated between different interest groups as much as the ecological system allows (Zimmerer and Young 1998; Scoones 1999; Bassett and Koli Bi 2000; Zimmerer 2000; Leach et al. 2002).

BY TYING TOGETHER the diverse ideas outlined above with historical and current evidence from Madagascar, I will be able to demonstrate how a dominant ideology (at the state level) and ecologically and economically rooted livelihood practices (at the land user level) interact in a historical process of struggle over the access to and character of natural resources. I will argue that the state's antifire received wisdom is scientifically unsupportable and will show how the struggle it inspires has unfortunate social and political consequences. Finally, I show how the land users gain a stalemate in the conflict by exploiting a nonunified state as well as the biophysical character of the resource. I begin this argument in chapter 2. First, however, I introduce the reader to Madagascar in general and to the specific research setting.

RESEARCH SETTING

The island of Madagascar contains a diversity of landscapes that bewilders the visitor: palm-lined tropical beaches; hills terraced by rice paddies in various shades of green, yellow, and brown; vast plains of golden pasture; lush tropical rain forests; massive granitic inselbergs and mountain ranges; and even cobblestone-lined urban streets. The history of this island is equally colorful, blending legends of huge flightless birds and man-eating trees (Dahle 1884) with mystery-shrouded hints of people migrating from as far away as Indonesia and dramatic sagas of conquest and kingdom building. Below I summarize some of the key aspects of the island's nature, society, history, and political organization.

An Introduction to Madagascar

As the world's fourth largest island, Madagascar covers approximately 587,000 km^2 in the southwestern Indian Ocean. Discovered rather late by humans, the island was only settled some 1,500 years ago through successive waves of emigration from present-day Indonesia, as well as from coastal Africa and Arabia. The original populations expanded, interacted with Indian Ocean trade and later European contact, and now constitute about twenty ethnic identities, all speaking a dialect of the same Indo-Malayan tongue, with Bantu and Arabic influences. Different regional polities and kingdoms emerged over the centuries; these are the source of some of the ethnic identities. The Merina kingdom of the central highlands, centered in a large rice-irrigating plain, dominated much of the island in the nineteenth century and developed a modern state.

France conquered the island in 1895 and ruled it as a colony through 1960, promoting economic development, commercial agriculture, and forest exploitation. A major rebellion in 1947 was brutally crushed. After Independence in 1960, the First Republic, under Philibert Tsiranana, was marked by continued tight relations with France. After a period of upheaval, Didier Ratsiraka took power in 1975 as leader of the isolationist, socialist Second Republic, which was troubled by major economic and fiscal crises. Strikes in 1991–1992 led to Ratsiraka's downfall and the election of opposition leader Albert Zafy, who took office, initiating the Third Republic, in 1993. Well-meaning but ineffective, Zafy was impeached in 1996, and national elections reinstated Ratsiraka in 1997. While the economy surged on neoliberal reforms, Ratsiraka lost reelection in late 2001, leading to a violent and economically disastrous six-month dispute with winner Marc Ravalomanana.

The population has increased from over 2,000,000 at the turn of the last century to just over 12,000,000 at the last census (1993); in recent years the growth rate has hovered around 3 percent. Due to political upheavals, flirtation with state-run socialism, and a debt crisis, the standard of living has dropped 40 percent since 1971. Madagascar is now among the world's poorest countries, with a 2000 GNP per capita of $258. Three-quarters of the people live on under $1 per day. In 2002, the country ranked 147 of 173 in the United Nations Development Program's Human Development Index. Some 70 percent of the population is rural and involved in agriculture; two-thirds of agricultural production is subsistence oriented and not commercialized. More than 60 percent of Malagasy families are net rice buyers (rice being the predominant staple), and mild chronic undernutrition is widespread. Drought, cyclones, and other natural hazards cause occasional regional famines (MARA 1996; Barrett 1997; Projet MADIO 1997).

Malagasy society is highly stratified (Desjeux 1979; Kottak 1980; Pavageau 1981; Rakoto 1997), with divisions based on a latent precolonial hierarchy of nobles, commoners, and slaves, and reinforced by wealth, ethnicity, age, and gender. In the highland village of Afotsara (see next section), the richest 20 percent of households utilize 50 percent of the rice fields, and 70 percent of the cattle are in the hands of just 10 percent of the households.[11] One quarter of households are classified as *andevo* or *nirahina,* descendants of slaves, and this stigma influences village social and political life. Community leaders are all *andriana,* descendants of nobles.[12] Speaking generally for the island as a whole, rich farming families may possess several hectares of irrigated rice fields, over a dozen head of cattle, and capital goods such as an ox plow, an oxcart, a bicycle, a radio, wooden furniture, and metal- or tile-roofed brick homes (in the highlands; homes in coastal areas are often built of bamboo, wood, and palm fronds). Poor families are often land poor, work frequently as wage labor, own little livestock beyond chicken, and sleep on the floor in small thatch-roofed adobe homes.

11. This is typical for highland Madagascar. For comparison, see, e.g., the series of excellent village studies performed in the late 1960s and early 1970s, listed in Donque (1979), especially Marchal (1974) and Bonnemaison (1976).

12. While no ex-slaves were found among the richest 10 percent of the village, and ex-slaves averaged less cattle and rice fields per household than ex-nobles, this difference was *not* statistically significant due to the wide variation in ex-noble wealth.

For administrative purposes, rural Madagascar is divided into some 1,000 *communes rurales* and 10,000 *fokontany.* The *commune rurale* is currently the primary rural political unit.[13] Run by a mayor and councilors, with appointed assistants and secretaries, a *commune rurale* governs an average of 5,000 to 10,000 people. The mayor is elected by universal suffrage among independent candidates to a five-year term. A *délégué* appointed by provincial authorities also serves the communal level, sometimes competing with the mayor for influence (R. Marcus, personal communication). The *communes* receive a budget from the central government and administer voting, schools, development projects, markets, and health centers in conjunction with the national technical services.

Each *commune* is split into six to twelve *fokontany,* previously known as *collectivités rurales* or simply *villages.* The *fokontany* is the lowest official political unit and serves, on average, 400–1,200 citizens. This unit held greater powers during the Second Republic (1975–1993), when the *fokontany* president[14] was popularly elected. The mayor of the *commune rurale* now appoints the president, whose job includes signing documents relating to births, deaths, sales, and *famadihana* (reburial ceremonies).

Currently, the *communes rurales* are grouped into about a hundred districts (*fivondronana*), which in turn are grouped into six provinces (*faritany*). These administrative divisions have metamorphosed numerous times, both during the French colonial period and after Independence.[15] According to the federalist constitution of 1997, the six provinces are to gain more control over regional affairs.

Depending upon the region, traditional or informal structures of authority may operate parallel to these official structures, especially at the local level. Village elders (*ray-aman-dreny*) play key roles, especially in ritual matters and in conflict arbitration. In eastern parts of the island, a ritual leader,

13. In colonial days, the corresponding level of administrative hierarchy, equivalent to one or two of today's *communes rurales,* was the *canton.*

14. Often best known as the PCLS (Président du Comité Local de Sécurité).

15. In the 1910s, the island was divided into about thirty *circonscriptions,* later called *provinces.* By 1938, there were eight *régions* divided into some eighty *districts.* By 1951 the eight *régions* were regrouped into five *provinces.* In the 1960s, the system included 6 *provinces* (Diego Suarez was split off from Majunga), each divided into 2 or 3 *préfectures,* and a total of about 110 *sous-préfectures.* The Second Republic did away with *préfectures,* and *sous-préfectures* became *fivondronana.* In 1996, the government proposed redividing the country into 28 *régions* and 160 *départements.* While this proposal never got off the ground, some *fivondronana* (e.g., Ambatolampy) refer to themselves as *départements* now. In this book, for convenience I sometimes refer to the *fivondronana* as "districts."

the *tangalamena,* has an important say in community affairs, as do petty royalty in the western parts of the island. The *fokonolona,* or the community group, takes on different levels of importance in different areas. This term is used to refer to the collected citizens of a hamlet, *fokontany,* or *commune rurale,* or sometimes to a village council, or sometimes more abstractly to a collective will of the people (Kull 2000b). Finally, additional informal structures of authority include lineage-based descent groupings, voluntary associations, church groups, and soccer clubs, among others.

Land tenure in rural zones is based on a mix of customary and state-sanctioned rules. While most permanent crop fields are private property (either officially titled or through customary tenure agreements), claims to up land zones are complex. Tradition gives land ownership to the descendants of those who originally clear plots; legislation upholds this practice but requires permanent cultivation for official title. Legislation also provides for streamlined land titling when individuals reforest hilltops. In most areas, de facto land rights in fallow-field zones and adjacent pasture or forest are under lineage-based control, unless a richer villager or outsider has obtained a government-sanctioned land title. All untitled, uncultivated land is also technically state domain, and the state claims most remaining natural forest areas as classified forests, reserved forests, or protected areas.

Malagasy Regional Geography

The island's size—it stretches nearly 1,600 km from end to end—and its rugged topography make for strong regional differences in climate and vegetation. Due in part to these differences, as well as to millions of years of isolation, the flora and fauna are highly diverse and unique. Of the island's approximately 200,000 plant and animal species, three-quarters are found nowhere else, including 97 percent of nonavian wildlife (Bradt et al. 1996). Human land use has significantly altered the island's vegetation cover, reducing total forest cover to about 23 percent (figure 1.4; table 1.2) and endangering some of the wildlife. As the history and geography of this island continent vary significantly from region to region, I briefly introduce four major regions below.[16]

The *highlands* cover about one-fifth of the island. Elevated between 900 and 2,700 m.a.s.l., the physiography of this zone varies dramatically, including rolling hills, large plains, volcanic cones, granite massifs, and deeply incised valleys. Soils are largely lateritic, with some volcanic soils and alluvial

16. Based, in part, on AGM (1969) and Jenkins (1987).

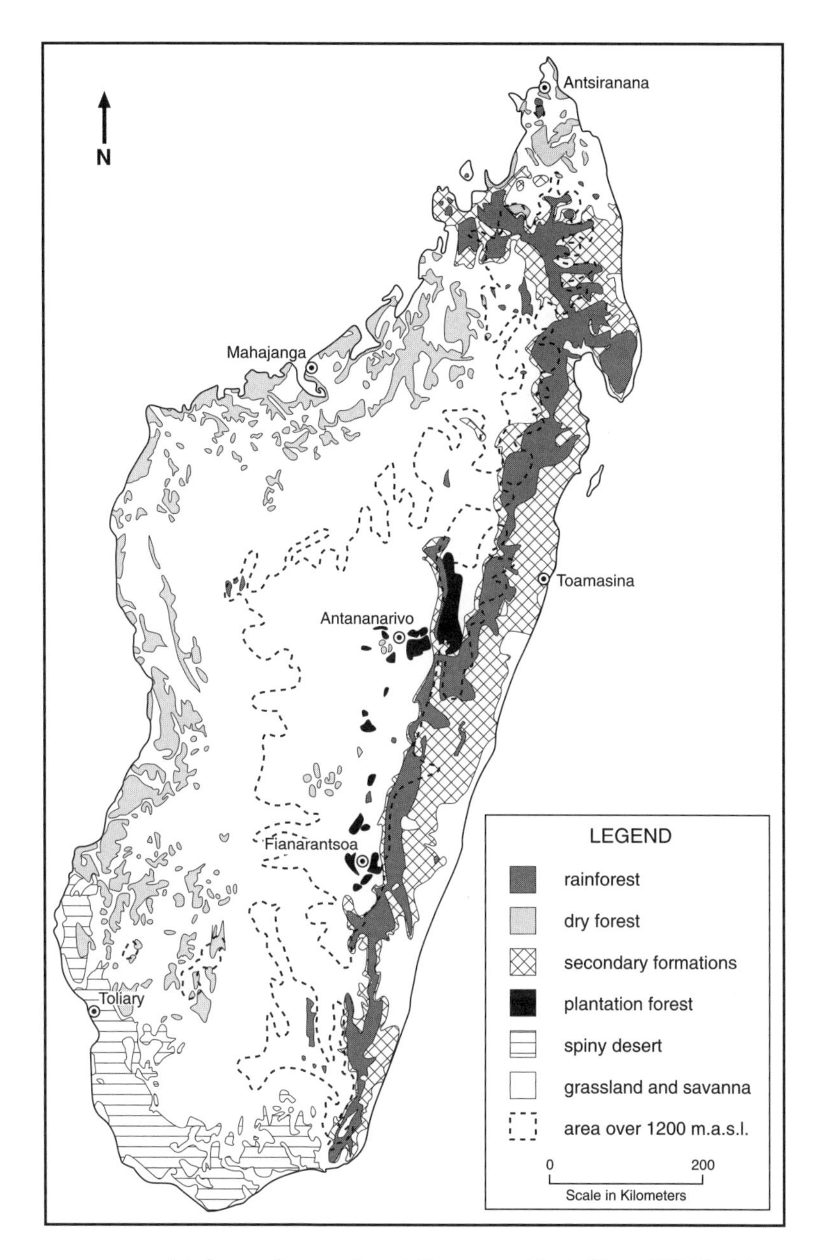

FIGURE 1.4. **Madagascar's current vegetation zones.** Adapted from AGM (1969), Green and Sussman (1990), and Conservation International et al. (ca. 1995). The category "secondary formations" includes cut forest and various stages of forest regrowth and brush, largely corresponding to areas of slash-and-burn farming.

TABLE 1.2. **Areal extent of different land covers (vegetation types) in Madagascar.**

Vegetation types	Area (km²)
dense humid forests*	57,210
sclerophyllous forests of the western slopes	2,600
montane forests and thickets	810
dense dry deciduous forests	49,900
xerophytic thicket (spiny brush)	14,440
mangroves	3,270
riparian and alluvial forests	1,210
exotic forest plantations (esp. pine and eucalyptus)	3,160
mixed land cover mosaic (crop fields, fallow fields, grasslands, forest islands, etc.)*	54,730
grasslands with woody plants*	200,190
grasslands without woody plants	171,390
permanent crop fields	16,750
other (marshland, water surfaces, open soil, built-up surfaces)	7,850
undetermined due to cloud cover	3,070
Total	586,580

* Includes *savoka* of different ages.
Source: DEF (1996).

soils in the valley bottoms and plains. The climate is moderated by the altitude, with cool, dry winters averaging 12 to 15 °C and warm, wet summers averaging 19 to 23 °C. Annual precipitation ranges from 1,200 to 1,400 mm, falling largely December through March. In the nineteenth century, grazed grasslands dominated the highlands, with irrigated rice in many valley bottoms. Land use today continues to intensify and now agriculture occupies most lowlands and many hillsides, while pine, eucalyptus, or wattle woodlots cover some hills. Grassland, however, still dominates vast areas. The highlands, home to the Merina, Betsileo, and Sihanaka ethnicities, are one of the most populous regions of Madagascar, with over six million people and local population densities exceeding 200 people per km² in some intensively cultivated valleys. The agricultural system is centered on intensive smallholder production of wet rice (the preferred staple) and a variety of dry land crops such as corn, cassava, sweet potatoes, beans, fruit trees, and market vegetables. Zebu cattle graze on pastures and fallow fields or are fed hand-cut grasses, and provide traction, manure, prestige, and capital.

The *humid eastern zone* stretches all along the eastern seaboard from Fort Dauphin on the southeastern tip to the Masoala peninsula in the northeast before hooking northwest towards the island of Nosy Be on the opposite coast. This region has a subequatorial humid climate, characterized by year-round rainfall (from 2,000 to over 3,000 mm/yr), warm temperatures averaging 19 to 27 °C (depending on altitude), and ferralitic soils. Some zones experience drier periods in September–October. The mountain escarpment squeezes precipitation out of the moisture-laden prevailing Indian Ocean winds, so rain forest is the natural vegetation. Two-thirds of the rain forest has now been converted to crop fields, secondary brush, or fire-maintained grassland, and deforestation continues at a rapid pace (Green and Sussman 1990). This region hosts the second largest concentration of human population, especially in the southeast (home to the Antaisaka, Antaimoro, Tanala, Antaifasy, and Antambahoaka ethnicities) and the center-east (home to the Betsimisaraka). People live off of irrigated rice production in the lowlands, slash-and-burn rice in the hills, and tropical market crops including coffee, vanilla, litchi, banana, citrus, and cloves.

The drier, hotter *western zone* sits in the rain shadow of the highlands, experiencing a lengthy dry season (March to December) shaped by the down-slope Föhn winds. Summer monsoon rains still drench much of this region, totaling over 1,500 mm/yr in the north (around Mahajanga) but decreasing and more variable towards the south, between 500 and 1,500 mm/yr. The average temperature of the coldest month remains above 20 °C; the annual average is 27 °C. The topography is gentler here, based on sedimentary relief, and soils are rather diverse but not lateritic. Vegetation includes extensive grasslands and some 50,000 km^2 of dry forests and woodlands. These forests remain largely stable in the face of frequent pasture fires in adjacent grasslands, but have been severely cut in certain periods for logging and maize cultivation.[17] The Sakalava people, united under several historically powerful dynasties, occupy much of this zone; other regional ethnicities include the Bara pastoralists of the south center, the Tsimihety rice-cultivators of the

17. The boundaries of many western and highland forests have remained stable, contradicting assertions that pasture fires eat away at the forest edge (e.g., Chauvet 1972; Roffet 1995). Philippe Morat studied 400 km of forest edge near Ankazoabo, between 1949 and 1970, and found only three cases of forest decline, the largest of which covered a mere 32 ha (Koechlin et al. 1974). Bertrand and Sourdat (1998, 137) report a study in the Menabe region that documents only minor forest reductions in the past half-century. A more recent satellite-based study in the northwest also documents forest stability (Andrianarivo 1990). There is disagreement, however, over the state of the Ambohitantely forest in the northwestern highlands. Some see clear evidence of forest loss due to fire (Ratsirarson and Goodman 2000); others document relative stability (Edwards et al. 2002). There is agreement, however, about forest loss in some western forests cleared for export maize cultivation or logging (Genini 1996; Raonintsoa 1996; Réau 1996; Smith 1997; Seddon et al. 2000).

north center, and the Antankaranana of the far north. Aside from denser settlements around coastal river floodplains (e.g., the Boina or Menabe zones) population densities are low: in the vast grasslands of the interior west, they average 2 to 5 people/km^2. The economy is strongly agricultural and pastoral: the Boina basin around Mahajanga produces significant rice, cotton, and tobacco, while the interior grasslands are cattle-raising zones.

Finally, the *arid south* is a semiarid lowland zone marked in many places by a unique vegetation formation often called the "spiny forest." Precipitation is erratic and averages under 600 mm/yr; average annual temperature is around 25 °C. This is a smaller zone, covering less than one-twentieth of the island, and also less densely populated. The principal people of this zone are the Antandroy, historically nomadic pastoralists (sheep, goats, and cattle), and the Mahafaly farmers (cassava, sorghum, beans, but little rice).

Research Approach and Case Study Sites

There are many ways one could investigate the fire problem on this diverse island. I needed to understand how and why rural Malagasy used fire in different regional contexts, and how this affected the land and vegetation. Also, I had to trace the government's antifire views and their roots. Finally, I needed to see how the conflict over fire played out both historically and in present-day, on-the-ground contexts. To address such a variety of objectives required an array of sources and methods, ranging from sifting through colonial archives, to supplementing the scientific literature on fire with field observations, to interviewing fire setters and participating firsthand in their daily lives.

As a result, this book is based on a wide range of sources, from village-level case studies (combining participant observation, household surveys, and interviews with fire mapping and the analysis of air photos and repeat landscape photos), to interviews of government and international agency staff (including twenty-two Forest Service agents in ten offices), to detailed archival and library research.[18] Each type of information served to cross-check against potential biases inherent in other methods. For example, the colonial archives privilege the voices of the Forest Service over those of rural fire setters, so oral history interviews became central to balancing the story.

18. I relied on libraries and archives in North America (Yale, Berkeley, Northwestern, McGill), France (the overseas colonial archives in Aix-en-Provence, and CIRAD in Montpellier), Norway (Norwegian Mission Society in Stavanger), and, most importantly, in Madagascar (National Archives as well as Académie Malgache, Musée d'Art et Archéologie, and CIDST, among others).

In my seven visits to Madagascar from 1992 through 2003, I spent a total of twenty-two months in different corners of the island. While this book addresses fire on the island as a whole, it highlights one major and seven minor case studies from my fieldwork (figure 1.5). I lived in the village of Afotsara, the major case study, from March through December 1998 and have visited four times since. I chose a long presence in this study site for two main reasons. First, I wanted to monitor an entire fire season to avoid a snapshot impression. Second, villagers are cautious when talking to outsiders about fire, as it is a very politicized issue. They may seek to avoid the subject, avoid casting their community in a bad light, or may feed a researcher lines borrowed from state antifire propaganda. A lengthy stay in Afotsara allowed me to gain the confidence of my informants, to be able to compare words with actions, to triangulate among sources, and to gain experience in cultural nuances for the other case studies. While I cannot claim complete authority in my interpretation of these interactions, I do hope to have gotten the best information possible.

Seven other case study sites figure prominently in this book. Some were chosen due to a history of my own or others' research in the place (see, e.g., the list of research in Donque 1979); others were chosen for their interesting characteristics; yet others grew out of invitations to collaborate with extant projects. The sample is arbitrary but reflects a diversity of cases around the highlands and eastern Madagascar.

I should note that as the use of fire remains a politicized issue in Madagascar, I give pseudonyms to those case study sites that are specific villages (as opposed to broader regions). I do not specify the location of these case study sites more precisely than a region of approximately twenty kilometers radius. I describe each of the case study sites below; they are described in further detail in Part 2.

The primary research site, Afotsara, was chosen because it included not only ample frequently burned grasslands, but also significant *tapia* woodlands. My hypothesis was that the existence of these woodlands, rare examples of native forest vegetation in the Malagasy highlands, indicated the presence of some sort of community-level fire protection arrangement that could provide a basis for islandwide fire policy. As we will see in chapter 4, the hypothesis was partially mistaken. Management does exist, but these pyrophytic woodlands are actively managed *with* fire, not protected from fire.

Afotsara lies in the 30 km long, straight valley formed by the Manandona and Sahatsiho Rivers south of the highland city of Antsirabe (Razanadravao 1990). The valley is bounded by the rocky heights of the quartzitic

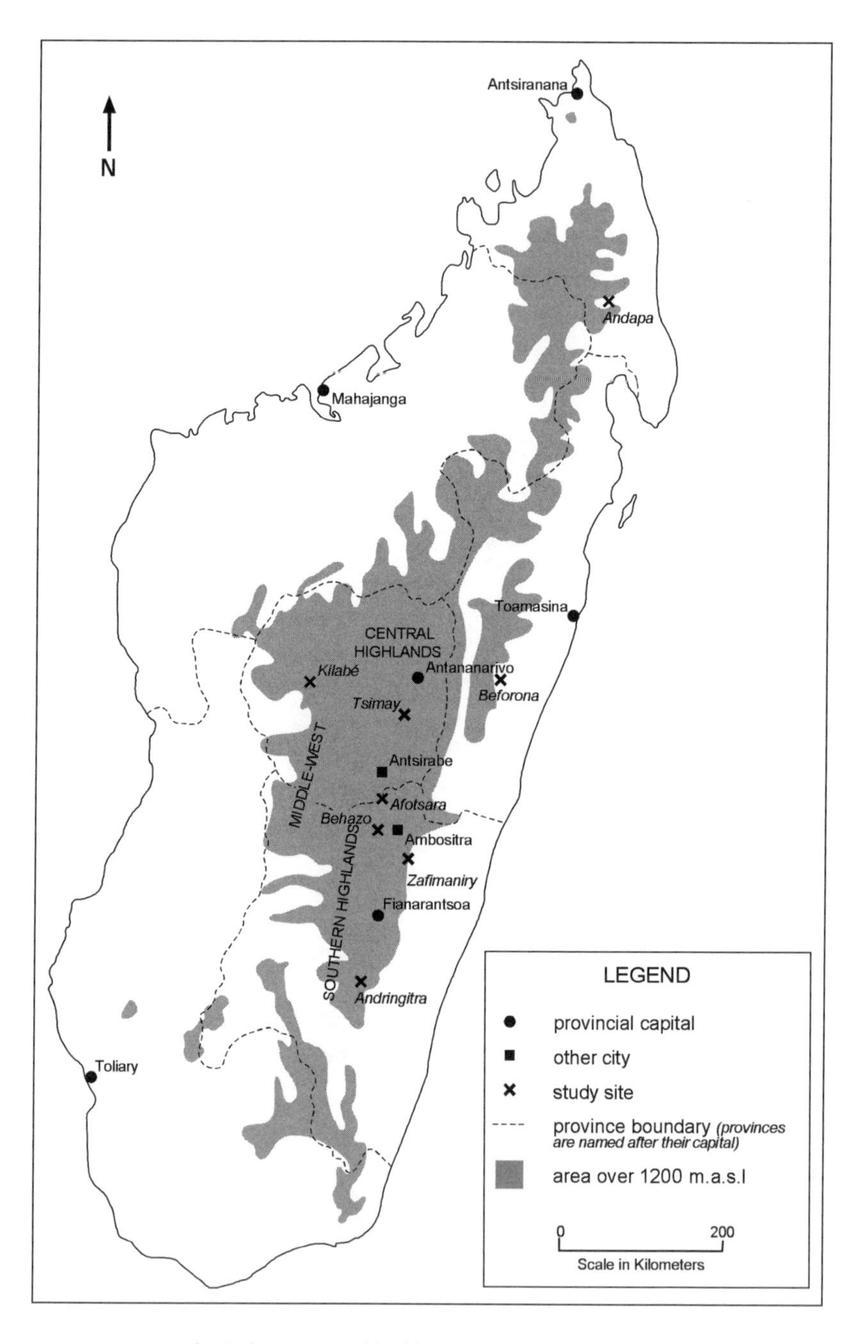

FIGURE 1.5. Study site locations and highland areas in Madagascar.

Ibity massif to the west, reaching 2,200 m.a.s.l., and the undulating, cool 1,700 m.a.s.l. Sahanivotry plateau to the east. Flowing at about 1,350 m.a.s.l., the Manandona River runs south until it is joined by the Sahatsiho, which arrives from the south, and together they breach the Ibity massif to the west in a dramatic gorge. Settled for at least 500 years (Raharijaona 1994), the valley spans a variety of landscapes with contrasting fire regimes. The valley bottom features intensive irrigated cultivation, while nearby hillsides and mountains support open pasturelands, rain-fed crops, private and government-sponsored pine plantations, and the aforementioned *tapia* woodlands.

The Manandona-Sahatsiho valley is bisected by the boundary between Fianarantsoa Province to the south and Antananarivo Province to the north, and includes three *communes rurales:* Manandona, Sahanivotry, and Ambohimanjaka-Sahatsiho. Together, the three communes are inhabited by 19,912 people, distributed over 564 km^2 (1993 census). Overall population density, including mountainous and outlying areas, is 35 people/km^2, yet in the densely populated central valley areas, density sometimes exceeds 100 people/km^2. Afotsara itself includes some 1,000 residents. They are distributed between a tiny central village, widely scattered hamlets, and isolated individual homesites.

Kilabé, the second field site, which I visited in 1994 and 1999, is located within a long day's walk southeast of the administrative town of Tsiroanomandidy, home to the country's largest cattle market. This region of the highlands, the Middle West, is known as the island's principal ranching zone, but also as a zone of agricultural colonization (AGM 1969; Raison 1984; Randrianatoandro 1990; Kull 1995). Long disputed between the highland Merina and coastal Sakalava royalties, this region became the grazing lands of the Merina royal herds in the 1800s.[19] During French colonization the area received an influx of settlers from around the island and several colonial beef cattle ranches were established, including the 1,000 km^2 Rochefortaise concession. Most of the ranches disbanded after Independence, though part of the Rochefortaise concession served as the Ferme d'Etat Omby, a state model ranch, from 1968 through 1999.[20]

Today, the Kilabé region consists of a scattering of villages founded by former employees of the concessions and filled by in-migration from across

19. J. P. Raison, personal communication, 1 Apr. 1999.

20. The Ferme d'Etat Omby was dysfunctional for most of the 1990s. Since its establishment, people have farmed and disputed land tenure within the Ferme's 17,000 hectares. The fences were stolen by the 1980s, and the last cattle sold off in 1999.

the island.[21] The isolated villages are tight clusters of houses found on the flat hilltops. Each village is surrounded by a few mango and eucalyptus trees. From a distance, the impression is of a vast plain of grass, only occasionally interrupted by villages and nearby hilltop fields of cassava or corn. The dense, dendritic network of stream valleys that cuts through this plain is hidden from view. These streams are lined with thin riparian forests; the floodplains and side valleys are increasingly developed as rice fields. Altitudes are moderate—only 800 to 1,000 m.a.s.l.—and the climate is warmer and sunnier than in the central and southern highlands. Population density is low, only 12.5 people/km^2.

Agriculture and ranching share the top subsistence and economic positions for regional households. Excess rice, corn, and cassava are sold for cash. Kilabé's population of 300 oversees about 300 cattle, though the distribution is very skewed as elsewhere in Madagascar: only one-third of households hold any cattle. Half of these practice a form of contracted cattle raising called *dabokandro*. An investor in Tsiroanomandidy buys a herd of twenty-five to thirty young steers from the breeding zones west of the Bongolava mountains and contracts with locals to fatten the cattle. The contractee benefits from the manure and trampling labor (for rice field preparation) and receives half the profits when the full-grown steers are sold (see also Cori and Trama 1979). Due to the importance of extensive cattle raising in the Kilabé region, pasture fires are a pervasive feature of the landscape.

Tsimay, in contrast, is located in one of the least burned regions of the island. Located within fifteen kilometers of Ambatolampy, a city of 20,000, the village and its scattered hamlets sit in a hilly landscape with irrigated rice fields in all drainages. I chose this case study because of the unusual condition of the hillsides: instead of being vegetated by grasses as elsewhere, they are covered by a brushy mix of shrubs like *rambiazina* and mimosa (wattle) saplings, as well as woodlots of coppiced eucalypts, indicating a relative lack of fires. This woody vegetation cover began to replace grasslands in the middle of the twentieth century, due to growing demand for wood fuel (see chapter 3).[22]

21. In the *commune rurale* of Kilabé, the population represented a dozen ethnicities, most significantly 25 percent Betsileo, 25 percent Merina, and 10 percent Tandroy, as well as Betsimisaraka, Antaisaka, and Antaimoro.

22. Late nineteenth-century writers (e.g., Girod-Genet 1898) describe Tsimay as a region of grassland with occasional forest islands, while air photos show that in 1950, the area was about 80 percent pasture, 10 percent eucalyptus, and 10 percent unclassified (air photos consulted at FTM, Antananarivo). By the 1940s, however, mimosa already covered some hills in the region, like

The accentuated hills of the region reach from 1,500 to 1,700 m.a.s.l., and the 2,600 m.a.s.l. Ankaratra Mountains to the west block the region from warmer western climates. As a result, the dry season is moderated by cool temperatures, occasional drizzle, and winter fog. All the same, annual precipitation there is similar to that elsewhere in the highlands, about 1,500 mm/yr. Population density is high, reaching 135 people/km^2. The rural economy is based on diverse, intensive smallholder farms, including rice-centered agriculture (and potatoes in higher elevations), wood fuel and charcoal production, dairy production, and animal husbandry. Fieldwork in this zone took place in 1999.

The next field location, Behazo, located in the Andina valley 20 km west of the city of Ambositra, is one of the most densely settled places in rural Madagascar (approximately 200 people/km^2). Behazo spans from 1,250 and 1,700 m.a.s.l. and has a typical highland climate. A long process of agricultural intensification has created an intricate smallholding farm landscape (Raison 1970; Kull 1998). Rice terraces climb the hillsides, fed by an intricate network of canals. The hillsides are a kaleidoscope of different uses, flush with woodlots, orange orchards, cornfields, and other agricultural lands. Hedges along paths and fields, relatively rare in the highlands, here present a diversity of plants from raspberry, blackberry, and mulberry to sisal, *ambiaty*, eucalypts, mimosa, guava, and even some *tapia* (*Uapaca bojeri*) spread by perambulating snackers. The economy is a diversified mix of field crops (grains, vegetables), fruits (especially oranges), and livestock. I include this case study because, like Tsimay, fire plays little role in this landscape. Unlike Tsimay, where market incentives led to a reduction in fire, Behazo's lack of fire is more directly related to demographic pressure. Research here is based on four visits from 1994 through 1999.

Only some fifty kilometers southeast of Behazo, the Zafimaniry region represents a completely different type of landscape. Straddling the dramatic escarpment between the Betsileo and Tanala people, the hundred-odd Zafimaniry villages practice slash-and-burn cultivation in remaining rain forests and on hillsides covered with *savoka* secondary vegetation. This humid region is cool because of its altitude (between 1,000 and 1,800 m) and its persistent orographically caused cloud cover. It receives 2,000 to 3,000 mm of

---

around Anosibe further to the northeast of Ambatolampy (ANM D81s). Topographic maps, based on air photos from 1959 and 1968, show a band of forest islands (probably eucalypts) and wooded savanna (probably mimosa and *rambiazina*) in a semicircle 15 km south, west, and north of Ambatolampy.

rain per year, with only about 100 days free of rain annually. The Zafimaniry once grew only maize and beans on their forest swiddens; now they also cultivate some irrigated rice due to the near exhaustion of available forest and *savoka.* This case study, which illustrates the use of slash-and-burn *tavy* fires and their effects on vegetation cover, is based on the work of Coulaud (1973; see also Bloch 1995) and a field visit in 2001.

The Andringitra case study represents an oft-burned cattle-raising zone like Kilabé, but with higher altitudes and, crucially, with the presence of a national park and an associated integrated conservation and development project. The dramatic Andringitra mountains, culminating at 2,658 m.a.s.l., form a granitic promontory at the southern terminus of the central highlands. Here, the highlands narrow to a 20 km wedge between the humid forests of the east and the dry, warm rangelands of the west. The colonial government decreed 31,160 ha in the mountain range a Strict Nature Reserve in 1927; it assigned a team of four Forest Service agents to oversee the reserve and enforce prohibitions on entry, burning, deforestation, and cattle grazing. Enforcement in this vast, mountainous, and remote area was difficult and necessitated compromises with the local people (Ratsirarson 1997).

In 1994, a multimillion dollar integrated conservation and development project, funded by Germany and administered by WWF, began to operate in and around the reserve. In 1998, the project achieved its goal of changing the statute of the reserve to national park, which would allow tourism, grazing, and controlled burns in part of the park. The project also worked closely with local communities, for example establishing a pact regulating fire use and creating pasture zones with management plans in the park periphery (Rabetaliana et al. 1999; Rabetaliana and Schachenmann 1999).

In multiple visits from 1994 through 1999, I researched a transect extending north from within the park, through the farming villages of the Namoly basin, to the broad upland pastures of the Fivanonana plateau. In this zone ranging from 1,400 to 1,700 m.a.s.l., with pastures up to 2,000 m.a.s.l., farmers combine agriculture and cattle raising in their quest for livelihoods. While overall population density is low, at 28 people/km^2, most people are concentrated in the Namoly basin. Here, valley bottoms are cultivated in rice, and *tanety* hillsides are sprinkled with dryland crops, woodlots (private and communal), and pastures. Upland areas—such as the Fivanonana plateau, the Andohariana plateau within the park, or the Ioramaro plateau between the forested ridges to the east—provide vast areas for pasture. This region contributes to the important cattle market at Ambalavao, 50 km to the north. Pasture maintenance fires are crucial to the extensive cattle-raising economy, yet the presence of the park has necessitated some interesting negotiations, conflicts, and compromises (see chapter 3 for details).

The Beforona case study is based completely on the work of a Swiss-sponsored multidisciplinary research team (Brand 1999; Brand and Pfund 1998; Messerli and Pfund 1999; Messerli 2000; Pfund 2000). Their detailed study followed a transect along the Antananarivo to Toamasina highway from the mountainous rain forest frontier (around 1,000 m.a.s.l.) down to the denuded low hills closer to the coast (around 300 m.a.s.l.). The tropical humid climate here is characterized by annual temperatures averaging 19 to 23 °C and year-round rainfall totaling 2,000 to 3,500 mm. Betsimisaraka farmers live in this region; population density ranges from 26 people/km^2 in the forest frontier area down to 19 people/km^2 in the longer-settled cleared areas. Farmers here practice *tavy* slash-and-burn cultivation and supplement this strategy—especially in areas where good forest or *savoka* no longer exist—with irrigated rice and market crops such as coffee and bananas. I include this case as an interesting window into the logic and consequences of *tavy* (see chapter 5).

Finally, the Andapa case study investigates rain forest cultivators and fire setters in the island's north, in the so-called "vanilla triangle." This case is based on a field visit in 1994 and on the literature (Neuvy 1979, 1989; Sakaël 1994; Laney 1999, 2002; Garreau et al. 2001). The humid Andapa basin, at around 500 m.a.s.l. and fully developed for irrigated rice cultivation, is surrounded by high, forested mountains up to 2,100 m.a.s.l. that have been declared protected areas (Marojejy National Park and Anjanaharibe-Sud Special Reserve). While the whole area has a tropical humid climate, rainfall and cloud cover vary significantly by location, due to both orographic uplift and rain shadow effects. In the basin itself, rain is frequent, falling about 270 days per year and totaling about 2,000 mm/yr; while just east of Marojejy mountains, there is less constant cloud cover but higher rainfall. The area was settled largely by Tsimihety migrants, beginning especially in the 1920s and surging after a road was built in the 1970s. Current population densities vary from 50 to 100 people/km^2 in the Andapa basin to under 10 people/km^2 in some surrounding areas. Smallholder production of vanilla, and to a lesser extent of coffee, dominates many people's livelihood strategies—the export economy is far more developed here than in many other areas of the island. People also cultivate irrigated rice in the basin and—until a 1994 crackdown by foresters and conservation agents—burned *tavy* clearings deep into the forest.

EACH OF THESE eight case studies illustrates a different aspect of Madagascar's fire problem. Drawing on these case studies, as well as a plethora of archival, interview, and documentary sources, I will make my case over the next seven chapters that the fire problem has been misunderstood. I will demonstrate how the government's views of fire were misguided and led to a century of conflict. By looking at the landscape from a rural farmer's perspective, I will show the many instances in which fire plays a crucial role in both seeking a livelihood and in environmental management. Finally, I will analyze the conflict over fire in detail, discovering why it has persisted so long and suggesting possible solutions. In the next chapter, I continue to lay the groundwork for understanding the fire problem by investigating in detail the nature of fire and differing views of it, introducing both the antifire received wisdom and fire-friendly modern ecological theories.

2 [THE NATURE OF FIRE

Bad Fire, Good Fire, Complex Fire

I know little more captivatingly beautiful than to see a Malagasy landscape, mile after mile, decorated with endless gold threads in all different patterns. Where during the day one sees mountains, now one sees crescents of fire, apparently rising up in the air, some standing on their legs, others lying on their backs. The plains and the hills are dressed up with rings, spirals, and wreaths often as far as the eye can see. And all of it continuously changing and moving, each moment new patterns and unexpected discoveries. The brightness varies with the wind speed and the grass's texture. A quiet, still fleck of light can suddenly burst up in the air like a wave hitting a rocky shore. . . . Night after night, week after week, we can enjoy this play, our national theater.

J. EINREM, NORWEGIAN MISSIONARY, 1912, 71

WHAT IS THE NATURE of fire? Is it an evil, primitive, destructive force, ruining lands and degrading environments, or is it a natural and well-adapted resource management tool used successfully around the globe? This chapter looks closer at the nature of fire, both in general and in the context of Madagascar. It examines why many colonial leaders disapproved of fires and details the challenges to their views that have come from range managers, ethnographers, and ecologists, among others. It also analyzes the specific characteristics of fire, for example its self-propagation. I combine these to argue that fire is certainly not a universal "bad" as often claimed by some colonial administrators and naturalists. However, far from always being a clear "good," fire is a complex and multifaceted process and tool with significant environmental consequences. It is self-propagating and unpredictable, and its effects and meanings are context dependent; it is easily used anonymously and can serve multiple purposes; it is highly visible and destructive, yet simultaneously constructive.

This chapter sets the stage for the book's argument in two ways. First, it highlights the ideological and material underpinnings of both the prevalent antifire received wisdom and its critiques. This background is crucial to understanding why the state has long criminalized fire and why rural land users resist such controls. Second, the chapter's investigation into the specific characteristics of fire lays the foundation for my later assertion that fire's ambivalent agency, easy anonymity, and complex nature helped the peasants resist state control.

Let me begin with a note on terminology. I use the term *fire* as shorthand to refer to fires occurring in the standing or cut vegetation and litter of croplands, grasslands, and forests. It includes pasture burning, forest fires, wildfires, crop field burning, and so on, and I will use these more specific terms when appropriate. I also employ the Malagasy term for slash-and-burn agriculture, *tavy*. The French often use the label *feu de brousse*, literally "brush fires," to denote practices as varied as grassland burning and rain forest swidden cultivation. The current frequently used American term "wildland fire" makes little sense in the case of Malagasy grassland and woodlands burning, since few, if any, of the lands in question are truly wild.

HUMANS BEGAN TO master fire some one million years ago, first by simply controlling and guiding natural fires, then by harnessing and carrying these fires, and finally by discovering how to light their own fires (Stewart 1956; Sauer 1975; Schüle 1990). They observed that the animals they hunted congregated on the flush of new grass after lightning-strike fires, and they mimicked this process. Ever since, throughout the world, throughout history, humans have relied on fire as a simple, effective, and easy tool to manage their environment (Thomas 1956; Batchelder and Hirt 1966; Pyne 1995). "Where vegetation would burn, people burned it" (Myers and Peroni 1983, 217). "When unrestrained," Denevan (1992, 372) notes, people everywhere "burn frequently and for many reasons." Reasons vary from wildfire prevention in California's mountain forests to swidden crop field preparation in the rugged hills of Laos.

Perhaps the most famous fire setters are the Aborigines, whose fire-stick farming manages most of the landscape of interior Australia. They burn about two-thirds of the land in the Northern Territories annually, avoiding fuel buildup and improving hunting and gathering resources (Gill et al. 1981;

Lewis 1989; Braithwaite 1991; Pyne 1995; Head 2000). But fire is not restricted to Australia. In historical times, English aristocrats burned moorlands regularly to manage grouse habitat (Maltby et al. 1990), eighteenth-century settlers in the Appalachian region relied on slash-and-burn agriculture (Otto and Anderson 1982), while Native Americans burned from California to the Great Plains to improve ungulate grazing, to clear underbrush and protect villages, to manage forests (e.g., redwoods, Douglas firs, and oaks), to create forest openings where plants useful for medicine and basketry could grow, and to improve hunting (Huntsinger and McCaffrey 1995; Stewart 2002; Vale 2002). In California, up to 13 percent of all nondesert lands—including forests—may have been burned yearly before white settlement (Martin and Sapsis 1991).

Diverse burning practices continue today, as ranchers and pastoralists in Kansas, South Africa, and Tanzania burn to manage their rangelands, as foresters in the southern U.S. or in India burn to encourage certain forest habitats, or as Brazilian peasants burn to improve pasture and prepare crop fields (Homewood and Rodgers 1991; Cahoon et al. 1992; Mistry 1998; Hill et al. 2000; Mbow et al. 2000; Ramos-Neto and Pivello 2000; Suyanto et al. 2001). The Gitksan and Wet'suwet'en peoples of northwest British Columbia even burn patches of wild huckleberry and blueberry to stimulate new growth and berry production and to impede conifer invasion (Gottesfeld 1994). Satellite analyses confirm that both anthropogenic and natural fires burn all over the globe each year (Dwyer et al. 1999).

Fire is a complex process, with diverse causes, purposes, and effects. Some effects are predictable, others not; some effects are intended, others not. Fires cause a transition in vegetation communities, they volatilize some nutrients and liberate others, and they can destroy homes (think of southern France or Colorado) and foul the air (think of Indonesia).

Figure 2.1 attempts to classify types of fire. By avoiding simplistic categories, this diagram emphasizes fire's complexity in three ways. First, agency is ambiguous: the cause of fire can be human or nonhuman, and human-ignited fires can be purposeful or accidental. Second, fire can have a variety of effects, which are good or bad depending upon one's interests. Third, the link between cause and effect is neither straightforward nor always predictable. To illustrate, a green flush of new grass in a rangeland may be caused by lightning, a planned prairie fire, or an escaped campfire. Likewise, any particular cause of fire—whether lightning, a prescribed burn, or an accidental ignition—can have multiple effects, from destroying woodlots, to renewing pastures, to incinerating a house.

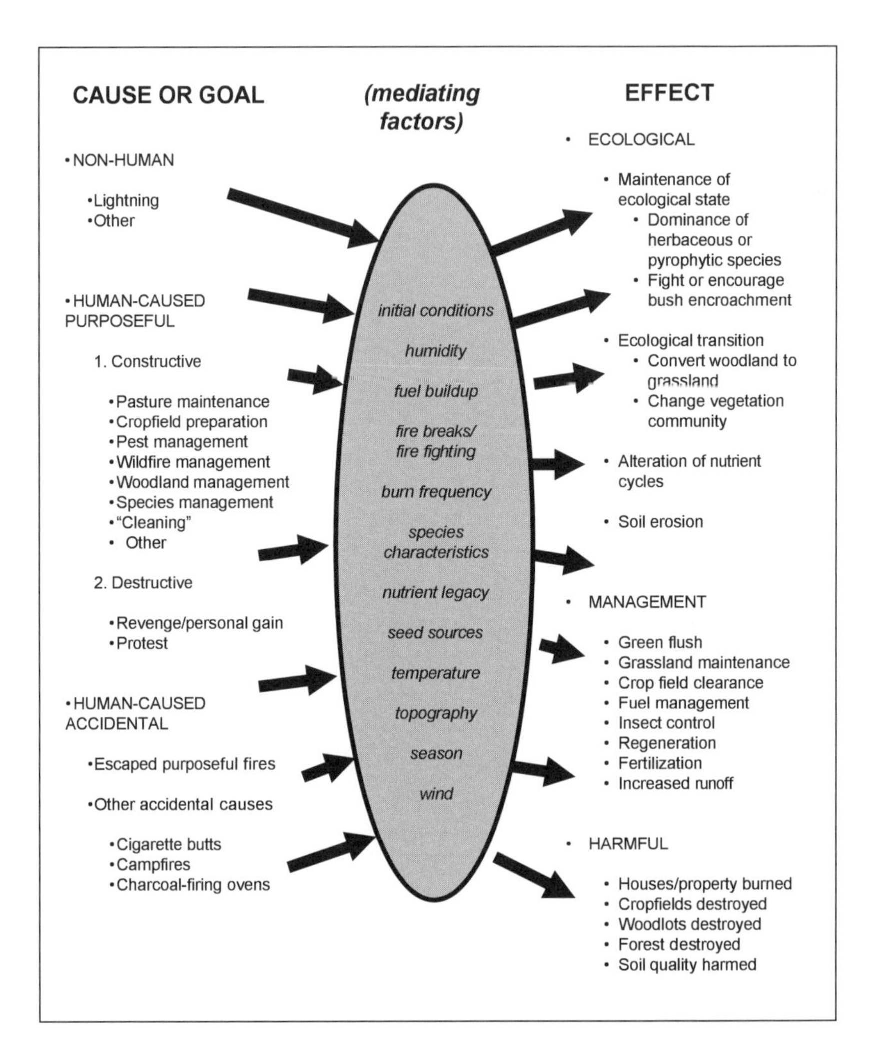

FIGURE 2.1. **Classification of fire types.** Fires can be classified by cause or by effect. Effects do not directly relate to causes, but instead to the physical initial conditions and mediating factors. People who light purposeful fires use their understanding (based on experience or research) of the initial conditions and of the resulting likely outcomes to predict the effects of their fires.

BAD FIRE

Despite this complexity, fire has often been seen as simply bad. In particular, as Europeans set out to conquer the rest of the world, they (or at least their intellectual, urban-based ruling classes) generally saw the fires used by farmers, hunters, and herders as destructive (Pyne 1997). To the state and elites, fire was a primitive, wasteful technique and a threat to economic assets and infrastructure like exploitable timber, cash-crop plantations, or buildings. In this section, I trace the origins of these pervasive negative views of fire.

Six factors underlie what Pyne (1995, 312) calls "Europe's pyrophobia." First, global power over the past two centuries was primarily centered in north-central Europe and eastern North America. Paris, Berlin, London, and New York are all located in subhumid temperate climates, where natural fire is quite rare and pastures remain green and nutritious throughout the growing season. In these densely settled regions, established agriculture relied on plows and labor, not fire. Land use had become intensive and industrial, leaving agricultural fire to frontier regions. As a result, these areas only knew destructive fires and saw agropastoral fires elsewhere as errant techniques.

Second, these same countries developed self-serving ideas of the stages of civilization. They observed that "civilized man" in the "advanced" countries did not use fire, while the "natives" of Africa, the Americas, and tropical Asia did use fire. From this, the seemingly logical conclusion arose that the use of fire was a "primitive," backward technique (Bartlett 1956).

Third, the negative view of fire came out of the urbanity of most leaders, academics and policy makers. Not only were they removed from the realities of rural life, but the only fires they did see, beyond cooking, heating, and blacksmithing fires, were dangerous house fires or even citywide conflagrations (Pyne 1997). As rural landscapes became more and more romanticized into aesthetic resources, black charred slopes were easily condemnable as scars in the landscape. This continues today, for example in the suburbs or urban-wildland fringe of California, where wildfires are extreme threats to houses yet residents oppose prescribed preventive burning as unaesthetic, damaging to property values, and potentially dangerous (Davis 1998).

Fourth, contemporary ecological thinking supported these ideas. The field of ecology in the early and mid twentieth century was based on notions of orderly succession, climax, and equilibrium. As a result, disturbances like fire were seen as negative, external shocks. Fire blocked "normal" succession towards vegetation climax and was therefore bad (Zimmerer and Young 1998; Scoones 1996, 1999).

Fifth, bad fire originated in the imperatives of states. In order to exert political control, states gain advantages from straight lines, single purposes, simple logics, and legibility (Cline-Cole 1996; Scoones 1996; Scott 1998). For economic advantage, states prefer large-scale, intensive, revenue-oriented land use. Both these political and economic preferences work against the use of complex fire, which necessitates opportunistic, not centrally planned management, and which, aside from sugar cane plantations, is employed only in more extensive land use systems.

Finally, as elites profiting from the natural resources of the colonies, Europeans saw fires wasting the wealth of the colonies—the soils and the timber. Such concerns grew out of domestic experiences. Government foresters saw trees as valuable economic resources and as tools against environmental degradation and thus tried to stop all fires. France in particular had a long tradition of state forest control, beginning with the Code Colbert of 1669. The French forest code of 1827 and later laws rested on the principles of state regulation of all forests for timber revenue production and for protection from devastating Alpine floods, shifting Atlantic coastal dunes, and Mediterranean fires (Kalaora and Savoye 1986; Prochaska 1986; Bergeret 1993; Sahlins 1994; Pyne 1997). While some local foresters in the fire-prone Mediterranean protected their forests from wildfires by systematically burning the undergrowth with *petits feux*, the forest bureaucracy, trained at the Nancy forestry school, only saw fire as destructive. As Pyne notes (1997, 115): "Obviously the practice [*petits feux*] had evolved out of agricultural traditions, and inevitably intellectuals condemned it."

If forests were economic profit and salvation from soil degradation, the logic went, then fire, which destroys forests, must be stopped. The French perfected measures of fire control, including severe regulations, a paramilitary forest service, firebreaks, hilltop lookouts, patrols, and fire fighting (Pyne 1997). It is this ideology of bad fire and of the need for fire suppression that the French colonial administration carried overseas. British foresters—trained in German and French forestry schools and hardened by inflammatory experiences in India—carried similar antifire ideas from colony to colony. Fire was their chief enemy, and fire control their most important task (Pyne 1997; Sivaramakrishnan 1996).

For these varied reasons, urban intellectuals came to fear fire and decry the use of this primitive technique. As colonial scientists observed environmental transformations such as deforestation and desert expansion, they elaborated increasingly detailed ideas about the precise role of fires in the degradation. For example, E. P. Stebbing, from the Indian forest service, toured West Africa in the 1930s and concluded that overgrazing and fire were causing dramatic levels of desertification (Swift 1996). Likewise, slash-and-burn agriculture was denounced as "wasteful," "barbaric," or "irresponsible" forest management leading to deforestation, erosion, and flooding (Bryant 1998). Across the board, scientists accused fire of impeding forests that were seen as the "natural" or climax vegetation, of nibbling at existing forest edges, of degrading rangeland quality, of causing erosion and soil degradation, and even of causing climate desiccation and reducing water supplies (Humbert 1927; Kuhnholtz-Lordat 1938; Aubréville 1952). This was the basis of an antifire received wisdom that held wide currency throughout the century in many parts of the world, like Côte d'Ivoire (Bassett and Koli Bi 2000), Costa Rica (Horn 1998), Ghana (Amanor 2002), and elsewhere (Pyne 1997).

"BAD FIRE" IN MADAGASCAR

In Madagascar, perhaps more than in any other tropical nation, this bad fire view became an unchallenged orthodoxy. While fires also burn in places as varied as Kenya, Guinea, and Haiti, they have always been particularly forcefully disparaged in Madagascar.[1] Perhaps, as Grove (1995) noted, tropical island edens are crucibles for the development of environmental ideas. The criticism of fires in Madagascar has even become central to political discourse (figure 2.2). The development of this antifire received wisdom is closely linked to ideas about the island's environmental history and its current environmental predicament; indeed it both contributed to and gained strength from those ideas. This section traces those ideas.

The writings of some explorers and missionaries show the roots of the received wisdom. While some early visitors matter-of-factly described indigenous agricultural techniques (e.g., Flacourt 1658; Ellis 1838, 1859; Oliver ca. 1863; Mullens 1875; Moss 1876), others marveled at the biological wonders of the island but criticized native burning practices. For example, missionary Baron wrote, "It is grievous to relate, however, that the forests of Madagascar are being destroyed in the most ruthless and wholesale manner by the natives" (Baron 1891, 324; see also Baron 1890; Elliot 1892). Others lamented the treeless "barren hills" or "desert" of the oft-burned highlands (Sibree 1870; Price 1989).

1. The recent strategic document for the third phase of the island's environmental action plan states "in Madagascar, the removal of vegetation is of a much more serious nature than most parts of the world." It goes on to blame the soils for this unique susceptibility to "total forest destruction," citing Humbert (1927) as its source (RDM 2002, 5n).

FIGURE 2.2. **Political cartoon of current issues, including fire.** The pilot of the plane crossing Madagascar says, "I can't see anything due to the greenhouse effect!" Signs depict the issues: a bubonic plague epidemic, the economic crisis, embarrassment over election results, and, of course, bushfires. From *L'Express de Madagascar* (23 Nov. 1996, 3).

The French colonizers took up this criticism of indigenous burning practices.[2] Their reaction was not only aesthetic, cultural, and moral, but also material. Colonial administrators were charged with creating a profitable colony, and they saw fires threatening the material riches of the colony: exploitable forests and rich grasslands. Colonial commentators in particular deplored *tavy* as an affront on the colony's wealth. Some probably accurately saw *tavy* as a means by which peasants asserted their control of forests in the face of state appropriation (Coulaud 1973; Jarosz 1996). Forester Saboureau (1950, 1709) argued for state-run sustainable forest exploitation in order to protect the national patrimony and that *tavy* was not necessary for farmers' livelihoods in forest areas. In a similar vein, Caltie (1914, quoted in Dandoy 1973, 29) wrote:

2. See especially the articles in the journal *Colonie de Madagascar: Notes, Reconnaissances, Explorations*, e.g., Chapotte (1898), Charon (1897), Girod-Genet (1899), Lafort (1897), Michel (1897), and Thuy (1898). These articles, as well as Catat (ca. 1895), are summarized in Bartlett (1955).

> I believe we should completely ban *tavy* and repress it with all possible means. Without a doubt, the natives' work . . . is antieconomic. The timber and secondary forest products have a much higher value than the few baskets of rice harvested by the Betsimisaraka.

Some critics of burning practices stretched these arguments of inefficiency or wastefulness into condemnations of indigenous attitudes and cultures. Petit (1934), for example, argued that *tavy* was not only the ruin of the forest, but also of society; he implied that *tavy* was a backwards way of life leading to poverty and starvation. The head of the Forest Service in the late 1950s, Kiener, painted *tavy* as a miserable livelihood, with negative effects on nutrition, physiology, and social development (Bertrand and Sourdat 1998, 36). This argument was repeated by Ratovoson (1979), who said that the isolation caused by *tavy* engendered illiteracy, ignorance, and the perpetuation of sorcery. Worse yet, Saboureau (1950, 1709) stated clearly that "*tavy* . . . is more than anything else the result of laziness." The biases of European or urban leaders for permanent, intensive agriculture and market-oriented activities, or simply anti-indigenous racist attitudes, contributed to this strand of thought.

These antifire ideas were given additional credence by the work of naturalists who asserted that fire had already destroyed most of the island's native forest vegetation. These influential ideas, most clearly and convincingly presented in the 1920s by colonial naturalists Perrier (1921, 1927, 1936) and Humbert (1927, 1949, 1955), assert that:

> A wide variety of forests covered nearly all of Madagascar before human settlement . . . Now only a few remnants of various sizes remain. By studying these . . . one can see the alarming progress of deforestation, . . . which is caused by *tavy*, logging, and grassland fires . . . [Instead of establishing permanent plots and irrigated rice fields], the natives cut and burn the remaining forest in order to cultivate temporary crops. . . . Shortsighted commercial logging is no less harmful, impeding regeneration just like *tavy*. . . . Deforestation was begun by *tavy* in the humid regions and advanced by rapid-spread fires into the easily flammable forests of the drier regions. The destroyed forest was replaced . . . by grasses, which the natives burn annually to renew pastures for their cattle. These fires cause a retreat of forest edges and lead inescapably to the sterilization of immense areas. . . . The prairie's lateritic soils are slowly impoverished, especially in the interior . . . where the scouring of erosion is rapid after deforestation. The degradation of old pastures pushes the natives to create new ones at the expense of the remaining forest. (Humbert 1927, 77–78)

A number of assumptions underlie the above interpretation of Madagascar's environmental history. The key assumption is the idea, known as the Perrier-Humbert hypothesis, that forests and other woody vegetation forms once covered nearly the entire island.[3] Observations of *tavy* carving fire-climax grasslands out of the eastern rain forests led the naturalists to conclude by analogy that the highlands and west must once have been forested, but had become grassland sooner due to their drier, more vulnerable location. It was just assumed, based on the ecological theories of the day, that forests were the climax vegetation (e.g., Humbert 1938). The naturalists also pointed to forest islands and toponyms in grassland zones, to the legend of a "great fire" that seared highland forests, to a presumed lack of native grazing animals, to biogeographical plant and animal distribution patterns, to stratigraphic analyses of fossil beds, and to the supposed paucity of primary colonizing plant species. Based on such evidence, researchers and foresters claimed the island was once blanketed by forest. The claims ranged from cautious to bold. On the cautious side, Perrier (1921, 61) made sure to note that he meant not a pure dense forest across the island, but an islandwide forest flora including tall shrublands and xerophyllous plants. At the other extreme, Lavauden (1931) imagined dense evergreen humid forest from coast to coast.

The Perrier-Humbert hypothesis that forests once covered most of a fire-free, prehuman island from coast-to-coast gave all the more urgency to the bad fire received wisdom. As contemporary evidence confirms, the settlers of the island and the fires they lit significantly decreased forest cover and expanded herbaceous vegetation (but the extent and nature of the prehuman forest cover is still debated—see section on deforestation later this chapter). In the colonial view, if fire was the cause for most of the deforestation, then fire needed to be stopped to protect the forests—which were, after all, in the words of one administrator after another, one of the colony's main riches.[4] As Pyne (1997, 442) astutely notes, the islandwide forest hypothesis probably "had more to say about France than about Madagascar." Perrier and Humbert's hypothesis was widely reproduced, and its conceptual influence great. Where the Malagasy saw pastures, the French imagined once lush forests.

Administrators and scientists blamed fire not just for the disappearance

3. For detailed discussions, see Gade (1996), Burney (1996, 1997), Lowry et al. (1997), Kull (2000a), and Ratsirarson and Goodman (2000).

4. See circulars by Gov. Augagneur (JOM 12 June 1909, p. 399) and Gov. Picquié (JOM 10 May 1913, p. 549).

of a primeval islandwide forest, but also for soil erosion, range degradation, and desiccation. They asserted that fires reduce the amount of vegetation covering the soil, cause the formation of lateritic hardpan surfaces, and thus lead to higher rates of erosion. The result was said to be the desertification of vast areas, the loss of massive amounts of soil, organic matter, and soil nutrients, the siltation of rice fields and rivers (causing flooding and port closures), and the scarring of the land by large erosion gullies called *lavaka* (Perrier 1921; Humbert 1949; Tricart 1953; Goujon et al. 1968; Le Bourdiec 1972; Rossi 1979; Randrianarijaona 1983; Paulian 1984). Second, many called fire an irrational pasture management tool that impoverishes the quality and quantity of forage and degrades diverse tall-grass pastures (Perrier 1921; Bosser 1954; Dez 1966; Cori and Trama 1979; Neuvy 1986; Rakatonindrina 1989; Rajaonson et al. 1995). Third, some linked fire to climatic and hydrologic desiccation. Fires were accused of drying up water sources and delaying the arrival of the rains (RDM 1980; Rajosefasolo 1992; Andriamampionona 1992; Rajaonson et al. 1995; Schnyder and Serretti 1997) or of causing drought and a colder climate with damaging frosts (Parrot 1924).

As we will see, some of these concerns are legitimate, others exaggerated or misleading (Kull 2000a). Together with the Perrier-Humbert hypothesis, however, these ideas contributed to the entrenchment of a strong antifire received wisdom, a kind of ideology that provided the justification for years of antifire politics and that continues to block consideration of fire as a legitimate resource management tool. For these reasons, fire was seen as a scourge that "handicaps the progress of agriculture, the rapid economic development of the nation, and the security of the population" (RDM 1980, 1). Despite some of the scientific uncertainty accompanying this antifire received wisdom, it persisted and gained strength in ways that are consistent with other environmental orthodoxies elsewhere, like those regarding desertification or overgrazing (Beinart 1996; Leach and Mearns 1996; Swift 1996). It persisted because it was plausible to those educated in scientific forestry and equilibrium ecology, because it was rooted in colonial economic imperatives, because it became institutionalized in the state bureaucracy, and because it helped to justify external intervention in resource management, perpetuating views of locals as incapable. This is the case in Madagascar (see Bergeret 1993).

The antifire received wisdom in Madagascar, arm in arm with the Perrier-Humbert hypothesis, had a great influence on contemporary public and scientific views. Among colonial scientists, Madagascar was considered a textbook example of the destruction of indigenous flora by fire and shifting agriculture (Kuhnholtz-Lordat 1938; Bartlett 1955). More recently, during the 1980s–1990s boom in conservation efforts, agencies frequently reiterated the long-established ideas of Perrier and Humbert. These ideas fit directly into the early proposals, reports, and policies of this era, for they corresponded closely to the dominant narrative of environmental crisis central to the legitimacy and funding of the conservation boom.

While the tone of scientific work and some other writing has recently changed, the antifire received wisdom continues to be repeated nearly word for word, in popular publications (e.g., Swaney and Wilcox 1994; Bradt et al. 1996; Holmes 1997) and by development and environmental organizations (e.g., World Bank 1988; Falloux and Talbot 1993; ONE/Instat 1994; USAID 1997a, 1997b). In the media, journalists use artistic license to dramatize Madagascar's environmental degradation, relying on images of red rivers "bleeding" into the ocean, visible even from space (Helfert and Wood 1986; Hoeltgen 1994; Apt 1996; Holmes 1997; Gallegos 1997; Morell 1999; Mitchell 2000) and on descriptions of the "gangrenous wounds" of *lavaka* (Murphy 1985, 68), also described as "gaping amphitheatres gouged from barren hills once draped with lush soil-preserving vegetation" (Swaney and Wilcox 1994, 10). These culminate in statements such as "ravaged by fire, overgrazing, and erosion, the mountains are often entirely devoid of vegetation" (Allen 1995, 4) and "more than a millennium of slash-and-burn agriculture has produced 100,000 square miles of virtually useless space where people once encountered forest" (Allen 1995, 13).[5]

Today, discussions of fire among policy makers are usually much more nuanced. Most still argue strongly for fire control, yet the argument no longer revolves around "native ignorance" but instead around the imperatives for nature conservation and economic development. Many policy makers recognize what is summarized in the conclusions of a report from the World Bank–sponsored Multi Donor Secretariat in Madagascar: *tavy* and pasture fires "are totally understandable strategies in rural societies in a phase of land colonization. They are economically efficient in terms of security and profiting from labor input. . . . but there are numerous consequences."[6] Gervais Andrianirina, research director at FOFIFA, the Malagasy agricultural research agency, opened a seminar on *tavy* with the following words:

5. For an interesting piece on how Malagasy nature was conveyed to foreign audiences through a variety of images, see Feeley-Harnik (2001).

6. From Note de Réflexion du Groupe des Bailleurs de Fonds concernant la question du *tavy* et des feux de brousse, written after a workshop held 4–5 Apr. 2001.

> *Tavy* has always been a problem—I would even say a real squaring of circle—for the Malagasy agricultural services . . . in the sense that, although considered a scourge because of its backwardness and its negative impacts on the environment, it remained almost a "necessary evil" for the peasants, who find it the most suitable solution, more secure and easier to execute in their geographical and agroeconomic situation. (Kistler et al. 2001, ix)

All the same, fire is still considered by many a grave problem not just in Madagascar (e.g., Andriamampianina 1998), but also globally (e.g., Myers 1999), and a deeper appreciation of fire as a "necessary evil" still leaves most of the bad fire argument intact. The recent strategic document introducing the PE3, the third five-year phase of the environmental action plan, states that "bushfires, lit for diverse agricultural reasons (pasture, *tavy* . . .), have not yet been eradicated" and stresses the "fight against bushfires" in its goals.[7] Likewise, new president Ravalomanana has made eradicating fire one of his priorities, stating in a speech on 28 September 2002 that fire is a chief obstacle to the development of the nation. As we shall see in chapter 8, these persistent views render difficult the solution to Madagascar's fire problem.

GOOD FIRE

While an antifire received wisdom shaped colonial policy making from Paris to India, its dominance was not total. There has always been an undercurrent of alternative ideas. For example, in the *sal* (*Shorea robusta*) forests of Bengal, the field-based foresters, local people, and the evidence of nature itself slowly convinced parts of the British colonial forest service that fire was central to *sal* regeneration (Sivaramakrishnan 1996; Bryant 1997). In fact, before the antifire view became entrenched in the late nineteenth century due to increasingly powerful interests in scientific forestry and state control, traveling Europeans were equally likely to accept fire as reject it. As mentioned above, a few early explorers of Madagascar, unfazed by the received wisdom, took at

7. RDM (2002, 5, 32). Goal number 1.2.4 of the PE3 Strategic Document is to reduce bushfires, with these specific subgoals (p. 32):

a) to sensitize and motivate the population to fight against bushfires
b) to strengthen the application of antifire laws
c) to improve the legislation for the fight against bushfires
d) to operationalize the fight against bushfires.

The proposed indicator of success is a 20 percent reduction in fires in key zones over the next 5 years (p 34).

face value the explanations of the "natives" regarding their burning practices. Rev. James Sibree (1870, 104) traveled around the island for the London Missionary Society, and wrote, "In many places great patches showed where the grass and ferns had been set on fire. This is done shortly before the rains come on, and the rank course grass is succeeded by a crop of fine short herbage, suitable for pasture."

In this section, I look at the roots of a number of profire ideas. Twentieth-century Western voices that challenged the bad fire view fit into four main categories: lower-level administrators caught between peasant needs and state views; resource managers in fire-prone ecosystems; anthropologists and geographers working with fire-starting "primitive" cultures; and "new ecologists" pushing ideas about vegetation change beyond equilibrium-based, deterministic theories.

First, lower-level administrators charged with implementing antifire policies were the first to hear peasant complaints; they sometimes conveyed peasant ideas defending fire back to the central government. Richards (1985, 29), for example, describes a period in West Africa during the 1930s when populist administrators and scientists softened earlier hard-line, top-down approaches to indigenous agriculture. They even learned "the value of bush burning" and other techniques from local experience. In Madagascar, district officers resoundingly defended the peasants' need to burn after the 1900 fire ban (anonymous 1904), and the administration (especially field officers) consistently dragged its feet in enforcing unpopular fire laws, such that exceptions became the rule (Lavauden 1934; see also chapters 6 and 7 of this volume). As Dandoy (1973, 29) noted, "in reality, slowing or stopping *tavy* was impossible because [the colonial administration] could not take away the peasant's means of subsistence."

The second group that pressed for the acceptance of useful fire was resource managers. Forest wardens and ranchers, faced with the pragmatic, on-the-ground task of growing trees or maintaining pasture in a variety of situations around the world, sometimes relied on fire as an effective tool. They were inspired by their own experiences elsewhere, by experimentation, or—perhaps most frequently—by the techniques of the locals. This occurred most often, of course, in fire-prone climates with lengthy dry seasons, such as Mediterranean regions or tropical savannas. Fire, after all, was "used in the management of grasslands and savannas in all tropical and temperate regions of the world" (Vogl 1974, 489). In the case of the Indian *sal* forests, both the testimony of local people and experiences in the forest convinced some British foresters of the need for fire, and these ideas spread to the managers of teak, pine, and eucalyptus forests, who were sometimes reluctant to adopt this viewpoint (Pyne 1997).

Managers of cattle ranches, stock farms, and game reserves in colonial Africa found burning a useful, if not indispensable, tool (Gillon 1983). Their views were supported by long-term fire experiments, for example in red oat grass pastures in the Natal (Geoffroy 1931) and in highland grasslands in southern Tanzania (van Rensburg 1952). At South Africa's Kruger National Park, burning old grass was considered essential to maintain ungulate habitat throughout the first half of the twentieth century. Wild ungulates, like cattle, prefer the short grasses produced by burning—the new green shoots are nutritious, and predators are easier to see in the absence of tall grass. In 1946, a new warden restricted fires (inspired, in part, by a new national policy on soil conservation), and as a result many game animals migrated away, bushes encroached, and catastrophic wildfires ensued. By 1954, park managers had reversed the restrictive policy, and new burn strategies were developed over the years in response to experience. Current fire management divides the park into some 400 burning blocks and rotates fires throughout the year in a cycle designed to mimic natural fires, following rainfall patterns and ecological needs (Gillon 1983; van Wilgen et al. 1990; Pyne 1995). In Australia, range managers have concluded, "as a generalization, there are relatively few arguments against burning the tropical and sub-tropical zones" (Leigh and Noble 1981, 472). As a result, Sandford (1983, 101) can write that while range scientists previously "tended to regard burning as a wholly undesirable practice, its useful role is now recognized by some of them both in Africa and elsewhere."

In the United States, especially outside of the temperate, consistently humid northeast, fire has often been seen with slightly different eyes than in Europe. The land is vast and, especially in the American West, undergoes yearly cycles of dryness that make it vulnerable to lightning and match. So while foresters trained in the French and German schools invented a Smokey Bear antifire policy for the national forests, land managers elsewhere harnessed fire for a variety of purposes. They burned prairies in Kansas and Wisconsin in the spring for improved range, burned crop stubble in Oregon and Idaho for field sanitation, burned pine woodlands in Louisiana and New Jersey to aid regeneration, improve range vegetation, reduce wildfire hazard, and kill scrub hardwoods, and burned public and private ranchlands in California for brush control. These land managers kept alive the idea of "good" fire, or at least useful fire, during the sharpest years of antifire policies (Kayll 1974; Biswell 1989; Pyne 1995).

In Madagascar, fire was likewise recognized as useful not only by the local people (part 2 of this book), but also by some European resource managers. Colonial cattle ranching establishments, such as the immense Rochefortaise concession in the western highlands, made use of fire to renew pastures.[8] In fact, Marcuse (1914, quoted in Bartlett 1955), an English traveler, condoned Malagasy pasture burning, for it created ideal feeding grounds and would lead to great profits in beef export through the canning industry. Colonial farmers burned to prepare crop fields even as the government cracked down on native fires.[9] The colonial military as well used fire to manage its terrains; a 1938 letter authorizes the burning of hundreds of hectares to avoid bush encroachment.[10] Finally, the colonial veterinary service reluctantly authorized a pasture fire to combat a tick epidemic in Ambovombe in 1953.[11]

A third group of voices defending fire came from researchers of an ethnographic bent who sympathized with their subjects' way of life; these researchers began to document the pervasiveness and logic of burning among global cultures. They sought to undo the damage done by colonial authorities who tried to bring about changes without understanding the environmental or cultural context (Netting 1968). Richards (1939), for example, looked at the *chitemene* woodland cultivation system of southern Africa, a practice considered backward and lazy by the authorities. She showed that this system—which involves cutting and burning tree branches and planting crops in the ashes—is a viable, and even the best, option. A few years later, Conklin (1954) began to dispel myths about slash-and-burn agriculture, noting that it is not haphazard, that swidden fires are not uncontrolled, that it is frequently not done in virgin forest, and that swidden cropping can actually protect the soil and mesh with natural regeneration cycles so that no permanent harm is done. In America at this time, anthropologist Omer C. Stewart (2002) was one of a few voices championing Native American burning practices (see also Sauer 1956). In short, these researchers all noted that people were not acting in a way that suggested they felt burning to be harmful to their long-term livelihoods.[12] Such research came to Madagascar in the

8. Interviews, Kilabé case study, 9 July 1999; see also Gilibert et al. (1974).

9. Files from Diego Suarez Province between 1935–1939 include burn permits for M. Wantz to burn tall grasses in his perfume plant concessions and for Mme. Tombo to burn surfaces for plowing in her concession (AOM mad ds//334).

10. Letter of authorization on 7 July 1938 (AOM mad ds//334).

11. ANM IV.D.73, subfolder 1.5.

12. Paraphrased from Stocking (1996, 151), who discusses erosion in these terms.

1960s and 1970s, when sociologists and geographers defended *tavy* and pasture burning as logical livelihood strategies—though also admitted an unease at the environmental consequences (Dez 1966, 1968; Coulaud 1973; Dandoy 1973; Ratovoson 1979).

Since that time, an academic literature has emerged that reevaluates and reappraises indigenous environmental practices in a better light. With regards to fire, the foremost author is clearly historian Stephen Pyne (e.g., 1995, 1997), who notes in stunning detail the reasons for each fire use, finding parallels across all continents (except, of course, Antarctica). He also captures the social and political dynamics that accompanied these practices through time. Pyne shows that, whether the purpose is to simplify hunting, to increase habitat for favored species, to prepare crop fields, or to increase grass production for domesticated ungulates, periodic burning was a well-established practice throughout history, across the globe (but see Vale 2002).

The fourth challenge to bad fire came from ecological researchers. Increasing doubts about the broadscale application of deterministic, equilibrium-based models for plant succession in land and resource management, and the development of alternative models, both probabilistic and chaotic, opened the way to consider fire as an integral part of vegetation community dynamics, rather than as a cause of "degradation." Traditional successional theory (e.g., Clements 1916; Odum 1969) held that each biome had a specific vegetal climax—frequently forest—and that vegetal change in a particular site followed a predictable, orderly succession towards a homeostatic climax. Fire, grazing, and other disturbances were seen as holding succession at pre-climax levels.

Experiences with fire and other disturbances, however, did not always match this theory. In the United States, some ecological researchers working in fire-prone western and southeastern ecosystems championed the natural place of fire in ecosystem maintenance, wildfire prevention, and range management (Phillips 1962; Biswell 1989). Others began to question deterministic theories of plant succession and began to include disturbances such as fire, drought, landslides, and even pests as normal, if somewhat unpredictable, factors in plant communities (e.g., Connell and Slatyer 1977; Reiners and Lang 1979).

To varying degrees, abiotic factors like fire and drought are thought to outweigh the competitive interactions that drive classic theories of vegetation development in many systems, rendering predictive models based on those classic theories of limited or no use. From this point of view, disturbance is not a cause of degradation, but simply another factor influencing the development of vegetation communities. Climax, or the end state of some linear successional development, can no longer be considered always achievable or necessarily desirable. Thus ecologists began viewing fires as normal parts of ecosystems, and different biomes were seen to have "normal" disturbance patterns, with expected fire recurrence intervals of tens or hundreds of years (e.g., Vale 2002, 29). This "new ecology" not only reshapes the way vegetation dynamics are understood, but also has consequences for the social sciences, in its emphasis on diversity, complexity, historicity, and uncertainty, and for applied conservation practice, in its implications such as adaptive management, learning by doing, and "flux (rather than fixity) and crossing over (rather than cordoning off)" (Zimmerer 2000, 364; see also Zimmerer and Young 1998; Turner 1998; Scoones 1999).

A current theory that incorporates disturbance, history, and multiplicity of pathways into ecological dynamics is the state-transition model, based on probabilistic, nonlinear, value-free notions of vegetation change (Westoby et al. 1989; Huntsinger and Bartolome 1992; Frost 1996; Dougill et al. 1999; Kepe and Scoones 1999). This theory suggests viewing a specific ecosystem condition as one of several possible "states"—discrete vegetation groupings that remain fairly stable for the duration of a similar management or disturbance regime. When management or disturbance regimes change, a "transition" moves the ecosystem into a new state. Some transitions are reversible, others irreversible. One cannot be assured of outcomes, only the probability of outcomes based on data.

State-transition models, developed especially with respect to rangeland ecosystems, are commendable for not theorizing "natural" nature separate from human intervention; they recognize management regimes as integral to ecological dynamics, not as outside disturbances (Kepe and Scoones 1999; Bassett and Koli Bi 2000). State-transition models remove the value judgments of traditional successional theory: the climax state (representing some sort of equilibrium) is not seen as inherently desirable, and fire is not considered inherently degrading. Deterministic plant succession based on competitive interactions remains an important pathway in many biomes, or one possible path in a state-transition model. In the wet-dry tropical grasslands, however, the wooded climax state may be achievable only with long-term fire exclusion and is hence highly unlikely.

As a result of these ideas that make room for good fire, fire as a management tool has become somewhat accepted in Western thought—at least in some places and contexts, away from the urban fringe. These ideas even surface occasionally in Madagascar, as part 3 of this book shows. Yet fire remains a powerful, feared force of destruction. The media reports massive incendiary disasters each year—Oakland, California, in 1991, Indonesia in 1997–1998, Greece in 1998, the western United States in 1999, 2000, and 2002, Australia in 2003—where fires destroy property, assets, and health. The Yellowstone fires of 1988, while forming an important event in public awareness about fire's ecological necessity, also highlighted the untrustworthy, uncontrollable, and catastrophic nature of fire due to the powerful imagery of failed firefighters and of a charred national treasure (Pyne 1995; Parfit 1996). For these reasons it is important to note that fire is neither simply good nor bad; instead it is best seen as complex.

COMPLEX FIRE

As the above discussion makes clear, fire is complex. On the one hand, it can contribute to property damage or environmental transformations, such as deforestation or soil loss, that people view as negative. On the other hand, fire is a legitimate, useful, and effective tool for land management and an essential ecosystem process. These contradictions are further complicated by the nature of fire as a natural and social process. The goals and causes of fire can be ambiguous, and the effects variable (figure 2.1). While a plow used in agriculture or an axe used in forest clearing have predictable and orderly effects, the impact of burning is less straightforward. A fire may or may not kill a certain tree. It might have resulted from lightning. If human lit, it can be quite unclear who started the fire, or why. The same match may burn a small patch or a large mountain. This section discusses these aspects of fire, aspects that Madagascar's peasants use to their advantage.

The Variability of Effects

The immediate and long-term impact of fire is complex and varies with a number of mediating factors, including frequency of burning, season, plant maturity or flowering stage, air temperature and moisture, soil moisture, vegetation type and density, litter amount and quality, wind, topography, time of day, direction of burn, historical legacy of land use and nutrients, and so on (Kozlowski and Ahlgren 1974; Sandford 1983; Gillon 1983; Biswell 1989; Goldammer 1990; Pyne et al. 1996). These factors determine fire's effect on standing vegetation and soil microfauna, as well as—in part—the successional changes that follow.

For example, the seasonality and the repeat interval of fires affect floral composition and forage quality in rangelands. In two decades of experimentation in the southern highlands of what is now Tanzania, from 1931 to 1951, researchers observed the differences between rainy season and early-, mid-, and late-dry season fires, between annual, two-year, and three-year fire cycles, and between burning, mowing, and protection. Each treatment had different results; the best for grazing was consistently those plots burned every second year in the mid dry season, that is, October (van Rensburg 1952). Sometimes, fire timing plays an extremely specific role: Bartlett (1956) observed that the *hariq* system of grassland firing in Sudan was specifically performed early in the wet season—exactly when grass regrowth had exhausted root- or seed-based energy reserves—in order to kill the grass to facilitate crop cultivation. In sum, similar fires can have different results, and postfire vegetation changes can vary based on a host of factors, including annual climatic variations, soil types, and seed availability.

Fire is complex not only in its results, but also in its spatial extent. When a logger sets out to cut fifty trees, the fifty selected trees will be cut. When a herder lights a pasture renewal fire, the final result is less predictable, for fire assumes its own agency—it is self-propagating. Even a tightly managed fire—surrounded by firebreaks and fire crews—may escape the bounds or not burn everything within bounds. Malagasy pasture fires, described in chapter 3, vary in exactly this way. Late dry season grassland fires tend to burn uniformly across a landscape. The fire front reaches all across a pasture, though it may burn at different intensities based on the vegetation type, density, moisture, and wind. It may stop at streams, trails, crop fields, or forests, but escaped flames or flying ashes can extend fire to new arenas. On the other hand, fires at the end of the rainy season are easily contained but give more patchy, unpredictable results of nonburned, poorly combusted, and burned zones within the planned burn zone.

The Ambiguity of Agency and Purpose

Fire's complexity is amplified by the ambiguity of agency and purpose in burning. Humans fully control neither the ignition nor the life span of a fire; the ignition of a fire can easily remain anonymous; purposes can be multiple and overlapping; and the effects of a fire do not necessarily correspond with the goals at ignition. The examples below—lightning fire, accidental fire, anonymous fire, and multipurpose fire—demonstrate these ambiguities.

First, humans have no monopoly on igniting fires; lightning also provides ample and effective sparks, with an estimated 6,000 lightning discharges occurring each minute across the globe. In the United States, lightning is the primary cause of wildfires. For example, in Forest Service lands in California, from 200 to 1,700 lightning fires occur each year (Biswell 1989). In southern Africa, lightning causes a significant portion of fires each year (van Wilgen et al. 1990), as it does in the Brazilian savanna (Ramos-Neto and Pivello 2000), but not in France (de Montgolfier 1991).

Perrier (1927) claimed that lightning does not start grass or brush fires in Madagascar. He was mistaken. There can be no doubt that lightning associated with convection cell activity at the onset and end of the rainy season caused many fires before human arrival (Burney 1996). Now, lightning fires are less common, for the Malagasy farmers tend to beat lightning to the task, removing susceptible fuels by dry season burning. Nonetheless, lightning fires occur each year, though these probably amount to less than five percent of all fires (Bloesch 1999). In the open plains of the western highlands, isolated trees around villages or along riparian corridors are struck every year, and some lead to wildfires. Few of these fires become large, however, as most grass has been burned by then.[13] In mountainous areas, fires are also common. Over three-quarters of the fires between 1943 and 1996 in the Andringitra mountain range were attributed to lightning.[14] In Afotsara, powerful lightning storms struck in September and October 1998, with thunder and lightning echoing off the mountain slopes. While no visible fires were ignited, three informants spoke of houses historically burned by lightning. In fact, two separate Norwegian missionaries stationed in the Afotsara region in 1878 wrote gripping accounts of lightning striking their houses, killing both cattle and people.[15]

Second, humans not only do not monopolize agency in the lighting of fires, but we also do not exert full control over fire once lit. Fire is self-propagating, and it can easily escape control. In the United States, an escaped forest service prescribed burn threatened the Los Alamos labs in summer 2000. In Madagascar, as elsewhere, such accidental fires are common. Rambeloarisoa (1995) labeled 25 percent of the fires in his study as accidental; in my tally of fires in Afotsara in 1998, 3 percent of the ignitions and 10 percent of the area burned were accidental. Commonly cited causes include escaped crop field fires, escaped flames from charcoal-creating ovens, dropped cigarette butts, and untended campfires (RDM 1980; Bloesch 1997; Rakatonind-

13. Interviews, Kilabé, 9 July 1999.

14. Bernardin Rasolonandrasana, personal communication, WWF Andringitra Project, Ambalavao, 1998.

15. Letters from P. A. Pedersen, Ambohiponana (Manandona), 21 Jan. 1878, and F. Bekker, Ilaka, 22 June 1878 (NMS Hjemmearkivet 1842–1919, Innkommende Brev, Boks 134, Legg 15-31/1878 and Legg 13-59/1879).

rina 1989; Rambeloarisoa 1995). Vignal (1956a), for example, described a dramatic escaped *tavy* fire in the Zafimaniry region that burned 3,520 ha of prairie and forest. During my October 1996 visit to the Andringitra region, a fire lit for agriculture escaped and burned 144 ha, crossing into the strict nature reserve. In Afotsara, the 22 August 1998 nighttime wildfire described in the introduction to part 1 had escaped from a canal clearance fire. Likewise, a pasture fire nine days later burned zigzag along the grassy corridor between some well-weeded cassava fields and ultimately torched some pine trees as well as a farmer's lean-to shelter (figure 2.3).

Third, in addition to the fact that fire is self-propagating, fire's very nature allows the identity of the igniter to easily remain ambiguous or anonymous to outsiders, at least for nonagricultural fires. Other land uses are not as easily anonymous. In a cultivated field, the cultivator must return at some point to harvest, while in a logged forest, somebody will be seen laboriously dragging a tree down the mountain or stockpiling wood behind their house. Fire, on the other hand, is quickly lit and easily anonymous, seriously complicating any attempts for outsiders to regulate its use. Fires can be ignited under the cover of darkness, or simply out-of-sight from authorities. Fires can even be lit after an arsonist is long gone: a variety of informants told me of a well-known technique for time-delayed ignition. One places a few matches and some dry grass over one end of a dried patty of cow dung, then ignites the opposite end. By the time this smoldering fire reaches the matches, the arsonist will be far away.[16] Frustrated colonial authorities, foresters, and even private tree growers have complained of the impossibility of apprehending fire starters for precisely these reasons (see chapter 6).[17]

Fourth, not only is agency ambiguous, but purpose is as well. As the stories in the introduction to part 1 show, different people frequently give different reasons for fires (e.g., the stories of the 25 June ridge top fire or the 21 July woodland blaze). People take advantage of the fact that fires may accomplish more than one task at the same time and that the purpose of a fire is not always clear to outsiders. So in a context of antifire repression (or simple criminality), this results in many scapegoats, half-truths, and excuses. For example, when locusts swarm (as they do once a decade or so), people frequently

16. The informants included an urban developer, a Forest Service agent, a rural mayor, a rural farmer (a former *fokontany* president), and a report (Andriamampionona 1992). This technique seems to have made its way into the popular imagination.

17. This applies chiefly to free-burning grassland or forest fires, and not to *tavy* fires. An illegal *tavy* cultivator, after all, has to return to the burned plot to sow and harvest, and his or her tenure of the plot is known.

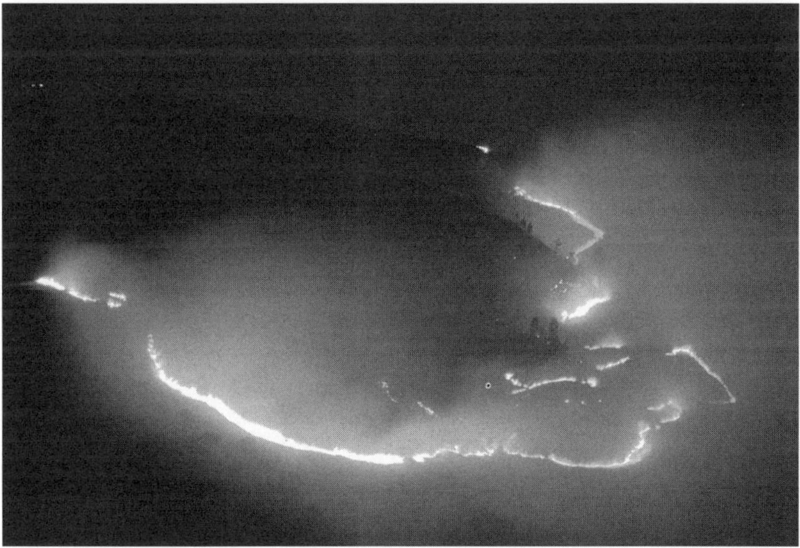

FIGURE 2.3. **Accidental fires in Afotsara.** (a) A cooking lean-to and storm shelter, as well as some pines, burned by a wandering pasture renewal fire on 31 August 1998. (b) The escaped canal clearance fire of 22 August 1998, described in the preface.

tell officials that they are burning lands to ward off the damaging locusts, when in truth the more central task of the fire is to renew a pasture (the fire can serve both purposes). Locals know that the state is less likely to enforce controls on antilocust fires, and people take advantage of this to burn unquestioned for pasture renewal goals.[18]

Criminal Fires and Fires for Conflict

As a result of the complex ambiguity of agency and purpose in fires, its anonymity, and the potential multiplicity of purposes, fire can clearly be used for both legitimate and illegitimate purposes. Fire has a long global history of criminal and destructive uses; characters from jealous neighbors in folktales to invading armies in real history all used fire to damage their enemies. Part 2 of this book describes in detail the legitimate, useful applications of fire; here I chronicle the use of fire in Madagascar for criminal and conflictual purposes. These manifestations of fire have clearly contributed at times to the antifire received wisdom and will always pose a challenge for fire management.

Sabotage and arson occur for purposes of revenge, jealousy, or personal gain. In interviews, several peasants mentioned malicious damage (*manao ankasoparana*) between rivals (*kakay*) as causes for fire—of fires lit near the home of the rival, of fires in the rival's corn fields, or especially of fires in pine woodlots. Jealousies over an inheritance led four peasants in Antsorindrana to burn the houses and granaries of local notables in 1936.[19] Similar interpersonal conflicts played a role in several fire-related arrests in the Antsirabe forestry district in the last two decades. For example, a fire that burned 694 trees in September 1995 arose out of love and blood brotherhoods gone awry,[20] while a fire in May 1992 that destroyed bean, corn, and cassava crop fields was clearly intentional arson.[21] Sharp differences in wealth or attitude can result in criminal

18. Multiple interviews, Afotsara, 1998. See also chapter 6. During the frequent locust passages in the Afotsara field site during 1998, not once did I see anybody move to put out an antilocust fire once the swarm had passed.

19. AOM mad pt//333: Cour d'appel de Tananarive, no. 9, 31 Aug. 1936.

20. File 16/95, Dossiers des Contentieux, DEF, Antsirabe. On 20 Sep. 1995, two residents of Laimbolo *fokontany* (Sahanivotry *commune rurale*) accused a man aged twenty-five years, resident of Ambatofinandrahana (50 km to the southwest), of intentionally burning eight hectares of land, including 694 privately owned eucalyptus, pine, and mimosa trees. The story involved love, blood brotherhood, jealousy, and revenge. The accused was spared prison, but fined 300,000 MGF.

21. File 02/92, Dossiers des Contentieux, DEF, Antsirabe. Two men aged forty-six and thirty-seven intentionally set fire to a hillside on 24 May 1992, in the *fokontany* of Laondany (Andrembesoa *commune rurale*). The fire burned 100 ha, including bean, corn, and cassava fields. A witness saw them light the fire and ran to the village to call for help in fighting it; the two accused refused

fires: for example, graduates of the Norwegian-sponsored Tombontsoa farm school who establish profitable market-oriented operations have been targeted by jealous neighbors with both house fires and crop field fires (EESR Lettres 1992; Jon Randby, personal communication 1998).

Woodlots and forests are especially contentious sites for conflictual fires. While the pines and eucalypts planted in Madagascar are fire tolerant, burning can destroy seedlings or force the salvage sale of adult trees.[22] Tree growers are uniformly frustrated by damaging fires; as one informant stated, "a reforestation never fails to be burned." In Afotsara during 1998, twenty-three hectares of pine woodlots were burned. At least one of these blazes, a 13 November fire that torched four hectares of pine saplings, was probably intentional and related to conflicts over water rights.[23] The frustrations of tree planters mask underlying class and resource-access tensions, for tree planting is a means used by richer farmers to establish legally defensible claims to portions of state land (Bertrand 1999). Poorer farmers, resenting this appropriation, may burn out of frustration. During the early days of the colony, Parrot (1924) documented a similar use of fire to contest resource control: the colonial state claimed all forested lands as its own; thus, by burning the forests the peasants could reclaim this land, void of forests, as their own.

Another form of illegitimate burning that takes advantage of fire's anonymity, ambiguity, and self-propagation relates to organized banditry. Western Madagascar has a long tradition—with parallels to the cattle cultures of East Africa—of intertribal cattle thieving (see Raison 1968; Hoerner 1982; Blanc-Pamard 1989; Aquaterre 1997). These bandits are known as *dahalo* or sometimes as *fahavalo* or *malaso.* Norwegian missionaries living in Manandona wrote of six *dahalo* raids between 1884 and 1891; the *dahalo* stole cattle, enslaved people, and repeatedly burned entire villages.[24] In 1884, mis-

---

to help put out the fire. This was perceived as a deliberate denigration of the honor of the community, especially considering the community's fear of *dahalo* bandits following a devastating 1982 raid (see discussion of *dahalo* in text). The Forest Service's recommended sentence was five years in prison and a fine of 300,000 MGF.

22. The notion that people burn to force salvage sales appeared originally in RDM (1980) and is repeated by Rakatonindrina (1989) and Andriamampionona (1992).

23. Interview, Afotsara, 20 Nov. 1998.

24. A particularly powerful raid in 1884 netted 550 cattle and 100 people and burned 45 houses. The remaining citizens, of course, fled. Letter, J. Nygaard to Hovedbestyrelsen, Manandona, Feb. 1884 (NMS Hjemmearkivet 1842–1919, Boks 136, Legg 6, J. Nr. 27/1884). Additional letters include ones mentioned from Nygaard cited below, as well as letters from Nilsen printed in the *Norsk Misjonstidende* in Feb. 1887 (42[3]: 44–50), Jan. 1889 (44[2]: 36–38), June 1890 (45[11]: 210–213) and Oct. 1891 (46[19]: 369–372).

sionary J. Nygaard asked for leave to return home, due to the *dahalo*.[25] He wrote, "People have seen *afo manjaka* (fires of power) west of the mountains, and this is taken as a sure sign that there is a gathering of bandits."[26]

Banditry continues today. The *dahalo*, however, are not a unified group. The term *dahalo* is a loose label for all perpetrators of banditry, whether by tribal groups from west of the highlands, by organized criminal gangs, or by local criminals. Banditry peaked in the mid-1980s, when a fiscal crisis paralyzed the state, but continues today, affecting many of the study sites.[27] *Dahalo* purportedly use fires to intimidate and threaten, but more specifically to hide traces of cattle rustling, to draw able-bodied men away from their homes (in that the men would fight a wildfire among crop fields or near houses while the *dahalo* pillage a village), or to create green shoots to attract wild cattle for theft.[28]

Finally, illegitimate fires include those for protest against the state. This is discussed in detail in chapter 6. Essentially, fire's anonymity, symbolism, and self-propagation make it a convenient vehicle for lodging protest. The paramount example of a highly symbolic fire linked tightly to national politics is the burning of the Rova on 6 November 1995. Built during the nineteenth century at the height of the Merina monarchy, the Rova is a collection of buildings on Antananarivo's highest summit. The fire destroyed the Queen's Palace, adjacent royal houses, and the historical collections within. The Rova symbolized the historical political and economic dominance of the Merina; the debate over the fire's causes led to intense speculation of arson symbolic of ethnic tensions. Other protest fires have erupted in the past century in protected areas or state forest plantations. However, I argue in chapter 6 that since fires can be ambiguously multipurpose, many of the fires perceived by leaders to be protest fires are more likely linked to peasant livelihood goals.

25. Letter, J. Nygaard to Hovedbestyrelsen, 3 Apr. 1884 (NMS Hjemmearkivet 1841–1919, Boks 136, Legg 6).

26. Letter, J. Nygaard to L. Dahle, 18 Apr. 1884 (NMS Hjemmearkivet 1841–1919, Boks 136, Legg 6, J. Nr. 35d/84).

27. In Afotsara, several herds of a dozen cattle were stolen and houses burned from 1985 through 1987. Additional cattle were stolen in 1990, and in 1998 some *dahalo* were captured west of the mountains after having stolen cattle and burned houses. In Behazo, as elsewhere, 1984 through 1986 were years of heavy *dahalo* thieving, and people sold off their cattle in fear of losing them. Finally, in Kilabé, over 100 head of cattle were stolen in 1982, while most recently, on 1 Apr. 1998, nine houses were burned down.

28. Interview, Forest Service officer, 6 Aug. 2001, as well as RDM (1980), Rakatonindrina (1989), Schnyder and Serretti (1997), and Andriamampionona (1992).

COMPLEX FIRE AND ENVIRONMENTAL DEGRADATION

We have seen that fire is by nature complex. While fire serves a useful role for resource management, it can also lead to environmental degradation. It has important consequences for forest-grassland dynamics, for soil and pasture quality, for biogeochemical cycling, and for air quality. In this section, I look at the complexity of fire from an environmental perspective. I show that while burning has serious environmental impacts, concerns over fire are sometimes overstated or oversimplified.

Deforestation

Anthropogenic fire is the proximate cause behind the loss of much of Madagascar's diverse, endemic presettlement vegetation. Since the arrival of humans on the island, the amount of fire has increased, and in many regions endemic woodland or forest landscapes have given way to herbaceous or grassland formations (Burney 1997). The pattern varies by region. In the rain forest zones of the east, fire accompanies the initial clearance of a forest for *tavy* agriculture. While forest would reclaim cleared land if permitted, repeated cycles of *tavy* agriculture covering an entire region lead to a long-term (circa 100–200 years) transition from forest to herbaceous vegetation communities. Ridge tops and hillsides become grasslands maintained by near-annual fires for pasture. At least two-thirds of eastern forest cover has been lost due to logging and *tavy* since settlement (see chapter 5).

In grassland zones of the east, highlands, and west, frequent fires serve to maintain a graminoid vegetation community (see chapter 3). In many of these regions, human-lit fires were responsible for the transition perhaps a thousand years ago from prehuman vegetation communities (whether forest, savanna, or heath) to grasslands and for later transitions from one kind of grassland to another (depending on soils, precipitation, and fire frequency). Grassland fires sometimes burn into adjacent forest areas, and if repeated (and depending on the type of forest) can expand the grassland at the expense of the forest.

Proponents of the antifire received wisdom have sometimes exaggerated, misconstrued, or oversimplified these effects of fire on deforestation. First, the Perrier-Humbert hypothesis exaggerated the extent of prehuman forest destroyed by anthropogenic fires. Critics challenged the hypothesis by noting the role of climatic desiccation in land cover change, by presenting biogeographic evidence of long-term forest patchiness or fire tolerance, and by positing that the diverse subfossil fauna suggested a variety of habitat types (e.g., Gautier 1902; Bartlett 1955; Dez 1970; Battistini and Vérin 1972; Bourgeat and Aubert 1972; Dewar 1984; see also Kull 2000a).

More recently, analyses of pollen and charcoal in lake sediments around Madagascar indicate that the presettlement highlands and west were never all forest (Burney 1996, 1997; Straka 1996). Instead, climate variations, lightning fires, volcanism, and now-extinct animals in these areas maintained a spatially and temporally varying mosaic of riparian forest, woodlands, heath, and grassland. These sedimentary records also show that the island's vegetation cover has always been changing; areas covered today by montane rain forest, like at Andasibe, were once heathlands during the last ice age (see also Gasse et al. 1994). There is no doubt that the amount of woody vegetation has decreased significantly since human arrival some 1,500 years ago, largely due to the use of fire. The received wisdom, however, exaggerates the amount and nature of the prehuman forests.

Second, proponents of the antifire received wisdom have occasionally conflated forest *tavy* fires with the pasture fires of the highlands and west.[29] Allen (1995) and Berg (1981), for example, blame *tavy* for historical highland deforestation, when in all likelihood repeated fires related to hunting or cattle raising converted the majority of the highland vegetation mosaic into grasslands. Others implicate pasture fires in the current deforestation of the eastern rain forest. The 1994 "State of the Environment" report from ONE/Instat, for example, includes an assessment of pasture fires within the section titled "deforestation." Ironically, the statistics presented in this assessment show that only 0.1 percent of the yearly fires, or 2,000 ha, occurred in forests.

Third, blaming eastern deforestation just on fire is far too simplistic. For one thing, a significant portion of forest loss in eastern Madagascar, in particular during the first thirty years of French rule, was due to colonial logging

29. The confusion is apparent in official documents. In 1914, Governor Hubert Garbit dedicated an entire circular to discussing the difference between "fires to renew pastures" and "bushfires in forested areas for the preparation of *tavy* swidden cultivation," due to the fact that his district officers often confused them (JOM 14 Nov. 1914). In 1938, Governor Léon Cayla defined *feux de brousse* (bushfires) as fires in woody vegetation (bushes, trees), and *feux de pâturage/prairie* (pasture fires) as burning grasslands to renew pasture (JOM 2 July 1938). Within a few years, his replacement, Governor A. Annet, essentially reversed the definitions (JOM 6 Dec. 1941), calling *feux de brousse* fires on denuded lands with only grasses and *incendie de forêt* (forest fire) fire in any land not used for agriculture or pasture, whether it is wooded, covered in ferns, or a savanna of widely scattered trees. An informant in Afotsara repeated the same complaint, that the government confuses the *tavy* fires of the east with the pasture and woodland fires of the highlands (interview, 27 Oct. 1998).

extraction. In this period, from 1 to 7 million ha of primary rain forest were opened to logging—though probably not clear-cut—out of an estimated 11 million ha. In addition, fire (together with cutting and cultivating) is just a proximate cause of deforestation. The expansion of agriculture is what drives much deforestation today, not simply fire, and this in turn is driven by a number of forces including population growth, government policies, and the search for livelihoods (for a detailed discussion of the causes of deforestation see chapter 5).

Fourth, Humbert (1927) blamed fires for nibbling at forest edges year after year (see also Coudreau 1937; Kuhnholtz-Lordat 1938; Guillermin 1947; and Gade 1996). This process is exaggerated or misunderstood. In humid forest zones, farmers occasionally allow *tavy* fires to burn a short distance into adjoining forests. They probably do so on purpose, for this starts the tough process of clearing new plots, and it absolves them of legal difficulties (authorization for *tavy* is next to impossible to obtain in primary forest, but not in secondary or burned forest; see chapter 5). In the highlands and west, pasture fires sometimes encounter forest edges, yet evidence for deforestation is slim. Confronted with humid forests, grassland fires usually stop due to moisture differences. In areas of dry forest, fires hardly affect forest edges, for these woodlands are often relatively fire tolerant.[30]

Soil Erosion

Burning accelerates soil erosion (Humbert 1949; Randrianarijaona 1983). Research in several areas of the highlands has demonstrated higher erosion rates in burned areas, at least in the first year (Bailly et al. 1967; Goujon et al. 1968; RDM 1980; Mottet 1988; Brand 1999). The consequences of erosion are locally quite severe. The irrigation works at Lake Alaotra, the nation's principal rice-producing area, fill with 30 to 60 million tons of soil each year. Siltation of the Ikopa River near the capital causes devastating floods about once per decade; siltation of the western port of Mahajanga proceeds at almost one meter per year, necessitating costly dredging.[31] In some spots, *lavaka* exceed 30 per km^2 and threaten roadways. More generally, fire-caused erosion is thought to lead to a long-term degradation of soil quality, often as a slow, almost imperceptible process (e.g., Tricart 1953; Granier and Serres 1966; Le

30. See chapter 1 note 17 for documentation, as well as chapter 4 on the stability of highland *tapia* woodlands.

31. *Tavy* is sometimes blamed for the siltation of rivers and bays (e.g., Roffet 1995; Apt 1996), yet the affected rivers are in the west and drain mostly non-*tavy* areas.

Bourdiec 1972; Rossi 1979; Randrianarijaona 1983; Paulian 1984; Olson 1988; Wells and Andriamihaja 1993; Brand 1999).

Despite these dire assessments, actual erosion may not always be as damaging as presumed. As noted in chapter 5, researchers analyzing *tavy* do not consider the erosive impact of *tavy* fires a priority issue (Brand 1999). Like deforestation or other vegetation transitions, erosion is often a matter of perspective: it benefits some while harming others. Soil eroded from one place can either benefit or harm downstream resource users, and this process is well understood by farmers. Farmers in the alluvial *baiboho* of northwestern Madagascar depend on the fertile soils eroded from the highlands (Rabearimanana 1994), and fieldwork in both the east and the highlands demonstrates that farmers actively manage *for* erosion in order to fertilize rice fields and improve swamp soils (Wells and Andriamihaja 1993; B. Locatelli, personal communication, 1996).

Concern over erosion often focuses on the visually compelling *lavaka*, but this can lead to overdramatization (cf. Stocking 1987, 1996). *Lavaka* are serious issues locally, but at a broad scale pale in significance compared to other forms of erosion. These gullies are natural parts of the landscape's evolution due to tectonism and climatic aridification. However, burning, forest clearance, erosion along paths, and irrigation canals have increased natural *lavaka* activity in some areas (Wells and Andriamihaja 1993).[32]

In grassland regions, erosion increases in the first year after a fire, yet regular burning does not necessarily increase long-term erosion (Biswell 1989; Pyne et al. 1996). The long-term effects of repeated burning on erosion depend upon a soil's tolerance for erosion, including soil formation rates and biogeochemical feedback cycles (Hurni 1983; Stocking 1996; Ojima et al. 1994), and vary from place to place.

Rangeland Degradation

Frequent burning in grassland regions can cause a transition from lush, nutritious grassland communities to relatively poor shortgrass pasture. While actual vegetation transitions in specific places depend upon soil types, seed availability, soil moisture, precipitation, and fire timing, rangeland specialists working in Madagascar have described a typical series of vegetation state transitions based on increasing fire frequency. The first fires replace forests with a savanna of high grasses (e.g., *Hyparrhenia cymbaria*) and then with a

32. Some sloppy authors (e.g., Helfert and Wood 1986; Gallegos 1997) imply that *tavy* leads directly to the creation of *lavaka*, yet most *lavaka* are found far outside the zone of *tavy*.

savanna of medium grasses including *Hyparrhenia rufa* and *Heteropogon contortus*. Further repetitive burning replaces this savanna with a degraded steppe of *Aristida* spp., which becomes less dense as fires increase to their maximum. The *Aristida* grasslands have the lowest nutritional quality of all the formations (Perrier 1921; Bosser 1954, 1969; Granier and Serres 1966).

Implied in this retrogressive model of range succession is a criticism of frequent burning as a pasture management technique. This assessment misses out on two important points. First, current theories of rangeland dynamics suggest that the amount and timing of precipitation may be at least as important as burning or grazing practices in determining grassland vegetation communities, especially in drier climates (Sandford 1983; Behnke and Scoones 1992). Second, the logic of Malagasy herders differs from that of western range managers (see chapter 3), especially with respect to the value of *Aristida* grasslands. Most herders rely on upland burned pastures for the critical period of September through November—this is when cattle are the hungriest and when they can no longer feed in fallow crop fields (for these are being plowed in anticipation of the rains). During this period, the cattle eat the resprouts of *Aristida* and other grasses. The young shoots are sufficiently nutritious to support the cattle until the rainy season arrives in full. Thus, while an *Aristida* pasture is far from ideal for year-round grazing, this is not what it is used for. Malagasy herders have just concluded, like ranchers and researchers elsewhere, that burning increases the nutritive value of the plants to their cattle at specific times of year (Gillon 1983).

Other Concerns: Nutrients, Climate, and Air Quality

Fires volatilize some nutrients, like nitrogen and carbon, and free up others for plant use. The effect of *tavy* on nutrients in the biomass and soils has been closely studied. After a fire, almost all of the carbon and nitrogen in the biomass are volatilized, yet ashes rich in calcium, magnesium, phosphorus, and potassium fertilize the soil. Biomass nutrients recover quickly after a fire, yet long-term continued *tavy* cycles lead to a reduction in overall nutrients (Brand and Pfund 1998; Brand 1999; see chapter 5).

The effect of pasture fires on nutrients has not been studied with the same level of detail in Madagascar. Elsewhere, however, Biswell (1989) argues that in Californian ecosystems, low-intensity annual burns volatilize little nitrogen. While some nitrogen is lost, other nitrogen is converted into forms more available to plants; additional nitrogen may be restored to the ecosystem by precipitation, dryfall, decomposition, and nitrogen fixation. In a detailed study of annually burned Kansas pastures, Ojima et al. (1994) determined that nitrogen loss due to annual burning has no harmful effect, since biological and biogeochemical feedback processes serve to fill the gaps and maintain grassland productivity indefinitely (see also Gillon 1983).

The emissions of biomass burning—e.g., carbon monoxide, methane, nitric oxide, hydrocarbons, and smoke particles—play a major role in climate and atmospheric chemistry and have attracted much research attention (e.g., Crutzen and Andreae 1990). Such emissions have been important ever since humans mastered and increased the amount of fire on the planet, and especially since we began burning not just vegetation but also fossil fuels. Much remains to be determined, however, about historical trends of emissions from natural and anthropogenic fires and about how these emissions and their impacts vary based on different frequencies and intensities of burning in different biomes.

These emissions from fire also affect air quality (Goldammer 1990). The health effects of smoke from the extensive fires in Indonesia in 1997–1998, for example, resulted in many days of front-page media coverage. In Madagascar, smoke from fires envelopes highland cities for a few days or weeks almost every year in October or November, aggravating lung-related diseases or even snarling traffic (Coulaud 1973; Andrianavosoa 1996).[33] However, as Biswell (1989) points out, fire smoke has been a natural part of the atmosphere forever and cannot be avoided. What is of concern is to control excessive amounts of smoke so as to avoid deleterious effects on public health and safety. Biswell suggests that frequent small prescribed fires, which release smoke in small amounts, are thus preferable to large, smoky wildfires.

IN SUM, fire is neither good nor bad, but complex. The ancient metaphor of the Phoenix is quite apt: fire is at once destructive and creative. The black waste of a burned field gives way to the lush greenery of fresh resprouts. In human hands, however, fire is even more complex. That is because one person's creation is another's destruction. The creation of pastures, rice fields, and livelihoods is the destruction of habitats, timber reserves, and biological diversity. This explains, perhaps, the vehemence of the conflict over fire. Fire is inevitable in a landscape of long dry seasons, and its physical characteristics give fire-setters a distinct advantage.

33. The particularly thick smoke that occurs annually in October or November is largely due to *tavy* in the forests of the eastern escarpment. The smoke is carried by the prevailing winds to the nearby highland cities (50 to 100 km). These *tavy* fires tend to occur later than pasture fires, which begin in June or July. When the smoke is thick, people say *mandoro ala ny taiva*—"the lowlanders are burning the forest."

Part Two

Landscape Burning and Livelihoods

WHY DO THE Malagasy burn their lands? What are the outcomes of these management techniques? The purpose of part 2 is to answer these questions. In the following three chapters I investigate the logic behind Malagasy burning practices. The conclusion is that rural farmers and herders utilize fire in a number of ways to manage vegetation, soils, and wildfire danger to their advantage. This management varies based on the character of the ecosystem and the social and economic context. In general, fire is a labor-saving tool that contributes to key components of many people's rural livelihoods. This explains why most rural Malagasy resisted the state criminalization of fire, as I will describe in part 3.

Chapter 3 investigates the most widespread uses of fire in the most fire-prone region: grassland fires in the highlands. I demonstrate the multiple uses and strategies of burning, and how these strategies are molded to fit different ecological, demographic, and economic contexts of land use. Chapter 4 then investigates the special case of the *tapia* woodlands, a rare type of endemic pyrophytic woodland found scattered about the highlands. Fire is a key contributor to the management of these woodlands, which are prized for their fruit, wild silk, and other products. Finally, chapter 5 addresses *tavy* slash-and-burn cultivation, in particular in the humid eastern regions. Fire is central to the agricultural rhythms of this region, both in repeated cycles of long-fallow cultivation and, once the land is exhausted, in pasture maintenance.

I should note that the chief focus is on highland and eastern zones. The fires of the drier lowland western zones are dealt with tangentially in chapter 3 (with respect to pasture management and other fire uses) and chapter 5 (which introduces some details on slash-and-burn cultivation in western forests). The spiny brush of the south sees fewer fires due to aridity and lack of flammable materials. All the same, this does not preclude fire practices, such as the burning of needles off of *raketa* cacti (*Opuntia* spp.) to create cattle fodder (see Middleton 1999; Kaufmann 2001). This region, the extreme south, is not specifically discussed in this book.

3 [GRASSLAND FIRE

The Agropastoral Logic of Fire across the Highlands

A pasture without fire is not a pasture.

RATSIRARSON (1997, 14)

THE EPICENTER OF Madagascar's fires lies in the vast grasslands of the central highlands and the interior west. Fire use in these regions—responsible for the paucity of woody vegetation—fits soundly into the agropastoral, ecological, and economic logic of peasant land use. While many policy makers see these fires as destructive, for rural farmers and herders burning can be an appropriate, efficient, sustainable, and inexpensive tool for resource management. In this chapter, I first describe the different types of grassland and farmland burning, documenting how they fit into the social, ecological, and economic context of rural life. Second, I compare fire use in five case studies across the highlands, showing that the peasants adapt their use of fire to socioeconomic and ecological contexts. While the lessons apply to the whole island, I focus on the highlands, where fire use is most widespread. Highland peasants harness fire for a wide variety of constructive goals, from pasture management to insect control (table 3.1). These uses reflect those of land users around the world (Kuhnholtz-Lordat 1938; Kozlowski and Ahlgren 1974; Sauer 1975; Sigaut 1975; Goldammer 1990; Homewood and Rodgers 1991; Bruzon 1994; Pyne 1995, 1997).

FIRE FOR PASTURE MANAGEMENT

The dominant role of fire in Madagascar is pasture management. Pasture fires affect by far the most surface area of all fires, burning about one-quarter

TABLE 3.1. **The purposeful and constructive uses of vegetation fire in grassland and farming areas.**

Sector	Purpose	Specific task or use
cattle raising	pasture maintenance	- fight bush encroachment - renew pastures (green shoots)
	pest control	- control tick populations
	cattle control	- facilitate cattle observation and mobility - attract free-ranging cattle to pasture
agriculture	field preparation	- clear brush for plowing/spade work - fertilize fields
	downhill transfers	- encourage erosion to fertilize downstream fields - encourage runoff to speed up irrigation
	cleaning	- clean out irrigation canals - remove pest habitat around fields (rats and birds)
other	wildfire prevention and control	- early burns for fuel management - burn fire breaks - backfires against wildfires
	pests	- ward off or collect locusts - control rats, ticks, and mosquitoes
	woodlots and woodfuel	- encourage pine regeneration - create dead branches for woodfuel collection
	travel	- clear trails and roadsides - light the way in the dark
	ground clearance	- expose mineral outcrops - expose wild tuber crops
	grass species management	- encourage *Aristida rufescens,* used for brooms and roofing
	celebration/spectacle	- provide natural firecrackers and entertainment

to one-half of all nonwooded, noncultivated zones each year. Early foreigners visiting Madagascar often commented upon these ubiquitous fires. The Reverend William Ellis (1838, 91) of the London Missionary Society wrote:

> In order to secure good pasture-land of the cattle, the inhabitants burn the grass which grows luxuriantly on the sides of the hills. They set fire to this about the close of the dry season; at which season of the year fires may be seen at an immense distance illuminating the horizon in a most splendid manner, for many

> miles in extent. As soon as the rains fall, the young and tender grass springs up, and a fine rich pasture is provided.

The situation is hardly different today. In the cattle-dominated Middle West (the sunnier western reaches of the highlands), multiple fires are visible in all directions from June through November. In Afotsara, fires in 1998 burned 38 percent of village lands, including 54 percent of pasture areas (table 3.2), largely for the purpose of pasture maintenance (table 3.3). Even in deforested zones of the east, large unsupervised fires burn across humid lowland pastures each year (Pfund 2000). Typical pasture fires cover anything from 0.5 ha to over 100 ha.

Fire is critical to extensive cattle raising. In intensive systems, where grasses are grazed to the ground, fire is difficult to light and unnecessary. In

TABLE 3.2. Area (in hectares) burned between March and December 1998 in an 18 km² surveillance zone in and around Afotsara.

Land use/land cover	Total area	Burned area	% burned
upland pasture zones	900	486	54
tapia woodlands	300	109	36
intensive zone*	645	101	16
TOTAL	1845	696	38

*Includes settled areas and nearby cropfields, woodlots, pastures, and mixed-use zones. Most of the area burned within the intensive zone was within small pastures (75 ha). The remainder was pine plantations (23 ha) and established crop fields and field or canal edges (3 ha).

TABLE 3.3. Causes of fires logged from March through December 1998 in Afotsara.

Cause	Number of fires	Area burned (ha)
pasture renewal and maintenance	18	342
locust invasions (including fires blamed on locusts but also achieving other goals, such as pasture renewal or woodland maintenance)	75	262
crop field preparation, rat control, and canal clearance	40	4
accidental	5	69
criminal	2	7
unknown cause	15	12
TOTAL	155	696

Classification based upon best information available.

extensive systems, fire plays two important roles: it prevents bush encroachment and removes old, unpalatable forage to stimulate protein-rich new growth. Below, I discuss these roles in detail; first, however, I introduce the grassland ecology and cattle economy of Madagascar.

Malagasy Grasslands and Cattle Raising

Malagasy grasslands, which cover some 40 million ha, are composed chiefly of pantropical perennial grasses (table 3.4). Typical grassland communities differ regionally (Bosser 1954, 1969). In the west, grass communities vary based on the underlying soils, with *Heteropogon contortus* or *Aristida barbicolis* on sandy soils, *A. rufescens* on eroded soils, *Bothriochloa glabra* in humid bottom soils, or *Trachypogon polymorphus* in gravelly soils. In the lower-elevation western parts of the highlands, an association of *Hyparrhenia rufa* and *H. contortus* is common, with significant populations of *A. rufescens.* Mid-level highland regions are commonly dominated by *A. rufescens,* but with a variety of communities dependent upon soils, local climate, and land-use history. Further east, and on the higher plateaus, where the dry season is shorter and winter drizzle possible, *Aristida* spp. mix with *Loudetia simplex, Ctenium concinnum,* and others. Finally, in the grasslands of the humid east, where a century or more of *tavy* has exhausted soils, the fire-maintained grasslands are dominated by *Imperata cylindrica, H. rufa,* and *A. rufescens.* The floristic poverty of these grasslands is striking (Perrier 1921; Bosser 1954).

In addition to such regional characteristics, grassland communities vary locally based on soils, land use, and microclimates. Afotsara illustrates this. In this case study site, one can recognize a number of major grass communities. A shortgrass prairie dominates on the cool, more humid 1,700 m.a.s.l. Sahanivotry plateau to the east, including *Aristida* spp., *Lasiorrhachis viguieri, bozamalemy* (unidentified), *C. concinnum,* and in fallow fields *Cynodon dactylon.* In the 1,300–1,500 m.a.s.l. central zone among settlements and crop fields, grass cover runs from monotone expanses of sparse *A. rufescens* to fallow-field pioneers (e.g., *C. dactylon, lavatanana* [unidentified], and *tsipipina* [unidentified grass with burr-like seeds]), to field and stream-edge communities (including *H. rufa, Pennisetum pseudotriticoides, Phragmites* spp., and *tsorodrotra* [unidentified]). To the west, *A. rufescens, A. similis,* and *C. concinnum* dominate the pastures above the *tapia* forest, from 1,500 to 1,700 m.a.s.l. Finally, the high-altitude (over 1,700 m.a.s.l.) pastures further to the west include the same species with the addition of *L. simplex.*

The established view of grassland dynamics in Madagascar is that frequent burning has replaced high quality, nutritious *H. rufa* and *H. contortus* pasture with low-value *Aristida* spp. (Perrier 1921; Kuhnholtz-Lordat 1938; Bosser 1954; Granier 1965; Cori and Trama 1979; Koechlin 1993; Herinivo 1994). However, in the most heavily burned regions of the island—in the Middle West—one continues to find plenty of *H. rufa* and *H. contortus*. This suggests that other factors, like climate or soils, also play a key role in grassland dynamics. Current range ecology theory suggests that grassland composition is nondeterministic and nonlinear. Transformations in Malagasy grassland communities probably follow multiple transitions between states, based on the seasonality and frequency of burning, grazing intensity, year-to-year precipitation and temperature variations, soil types, and historical legacies (see discussion of new ecology in chapter 2; Behnke and Scoones 1992; Kepe and Scoones 1999). Unfortunately, these dynamics remain poorly researched.[1]

The grasslands of Madagascar are central to the island's cattle economy. Some ten million head of cattle graze the grasslands.[2] Cattle raising contributes to both subsistence and monetary income (Raison 1968; Randrianarison 1976; Cori and Trama 1979; Ramamonjisoa 1994). Before colonization, Madagascar already exported meat to Mauritius and Reunion (Brown 1995). By the 1920s, each year at least 350,000 head of cattle were consumed locally and 30,000 exported. Six factories canned beef for export (Geoffroy 1923). In the last decade, beef continued to be a considerable yet erratic export earner,

1. The ecology of Malagasy grasslands has elicited relatively little research interest, due to the attraction of the forest (Koechlin 1972; Rakotovao et al. 1988) and to the widespread image of the grasslands as degraded pseudosteppe occupying once-forested lands. The primary sources are Bosser (1954, 1969), Granier (1965), and Granier and Serres (1966); Koechlin (1993) reviews the literature. Most effort by rangeland specialists focused on the introduction and cultivation of improved forage species (e.g., Razakaboana 1967). Research on grassland fire ecology is likewise in its infancy in Madagascar, owing to the long-established bias against fire as a legitimate range management strategy. Early efforts—not always related to rangeland management—included experiments at Kianjasoa in the Middle West in the 1950s (mentioned by Dez [1966] and Albignac [1989]), as well as research reported by Metzger (1951), Dommergues (1954), Bailly et al. (1967), Gilibert et al. (1974), and Rakotoarisetra (1997). A few encouraging recent efforts investigate fire ecology in or near protected areas (Razanajoelina 1993; Herinivo 1994; Bloesch 1997, 1999; Rakotoarisetra 1997; Schnyder and Serretti 1997; Rabetaliana et al. 1999).

2. The number of cattle, admittedly poorly known, has according to administrative counts hovered between 5 and 8 million between the 1920s and today; more realistic estimates by the Veterinary Service pin the number of cattle at approximately 10 million since at least 1980 (source: Direction des Services Vétérinaires, Ampandrianomby, Antananarivo). In some regions, such as Ambositra district, the bovine population has decreased in recent decades (MARA 1996; Aquaterre 1997; Kull 1998).

TABLE 3.4. Important grass species (family Poaceae). Includes only those mentioned in text.

Genus, Species	Local Names**	Comments
Aristida rufescens	horona (SV, NV), horombohitra (T), kofafavavy (A, N), horombavy, pepeka	widespread extremely fire-tolerant perennial; used to make grass roofs and brooms, little nutritious value when mature
A. similis	horombavy, kofafavavy, ahitsorohitra madinika (SV)	extremely fire-tolerant, perennial, hygrophile
**A. congesta*	kofafalahy (N), *three-awn*	smaller plant found in cooler, higher areas
* *Ctenium concinnum*	ahitsorohitra (SV), *sickle grass*	perennial, mediocre forage found in repeat-burned short-grass savannas > 900 m.a.s.l.; associated with *Aristida* and *Loudetia* spp.
* *Cynodon dactylon*	kindresy (N), fandrotrarana (SV), *couch grass, Bermuda grass*	creeping perennial grass, found worldwide in lawns and pasture; good forage; found near villages and on cool high pastures (> 1,600 m.a.s.l.)
* *Heteropogon contortus*	danga (SV, T), ahidambo (N), lefondambo, ahimoso, *spear grass*	perennial, common < 1,300 m.a.s.l. in association with *H. rufa;* moderately tolerant of fire; good forage when young
**Hyparrhenia cymbaria*	verobe, verotsanjy, vero, verovato, *thatching grass*	large perennial, intolerant of fire, found on good deep soils, at forest edges
* *H. rufa*	vero, verofotsy, veromena, fataka, *red thatching grass*	perennial, good forage (esp. in association with *H. contortus*), less tolerant of fire
* *Imperata cylindrica*	tenina (SV), manevika, fehena, antsoro (NV), *cottonwool grass*	perennial, rhizomous, < 2000 m.a.s.l.; in dry, well-drained spots, not on eroded soils
Lasiorrhachis viguieri	haravola (SV, NV)	used to make straw mats, baskets; cattle eat new shoots only; in cool humid highlands remains green through winter
**Loudetia simplex* (var. stipoides)	horona (A), horo (N), berambo, kilailay, kirodrotra, felika (SV), *common russet grass*	perennial, widespread, medium pastoral value, quite fire resistant, found on leached, poorly drained soils

TABLE 3.4. (*continued*)

Genus, Species	Local Names**	Comments
* *Panicum maximum*	verotsanga (T), *guinea grass*	forage species, in alluvium and colluvium
**Paspalum spp.*	gebona, gebo (N)	high pastoral value esp. March to May
Pennisetum pseudotriticoides	horompotsy (NV), diña (SV), tohiambazaha	endemic, 1,000–2,000 m.a.s.l., found on humid, cool, silty, or sandy alluvions
* *Phragmites spp.*	bararata	found in humid soils along streams

* Pantropical species.
** Regional variations on local names: SV = southern Vakinankaratra (i.e., Antsirabe); NV = northern Vakinankaratra (i.e. Ambatolampy); T = Tsiroanomandidy; A = Ambositra; N = Namoly-Andringitra. English names from South Africa.
Sources: field interviews, observations, and Bosser (1969), as well as Ratsimamanga (1968), Gade (1996), and Bloesch (1997).

with 2,000 tons/yr exported to the European Union before a 1998 ban. Some 1 million head of cattle were being slaughtered annually (EIU 1998, 2001).

While the colonial days saw some large-scale commercial ranching, cattle raising is largely a household affair today. In the central highlands, about half of households own cattle, averaging about four head per cattle-raising household. This amounts to about 0.1 to 0.3 cattle per person in these regions.[3] Dairying is as common as raising cattle for beef, traction, and savings. Elsewhere in the highlands and in the west, beef cattle form a greater part of the household economy. In these regions, there are upwards of 1.0 cattle per person, and individual herds can include hundreds of head of cattle. In some areas, investors practice a kind of contracted cattle raising called *dabokandro* (see description of Kilabé in chapter 2). In the eastern deforested regions, cattle raising is also common, but at lower densities than in the highlands—under a third of households have cattle. The net flow of cattle trading is from the west towards the highlands and east coast (AGM 1969; Projet MADIO 1997).

3. In the Antsirabe region, 69 percent of households own cattle, an average of 3.6 head per cattle-raising household. Some 71 percent of cattle owners sold cattle in 1997, an average of 1.2 head per seller, valued at 715,000 MGF, equal in 1997 to US$160 (Projet MADIO 1997). In Afotsara, 34 percent of households own cattle, averaging 4 head per household (ranging from 1 to 20), with a total of 130 cattle for 105 survey households.

Bush Encroachment

Malagasy cattle herders rely on fire to maintain their pastures. One of the key roles of repeated mid or late dry season fires in rangelands is to maintain grass dominance and avoid the encroachment of woody species. Without such fire, grass litter begins to accumulate and enrich the soil organic horizon. Depending upon the climate, seeds, and soils, woody plants and other nonherbaceous species may invade, and pasture is lost. Common Malagasy invaders include the leguminous *Sarcobotrya strigosa,* the guava bush *Psidium guyava, Erica* spp., and *Helichrysum* spp.[4] Various fire exclusion experiments in the Malagasy highlands—planned or unplanned all demonstrate some form of bush encroachment. At the Kianjasoa research station near Kilabé in the Middle West, forty years of exclusion resulted in impenetrable brush, extremely flammable and of no value to cattle (Albignac 1989). In the altimontane prairie of Andringitra National Park, fire exclusion led to invasion by ericaceous brush (especially *Erica* spp.) and reduced plant diversity, for example among orchids (Koechlin et al. 1974; Rabetaliana et al. 1999). In the field sites of Afotsara, Tsimay, and Behazo, in the central spine of the highlands, unburned pastures develop thickets of mimosa,[5] bracken fern, and shrubs like *Erica* spp. and *Helichrysum* spp. In the Kilabé case study, pasture encroachment by leguminous *rohy,*[6] once established, is difficult to control—its seeds are not killed by fire. Thus frequent burning is necessary to avoid its establishment in the first place. In all cases, bush encroachment on pasturelands reduces the grazing potential and is actively fought against by regular burning.

Pasture Renewal: The "Green Bite"

Pasture fires also renew grassland vegetation, physically removing old grasses, releasing their nutrients, and stimulating a flush of new growth. After the rainy growth season (December through March) most major grasses

4. Granier (1965), Cori and Trama (1979), Koechlin (1993); see Holland and Olson (1989) for a detailed analysis of Malagasy woody pioneers.

5. An exotic plant, mimosa grows spontaneously in many zones over about 1,200 m.a.s.l. and is known as an excellent nitrogen fixer. Mimosa is either *Acacia mearnsii* (black or tan wattle) or *Acacia dealbata* (silver wattle); these two species are very similar and the literature often does not distinguish between them (NAS 1980; see also Borie 1989). This Australian leguminous tree was introduced to Madagascar by the French—they planted mimosa along the Antsirabe train line in the 1920s to meet energy and tannin needs (AOM mad-ggm D/5(18)/9). The species is considered a noxious weed in Hawaii and has been called "green cancer" in South Africa (NAS 1980).

6. Interview, Kilabé, July 1999. Also called *tsilonavotra;* unidentified.

begin to lignify, or harden with age. As the dry season progresses—aided in higher areas by winter frosts—the stalks dry out and become poor in nutrition and largely unpalatable to cattle. At this point, herders pasture their cattle on crop stubble in fallow fields and on streamside vegetation, not in pastures. In some regions they also feed the cattle rice straw, cassava, or hand-cut grasses. The late dry season, September through November, however, is increasingly difficult for cattle, as crop stubble and rice resprouts in the valley bottoms are plowed under in preparation for the rains and the pasture grasses no longer have any nutritious value. With increasing temperatures and the first tentative rains (September or October), pasture grasses begin resprouting.

Burning plays several roles at this point. First, it removes the dry stalks of old grass that can impede the access of cattle to the small new shoots.[7] Second, it releases inaccessible nutrients in the old grass back to the soil, fertilizing the new growth. Third, burning overrides the competitive effects of selective grazing, giving favored forage species a better chance.[8] Fourth, by exposing the soil to the sun, in areas of sufficient moisture fire may accelerate the growth of the resprouts. In fact, burning provokes a resprout at any time of year, though the vigor and speed vary.[9]

This "green bite" (Jolly 1989) is critical to the health of the cattle, for the protein-rich resprouts—even of poor-quality forage species like *A. rufescens*—carry the cattle through the annual hungry season. When asked to justify pasture fires, informants' responses included "so that the cattle are full" and "what would the cattle eat without fire?"[10] The logic of the green bite also explains the hesitance of Malagasy farmers to engage in *feux tardifs* (fires very late in the dry season and into the rains, late November and December) and *feux précoces* (fires very early in the dry season, March and April). These alternative burning strategies were suggested by technical experts to mitigate

7. The removal of *danga* (*H. contortus*), also called "spear grass," is especially important, for its spiny seed heads are known to injure cattle, sometimes blinding them (Tachez 1996; interview, Ambatofinandrahana, 28 June 1999)

8. In light grazing systems with low numbers of cattle per unit area, cattle eat only their preferred forage grasses. This favors the growth of less nutritional grass species. Fire overrides these competitive effects by reducing all grasses to ashes. In essence, this creates a clean slate for all grass species (L. Huntsinger, personal communication).

9. Cori and Trama (1979); Mainguet (1997); interview, Afotsara, 2 Dec. 1998.

10. Multiple interviews. These quotes: Afotsara, 6 Dec. 1998, and Behazo, 25 June 1999. Blanc-Pamard and Rakoto Ramiarantsoa (2000, 76) even mention the collection of *A. rufescens* as valuable forage (though the vernacular name they assign to the plant, *horompotsy*, may indicate *Pennisitum pseudotriticoides*; see table 3.4).

the environmental consequences of overburning the island (Metzger 1951; Dez 1968). However, they do not provide the green bite necessary to cattle when they are most hungry.

Strategies and Logic of Pasture Fires

The goals of pasture burning are clear: defending rangelands from bush encroachment and assuring cattle survival by providing a nutritious green bite. As we will see below, they also reduce populations of bothersome ticks and protect the hills against unpredictable and destructive wildfires. Most burning is accomplished through an opportunistic strategy of temporal and spatial rotation. That is, people burn pastures in patches, and they burn these patches at different times, taking advantage of the best moments—both ecological and political—to do so. Below I describe these strategies and defend their overall logic.

Pasture burning follows an unseen, unwritten management plan. Management is opportunistic, in the sense that people exploit environmental opportunities (such as rainfall or vegetation growth; see Westoby et al. 1989 and Behnke and Scoones 1992) as well as social opportunities (like a reduced chance of enforcement, or happening to be passing through an area that needs burning) to ignite fires.[11] A father tells his son to light the grass on the way home from weeding a distant cassava field if the wind is right; a group of elders walking to a *famadihana* note an area that "should" burn and light it; or an escaped crop field fire headed towards an area that "needs" burning is allowed to continue. Such opportunistic management matches that of rural peoples elsewhere—the Australian aborigines set fire to the land when passing through, when fuel and dryness are sufficient, and when burning is necessary (Lewis 1989).

This opportunism in Malagasy fire management is often seen by observers as casual and careless, or even irrational and pyromaniac (Einrem 1912; Perrier 1921; RDM 1980). Indeed, "careless" is an epithet flung at fire users around the world: at African herdsmen, Californian sheepmen, Native American hunters and gatherers, and European peasants (Pyne 1995, 1997). This is due to the contrast between opportunistic strategies and Western, scientific, modern, or governmental expectations of written management plans, official meetings, permits, and fire trucks at the ready, or more broadly of standardized planning, uniform procedures, straight lines, and rigid zones (Scoones 1996; Blumler 1998; Scott 1998; Zimmerer 2000).

11. As most burning is illegal and could result in fines or imprisonment—though it rarely does—people burn when the authorities are distracted or weak, such as during locust invasions or elections. See chapter 6.

Despite the opportunism, the result demonstrates a distinct logic. Sufficient areas burn to maintain pasture, and spatial and temporal rotation ensures resource patchiness and fire control. About one-quarter to one-half of Malagasy pastures burn annually, so it can be deduced that an average plot burns about once every two to four years. This is consistent with examples of fire use in pasture management around the world, where fire frequencies vary from annual, to three out of every four years, to once every two to four years (Kozlowski and Ahlgren 1974; Gill et al. 1981; Gillon 1983; Ojima et al. 1994; Campbell 1996; Pyne et al. 1996; Mistry 1998). In general, areas of higher precipitation and less-intensive grazing allow more frequent burning.

People ignite fires throughout the year. My observations from Afotsara in 1998 demonstrate this (figure 3.1). In April and June, many fires were associated with locust invasions. These served to ward off the pests, but also to renew pastures and maintain woodlands without the risk of punishment. Winter fires in June and July began the normal cycle of pasture renewal. These areas provided the earliest green bite in late September, when the first cattle herds moved out of rice fields and up into the pastures. August, which marked the beginning of warmer spring days and reduced humidity, resulted in a great batch of pasture-renewing fires. By September and especially in the heat of October, burns were clearly destined to "fill in" unburned patches. In October and November, a large number of fires were small crop-clearance burns. The coming of the rains in November and December slowed the burning.[12]

As figure 3.1 illustrates, fires burn different areas in succession, filling in gaps as time goes on. Research shows that early dry season burns are better for woody species, while mid and late dry season burns benefit grasses (Kayll 1974; Campbell 1996; Bassett and Koli Bi 2000). The effects of a successive rotation of fires are multiple. First, early dry season fires in the woodland favor woodland maintenance, as shown in chapter 4. Second, the yearly burn rotation in grasslands results in a patchier grassland, which combines zones of unburned grass (also critical to collection of grass thatch for roofing material), early fire zones, midseason fire zones, and late fire zones. This provides cattle with grasses in different stages of development and different species assemblages. It gives cattle a place to go in September for the first resprouts and

12. In grassland zones of the humid east, fires also burn at a variety of times, taking advantage of drier moments. Most pastures are burned in September and October (Pfund, personal communication 2001; interviews, Fort Dauphin, August 2001) but fires are also seen in June through August (personal observation).

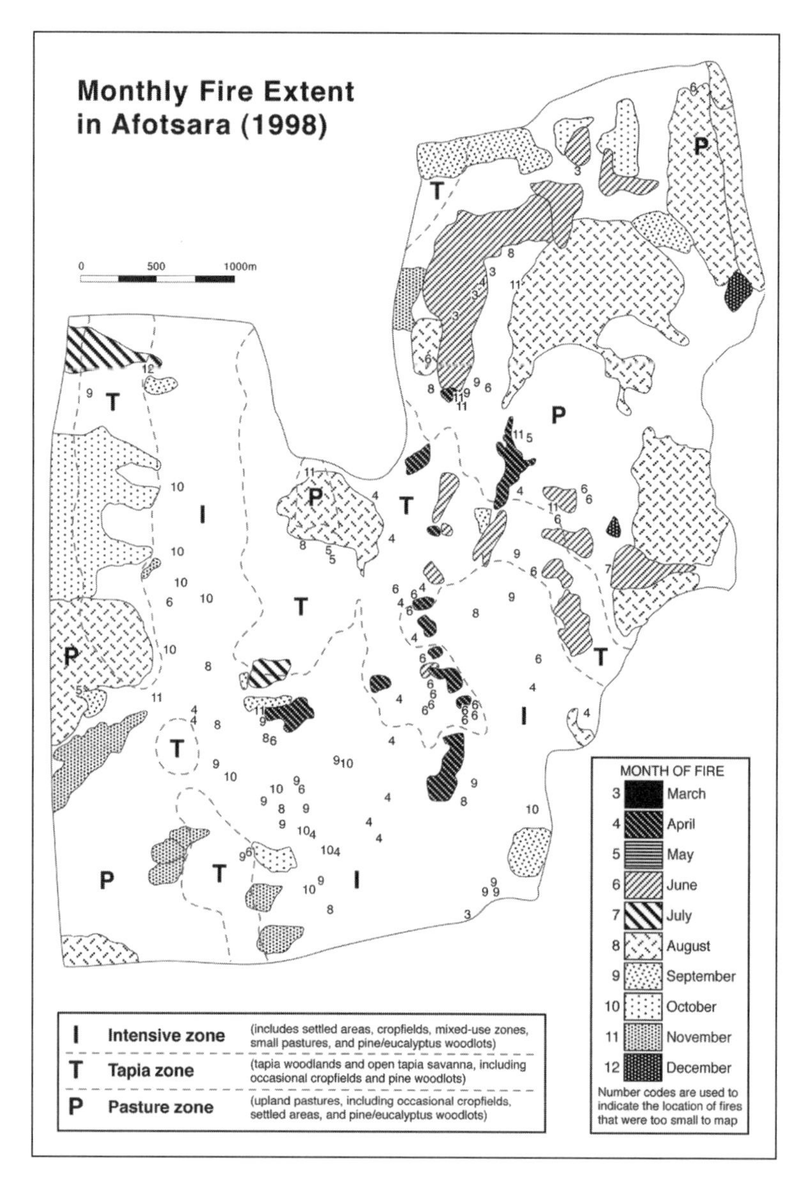

FIGURE 3.1. **Monthly fire extent in Afotsara, demonstrating the spatial and temporal mosaic of burning.** Map covers 18 km² surveillance area from March to December 1998.

other places for November resprouts.[13] This patchiness probably also serves as insurance against the vagaries of climate, such as early rains (which cut off the burn season).

Fire patch rotation also serves as a built-in protection against running wildfires. Fires lit at the beginning of the dry season and during cool, overcast, humid winter days tend to progress slowly and burn incompletely and are easily controlled. Burning piece by piece, month by month, creates a patchiness in the grassland landscape that later serves to check the spread of hotter, hard-to-control late dry season fires, including the inevitable accidental burns resulting from escaped crop fires. As Pyne (1997, 90) notes, fire spreads quickly to untended, overgrown sites, and the solution is "good housekeeping," burning the land in patches, and burning the patches at different times. Such sequencing of fire through the dry season is common across the tropics, in West Africa (Fairhead and Leach 1996; Laris 2002), Australia (Lewis 1989), and Kruger National Park in South Africa (van Wilgen et al. 1990).

Pastoralists and herders in Madagascar largely practice a form of extensive or semiextensive cattle raising, relying for at least a portion of the year on vast communally held pastures on the hills and ridges outside village farmlands and valley bottom rice fields. In extensive range management, land area is large, cattle numbers are low (limited by dry season forage supply), and capital and labor inputs are low. Low stocking densities result in *undergrazing* during the wet season,[14] thus necessitating some form of pasture management. Burning is the most efficient, cost-effective, and appropriate solution (Vogl 1974; Mistry 1998). Viewed from a logic of the Western, technical ideals of intensive production (including fenced grazing, mowing hay, stall feeding, fodder cultivation, and improved breeds), this extensive system seems illogical or wasteful, as it did to colonial specialists and administrators. Their efforts thus focused on "modernization" (intensification), stopping fires and perceived overgrazing (Scoones 1996). Clearly, however, for the

13. Interviews, Afotsara, 3 Dec. 1998; Behazo, 24 June 1999.

14. While many claim that Madagascar is overgrazed due to the pressure of ever-larger herds on smaller land reserves (Olson 1984; Koechlin 1993; Allen 1995; Wells and Andriamihaja 1997), herd size has not been increasing dramatically (note 2 above) and pasture areas remain vast. In fact, Malagasy pastures are *undergrazed*. At Ankazobe, for example, range specialists found the ideal stocking rate to be 4 to 8 ha per adult cow, while actual stocking rates only placed one head in each 15 to 17 ha (Schnyder and Serretti 1997). In the dry season, cattle graze in the fallow valley bottom fields—the rangeland being too dry—yet in the rainy season the rangeland grows much too fast for the cattle to keep up, resulting in undergrazing (Granier and Serres 1966; Cori and Trama 1979).

Malagasy farmers and herders, burning is a well-adapted, highly effective, economical, and logical tool for pasture management (Cori and Trama 1979).

AGRICULTURAL FIRES

Burning also plays a key role in agriculture, whether in humid forest zones (see chapter 5 on *tavy* in the east) or in the open, drier landscapes of the highlands and west. One-quarter of fires I logged during 1998 in Afotsara were specifically for crop-field preparation, though they accounted for only 0.5 percent of the total burned area (table 3.3). Typical crop field fires burned between 0.01 and 0.2 ha; these fires were most common in the months leading up to the rainy season, especially October and November.

In crop field fires outside of forest *tavy* plots, farmers burn the standing vegetation in the plot they intend to cultivate. The vegetation may be uncultivated grassland, long-fallow fields covered with grass, ferns, or bushes, or short-fallow fields covered with grass, weeds, and crop stubble. In Afotsara, agriculture has spread to ridge top grasslands. After burning a plot, a team of two to three farmers labors with spades to turn over large dry clods of the upper layer of soil, thus burying the nutritious ashes. Later, when rain has softened the clods, they break them up, level the field, and plant the crop—typically cassava, sweet potatoes, beans, or corn. Farmers recognize that preparing a field without burning is feasible, yet it requires more time and labor.[15] The laboring of unburned fields is practiced in the Behazo case study (see below), where dense settlement makes fire impractical.

In some cases, farmers collect additional fuel to burn in their fields. Most typically, piles of dry rice straw or cut grasses are burned in dry paddies destined to be flooded and used as rice nurseries. In the Tsimay case study, farmers cover future sweet potato fields with a shrub cut from nearby hillsides, *rambiazina* (*Helichrysum bracteiferum*) and then burn this shrub to provide a fertilizer input (Blanc-Pamard and Rakoto Ramiarantsoa 2000). This practice is similar to the much-documented *chitemene* system of southern Africa, where farmers collect woodland branches from around a field to burn before cultivation (Richards 1939; Bartlett 1956; Campbell 1996).

Other agricultural fires take place outside the actual crop field. First, fire in the catchment basin above rice paddies can encourage the erosion of important nutrients or soil particles into the paddies. Farmers thus manipulate fire and erosion to concentrate nutrients or soil particles where they will be most

15. Interview, Afotsara, 23 Nov. 1998.

productive. This has been noted in Sri Lanka (Stocking 1996) and in Madagascar (Dez 1966; Wells and Andriamihaja 1993; Rakoto Ramiarantsoa 1995a; B. Locatelli, personal communication, 1996). One farmer in Afotsara described this use of fire, stating however that it gives only a small advantage, for it can also lead to problems of siltation and hillside soil degradation.[16] Second, fire in the catchment basin above rice paddies can also facilitate quicker runoff during the first rains, aiding the all-important quest to fill one's rice paddies with water for transplanting (Borie 1989; Jolly 1990; Rakotomanana 1993; Schnyder and Serretti 1997). In Afotsara, for example, the start of the rainy season stalled for about two weeks in November 1998. Rice seedlings in the nursery fields were aging, for there was insufficient water to flood rice fields and transplant the seedlings. Tensions were high and accusations of water theft abounded. At this time, over a period of two weeks, someone systematically burned—night after night—the steep slopes above one irrigation zone. While the evidence is circumstantial that this was the burner's purpose, when the rains finally did come, the burning ensured the prompt delivery of the water to those rice fields, as bare ground reduces retention and infiltration and leads to faster runoff.[17]

Finally, farmers burn to clean irrigation canals and field edges. Just like Joe Mondragon's New Mexico father in *The Milagro Beanfield War* (Nichols 1974), farmers drop a match into irrigation ditches to clean them out before irrigating. At least two such fires escaped into wildfires in Afotsara in 1998. Field edge fires—such as along the berms between rice paddies—facilitate travel along the berms, contribute to a clean landscape, and, most importantly, help control pests.

PEST CONTROL

Fires help control a wide variety of pests harmful to crops, livestock, and humans (Grandidier 1918). Locust invasions, in particular, lead to the widespread use of fire. Locusts swarmed out of gregarious zones in southwestern Madagascar six times in the last century, each invasion lasting three to ten years (Olson 1987). When locusts arrive, people burn for two reasons: because smoke can ward off locust swarms in flight, and because burning where locusts rest at night kills many of them, facilitating collection.[18] Peo-

16. Interview, Afotsara, 19 Nov. 1998.

17. A focus group on 14 Aug. 2001 in Afotsara emphatically disapproved of this practice.

18. French colonial officers in Algeria recommended the use of fire to fight locusts in the late nineteenth century (Kunckel d'Herculais 1890); officers noted and approved the same practice in Madagascar (anonymous 1904).

ple also burn for a third reason: locust invasions serve as a convenient pretext for burning for other goals, such as pasture renewal, without the fear of state prosecution (see chapter 6).

The latest invasion began in late 1997 and captured public attention throughout 1998 and 1999. Locust swarms arrived in the Afotsara area at least fifteen times from April through September 1998. The cries of children echoed up and down the valley—"v*alala ôôôô! valala ôôôô!*"—each time a swarm of locusts appeared over the mountain ridges. The black clouds of millions of locusts, the rustling of their wings audible against the wind, are a formidable sight. Immediately after the alarm is sounded, smoke billows from a hundred points in the valley. People light straw piles along their rice fields and stand in ripening fields waving long sticks with plastic bags attached in order to ward off the pests. Others ignite the grassy slopes behind their homes, hoping to contribute to the smoke driving the locusts away. Once the swarm has passed, in the evening, the horizon is aglow with fire in the zone where the locusts settled down for the night. Before dawn, villagers depart with baskets and sacks, eager to collect locusts that have been burned as they rested. The bushels of locusts are eaten fried as a welcome supplemental protein source or dried and used to fatten up chicken and pigs.

The locusts caused much fear at first, but only isolated crop damage in Afotsara (a few individual farmers, with fields completely devoured, did suffer). For the average family, locusts may have been more good than bad, a welcome infusion of a "free" protein. By the end of 1998 in Afotsara, over a third of the area burned—or 75 of 155 fire scars—was blamed in the first instance on locust invasions (many fires probably served other purposes—like pasture and wildfire maintenance—at the same time).

Fires also control a variety of other pests. Small cleaning fires (*feux de nettoiage*) that remove brush and weeds from irrigation canals, crop field edges, and around residences also mitigate against rat populations. Around rice fields, burning serves to destroy the brush in which rice-thieving birds such as the *fody* (*Foudia madagascariensis*) perch between raids.[19] The burning of rice stalks after harvest can help control rice borer infestations (Zahner 1990). Fires also help control populations of ticks, a perennial annoyance of cattle.[20] In addition, fires may control mosquito habitat and populations,

19. The thefts by these birds are acutely felt; in November and December, those among Afotsara's farmers who had cultivated early season rice arranged around-the-clock watches to scare off *fody*. In Tsarahonenana, north of Antsirabe, locals noted with disdain the increase in *fody* populations linked to twenty years of hillside afforestation and fire control (Blanc-Pamard et al. 1997).

20. Humbert (1949); Bartlett (1956); Homewood and Rodgers (1991). In 1953, colonial au-

thereby reducing the spread of malaria (Einrem 1912; Grandidier 1918; Bartlett 1956). Finally, as we will see in chapter 4, people burn in *tapia* woodlands to protect wild silkworms from a parasitic ant.

FIRE TO FIGHT FIRE

The most effective means of avoiding uncontrollable or destructive wildfires is probably fire itself. There are three ways in which people use fire to fight fire. First, as discussed earlier, the rotation of fires in pasture management serves to reduce the danger of wildfires by reducing fuel loads. In fire-prone regions, the more biomass there is, the graver the danger of massive hot wildfires that are difficult to control and damaging to the soil and trees. Frequent low-intensity fires, in contrast, reduce fuel loads and are easier to control (Minnich 1989).[21] In fact, the Malagasy consider a landscape that is regularly burnt and covered with grasses as opposed to bushes as *madio* or "clean."[22]

A second fire-avoidance strategy is firebreaks. As described earlier, a rotation of pasture fires through the burn season facilitates the control of later fires by establishing a patchy landscape of informal firebreaks. Some people deliberately burn strips of land early in the dry season to establish firebreaks around settlements, woodlots, and parks. Only one man in Afotsara burned such firebreaks; he was a rich farmer settling new terrain among vast pastures, and he relied on firebreaks to protect his crop fields and pine plantations from pasture fires. Elsewhere in the highlands, firebreaks (burned or mowed) were visible around private woodlots or prescribed for national parks (e.g., Andringitra and Ankarafantsika). It is the responsibility of people with homes near oft-burned pastures to keep their yards free of dry weeds to avoid house fire danger.[23]

A third, last-minute use of fire against fire is backfires (French *contre-feu*), a technique whereby a wildfire is fought by lighting a controlled fire

thorities in Ambovombe reluctantly overlooked antifire prejudices and authorized a prescribed pasture fire to fight a devastating tick epidemic that had killed one-third of the cattle in the region (ANM IV.D.73, subfolder 1/5).

21. Minnich has also said, "Fuel, not ignition, causes fire. You can send an arsonist to Death Valley and he'll never be arrested" (cited by Davis 1998, 101).

22. Bloch (1995) writes that in Madagascar's forest zone, Zafimaniry people preferred post-*tavy* landscapes that were "clear" (*mazava*) and "open" (*malalaka*). A similar logic of preferring a clean, well-managed landscape is expressed in southern Mali (Laris 2002) and among the Aborigines of Australia (Head 2000). According to an Aborigine elder, a south Australian firestorm that killed seventy-two people was a crime, a "crime in allowing a country to get so dirty," that is, not keeping it clean with regular burning (Lewis 1989, 940).

23. Interview, Afotsara, 15 Oct. 1998.

from a defensible firebreak (taking account of wind and slope) in order to head off the wildfire and not allow it the chance to jump the firebreak (RDM 1980; Pyne et al. 1996). When fighting the 21 July flare-up in the *tapia* woodlands, the villagers in Afotsara demonstrated a clear understanding of the use of backfires (see introduction to part 1).

FIRE AND FORESTRY

Fire is also used for purposes of forestry. As chapter 4 details, burning plays a key role in the maintenance of *tapia* woodlands. Here I highlight the role of fire in the management of the pine and eucalyptus plantations found around the island.[24] Fire can be both useful and harmful in such forest plantations: useful in provoking regeneration and clearing out undergrowth to avoid wildfires, but harmful if seedlings or adult trees burn. Fire is either necessary or conducive to regeneration in a variety of forests from Californian sequoias to Indian *sal* (Pyne 1997). As one nineteenth-century observer wrote: "Impossible to raise a finer crop of pines than that which follows fire" (Pyne 1997, 115; see also Borie 1989). In Madagascar, the chief plantation species are pine (*Pinus patula,* from Mexico, and *P. kesiya,* from southeast Asia) and eucalyptus (*Eucalyptus grandis* and *E. robusta*). The pines benefit from fire to open the seed and facilitate regeneration, though fire is not necessary. The eucalypts can also be managed by fire: since they resist fire better than the endemic brush, annual burning allows these eucalypts to proliferate.[25]

An important side effect of fires in woodlots and forests is that people may freely collect as fuelwood the dead and downed wood that results from fire (Andriamampionona 1992). *Miteraka kitay ny afo,* people say: "Fire gives birth to fuelwood." As long as people have the right to collect dead wood from forests—whether a legal use right in the case of state forests or community traditions in the case of private or community forests—then fire may sometimes benefit people by creating firewood.[26]

OTHER USES OF FIRE

Fire has myriad other uses (table 3.1). A clean landscape facilitates the surveillance of cattle or—in times of conflict—of enemies. It eases foot travel

24. For more information on these plantations, see ANM D76s, Aubréville (1954); Vignal (1956b); Ramanantsoa (1968); Olson (1984); Gade and Perkins-Belgram (1986); Rahamalivony (1989); Raharivelo (1995); Rakoto Ramiarantsoa (1995a, 1995b); Bertrand (1999).

25. Interview, Projet PDFIV, Ambatolampy, 26 Apr. 1999; also Bertrand (1999).

26. Interviews, Afotsara, 12 Sep. 1998 and 9 Dec. 1998; also Favre (1996).

by reducing the annoyance of scratchy grasses and burrs and by making footpaths easily visible. The highway authorities sometimes burn the grasses along highway edges, increasing visibility. During the dark of the moon, fires can serve to light the way or serve as beacons.[27] Would-be mineral and precious stone exploiters profit from fires to easily survey surficial stone deposits (Dez 1966) or to increase erosion for hydraulic gold exploitation (Rajaonson et al. 1995). A certain rhythm of fire, followed by protection, encourages the growth of hard-stalked *Aristida rufescens* grasses, prized for their use in brooms and roofing material (Andriamampionona 1992).

Finally, fire is a spectacle for the senses, from its crackling roar to the flickering flames that light up the night. Fire serves as a distraction, as celebration, and as the most natural form of firecrackers. On 25 June, the eve of the national holiday,[28] families light bonfires beside their homes and children run up and down the paths carrying *iloilo* (burning tufts of hay attached to long poles). Others just light the natural canvas behind their homes to celebrate, but also perhaps to renew the pasture. The spectacle is not restricted to holidays: Coulaud (1973, 162) describes how the children watch *tavy* fires with a "festive ambiance." Perrier (1927) wrote of a Professor Lemoine, who out of sheer joy at the spectacle lit the summit area of Madagascar's highest mountain range, Tsaratanana.

WHERE FIRES BURN AND WHERE FIRE IS SQUEEZED OUT

We have seen that fire is a useful tool relied upon by the Malagasy peasants. However, fire use across the island is far from uniform. The vast, open grasslands of the Middle West burn much more than the dense agricultural lands of northern Betsileo country, for example. I argue that people adapt their use of fire to the local context. They know where and when to burn, and where and when not to burn. To show this, I compare five case studies across the highlands with very different fire patterns. The chief contextual factor shaping fire use is the intensity of land use—whether due to demographic or market forces; additional factors include the climate, location, history, and local politics. I survey the case studies (introduced in chapter 1) from north to south; the salient points of comparison are summarized in table 3.5 and discussed in the conclusion.

27. Moss (1876) writes of lighting fires for this purpose; also interview, Afotsara, 3 Dec. 1998.

28. Before Independence, the equivalent date was the French national holiday, 14 July; before colonization similar burning took place during the festival of the Royal Bath, Fandroana, a ceremony marking the New Year (Dez 1966).

TABLE 3.5. Comparison of the five highland case studies.

	Location	Biophysical Zone	Population	Cattle	Economy	Fire Use
	(a) name of region (b) *fivondronana* (administrative region) (c) specific location	(a) altitude (m.a.s.l.) (b) annual rainfall (mm) (c) landscape (d) climate	population density (pop/km²) for (a) *fivondronana*[a] (b) commune (c) study site	cattle owned per capita for (a) *fivondronana*[b] (b) commune (c) study site	character of the local production economy	(a) summary of *local* fire use (b) INFA[c] for *fivondronana* (c) INFA ranking among all 111 *fivondronana*
Kilabé	(a) Middle West (b) Tsiroanomandidy (c) within 30 km of Tsiroanomandidy	(a) 800–1000 (b) 1500–1600 (c) plain dissected by dendritic valleys (d) sunny, long dry season	low (a)19 (b)20 (c) na	high (a) 0.8 (b) 0.95 (c) na	cattle ranching and agricultural expansion	(a) very high (b) 2.86 (c) 9
Tsimay	(a) north Vakinankaratra (b) Ambatolampy (c) within 15 km of Ambatolampy	(a)1500–1700 (b)1400–2000 (c) hilly (d) chilly, humid winters	high (a)116 (b) 84 (c) na	low (a) 0.1 (b) 0.19 (c) na	diversified agriculture, livestock, dairy, fruit, and woodfuel	(a) no fire (b) 0.21 (c) 85

Afotsara	(a) south Vakinankaratra (b) Antsirabe II and Ambositra (c) Manandona-Sahatsiho valley	(a) 1300–1700 (b) 1500 (c) deep valley (d) transition zone	medium (a) 94 (b) 18 (c) 97	medium (a) 0.2 (b) 0.33 (c) 0.22	agriculture, livestock, and woodland products	(a) high (b) 1.40 (Antsirabe II); 2.51 (Ambositra) (c) 35 (Antsirabe II); 13 (Ambositra)
Behazo	(a) north Betsileo (b) Ambositra (c) Andina *commune* west of Ambositra	(a) 1250–1700 (b) 1400 (c) steep hills (d) typical	high (a) 56 (b) 63 (c) 200	medium (a) 0.1 (b) 0.3 to 0.4 (c) 0.31	diversified agriculture, livestock, and fruit	(a) low (except ridgetops) (b) 2.51 (c) 13
Andringitra	(a) south Betsileo (b) Ambalavao (c) northern periphery of Andringitra National Park	(a) 1400–2000 (b) 1000–2000 (c) mountain + plateau (d) east: cool + winter drizzle; west: sunny	low (a) 28 (b) 42 (c) 28	high (a) 0.6 (b) 1.20 (c) 0.87	agriculture, cattle ranching; eco tourism and presence of National Park Project	(a) high; some controlled (b) 2.12 (c) 20

Notes

na = data not available.

[a] Source: 1993 census.

[b] Source for cattle statistics: Direction des Services Vétérinaires.

[c] Index of Night Fire Activity. Based on DMSP satellite images, Aug.–Dec. 1997.

Source: PACT, Antananarivo (see figure 1.3).

Kilabé: Big Fires in Open Cattle Country

Big pasture fires burn annually across the Middle West, the lower western portion of the highlands. The districts (*fivondronana*) of this region all rank among the most incendiary 18 percent of the island's districts.[29] Kilabé, as described in chapter 1, is a village located near Tsiroanomandidy, an administrative town with the country's largest cattle market. Villagers in this warm, sunny, and lightly populated region depend upon agriculture and cattle raising for their livelihoods.

The pastures of Kilabé are critical to the livestock portion of the local economy (figure 3.2). The local tallgrass prairie grows over peoples' heads in the warm climate, despite annual fires. The prairie varies between nutritious *H. rufa–H. contortus* associations and less nutritious *A. rufescens,* as well as *Panicum maximum.* Introduced grasses or weeds escaped from the former state farm are sometimes found around villages or along streams, especially *Stylosanthes* and *Brachiaria* spp. Cattle rely on fallow rice fields and riparian vegetation during the dry season, and on the hillside grasslands (the *tanety*) for most of the year.

These *tanety* pastures are maintained through annual fires, as the cattle raising system is extensive. From June through November, fire is a constant presence. Smoke pillars grace the horizon each afternoon; flames glow on the horizon each evening. While Kilabé is clearly an extensive cattle-raising agropastoral system, it does show some signs of slow intensification: cattle are penned at night to collect manure and prevent rustling, some cattle receive cassava to supplement their diet, and crop fields are gently expanding into the *tanety* (AGM 1969; Raison 1984; Kull 1995).

Tsimay: No Fire for Milk and Wood Fuel

As chapter 1 noted, the village of Tsimay is notable for the way in which its hillsides are not covered by open grasslands, but by a brush of mimosa (wattle, or *Acacia* spp.; see note 5 this chapter) and *rambiazina* (*Helichrysum bracteiferum*) as well as woodlots of coppiced eucalypts. In contrast to Kilabé, the brushy *tanety* of Tsimay indicates that fire is rare. In fact, Ambatolampy is now one of the least-burned districts in Madagascar (table 3.5). All the same, fires of up to ten hectares do sometimes burn across the woodlots or in open pastures. More common are controlled crop fires: as described

29. Based on an index of night fire activity (INFA), a statistic averaged for the territory of each district from satellite images, August through December 1997 (see figure 1.3). I include as the Middle West Ankazobe, Betafo, Miarinarivo, Fenoarivo Centre, Tsiroanomandidy, Soavinandriana, and Faratsiho districts in Antananarivo Province; and Manandriana, Ambatofinandrahana, Ikalamavony, Ihosy, and Ambalavao districts in Fianarantsoa Province.

FIGURE 3.2. A view of the vast grasslands of the Kilabé case study.

earlier, a local innovation is to gather *rambiazina* shrubs and to burn them on fields destined for sweet potato cultivation.

Why does this region not burn like elsewhere? The primary reason is economic. In the densely populated periphery of Ambatolampy, it is simply more profitable to manage *tanety* lands for the urban wood fuel market, substituting extensive pasture-based cattle ranching with more intensive dairy production.[30] In a zone stretching 10 km north and south from the city, approaching the Ankaratra mountains to the west, and 5 km to the east (figure 3.3), there exists a periurban wood fuel zone. Each highland city has such wood fuel zones, e.g., Antananarivo's charcoal belt of coppiced eucalypts (Manjakatompo and Mantasoa), the pine-covered mountains north and east of Antsirabe (especially Vinaninony), and Fianarantsoa's wood fuel zone centered around Ambohimahasoa (Rakoto Ramiarantsoa 1993, 1995a, 1995b; Bertrand 1995, 1997, 1999).

In Ambatolampy, a small city of 20,000, the majority of the population relies upon charcoal and firewood for cooking. Official statistics for forest

30. Rough calculations confirm this statement. In extensive pastures, one head of cattle requires 4–8 ha (Schnyder and Serretti 1997). If cattle are sold every second year, for an optimistic 1,000,000 MGF, the gross proceeds (excluding vaccines, etc.) are 40,000–80,000 MGF/ha/yr. If the same hectare is used for mimosa fuelwood, gross proceeds are 180,000–360,000 MGF/ha/yr (assuming growth of 3 t/ha/yr, oxcart loads of 250–500 kg, and 30,000 MGF per load).

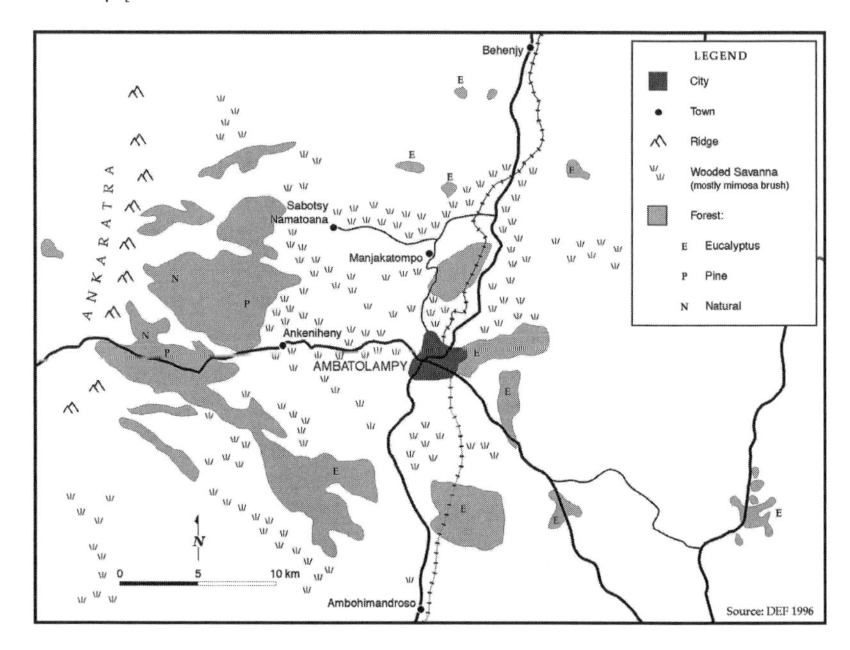

FIGURE 3.3. **The wood fuel zone of Tsimay, around Ambatolampy.** Source: DEF (1996).

exploitation in the 1,800 km² *cantonnement forestier* (table 3.6) show that more than half of charcoal production is "exported" to the capital, 67 km to the north, by truck. Meanwhile, 78 percent of firewood production is consumed locally. Typically, oxcart loads of mimosa firewood arrive at wholesale merchants in the city, who pay about 30,000 MGF (1999; US$5) per load. The firewood is resold to consumers in small bundles of seven to ten sticks of 1 cm diameter and 40 cm length, at a price of 50 MGF (1999; US$0.008). The most common use is for short-order cooking (not rice or beans) by streetside vendors. Charcoal, largely made from eucalypts, is sold for 5,000 to 10,000 MGF per 30 kg sack (1999; US$0.83 to US$1.67), or more commonly in small baskets or platefuls (see Blanc-Pamard 1998).

Only about one-third of lands in Tsimay are formally registered and titled with the state; these are typically large hillside plots of coppiced eucalyptus woodlots held by richer investors. The rest is governed by a traditional lineage-based tenure system. Formerly, open lands, while putatively belonging to a certain lineage, were open to all for grazing and shrub collection. Now, grazing and cutting of *rambiazina* is still open, but mimosa plots are all privately claimed due to their economic value.

Mimosa, introduced last century (see note 5), grows spontaneously on

TABLE 3.6. **Average annual wood product exploitation in the *cantonnement forestier* of Ambatolampy, 1996–1998.**

Product	Local use	Export (to capital)	Total
Construction wood	2,550 m^3	4,330 m^3	6,880 m^3
Firewood (largely mimosa)	3,000 m^3	867 m^3	3,867 m^3
Charcoal (largely eucalyptus; in sacks of 30 kg)	10,400 sacks	15,600 sacks	26,000 sacks

Source: DEF Ambatolampy.

unburned lands. Occasionally, people collect and spread mimosa seeds. Harvesting of this dense, high-quality firewood occurs in a very short rotation of two to six years; typical poles are from one to two meters long and from one to three centimeters in diameter (cf. NAS 1980). Eucalyptus woodlots are normally coppiced for charcoal production on a cycle of about seven years or when poles reach a diameter of about ten centimeters; each stump grows two to ten poles of five meters' length. Some people allow pine or eucalyptus woodlots to mature as a kind of savings account; when money is needed the wood is sold as construction lumber.

As a result of the wood fuel economy, the *tanety* in Tsimay is a mosaic of brushy mimosa, eucalyptus woodlots, and crop fields (figure 3.4). Mimosa is *zavatra sarobidy,* a precious commodity.[31] In addition to mimosa, common shrub species include *anjavidy* (*Erica* spp., consumed for firewood), *rambiazina,* and *tsipopoka* (*Nicandra physaloides*). Farmers harvest the leaves of *rambiazina, tsipopoka,* and an introduced hedge plant, *Tephrosia* spp., to make compost. The brush-covered *tanety* also hosts grasses. *Lasiorrhachis viguieri* dominates, along with *Aristida* spp., *tsivongo* (unidentified), *P. pseudotritricoides,* and *C. dactylon.*

The loss of open pastures, the density of the population, and the growing highland dairy industry have pushed a transition to intensive cattle husbandry. Rice straw is actively stocked and sold, and most cattle owners collect fodder plants—from spontaneous field edge grasses to specially cultivated oats, banana stems, or introduced fodder plants like *Pennisetum, Brachiaria, Chloris,* and *Setaria.* Most cattle are at least partly stall fed. At least one farmer in Tsimay village owns pure-breed European dairy cattle[32] and many local cattle are mixed zebu-dairy breeds. In the highland region as a whole,

31. Interview, Tsimay, 12 May 1999.

32. He owns a Pie Rouge Norvégienne bull, sponsored by the Norwegian funded organization FIFAMANOR, and a Pie Noire Française cow.

FIGURE 3.4. **A hillside of mimosa *(Acacia)* and other brush in the Tsimay region.**

milk production rose from 600,000 liters in 1960 to 10 million in 1990 (EESR Lettres 1992).[33] Dairy production was pushed by French colonists, Norwegian development projects, and a recently booming dairy industry (see Vea 1992; Ramamonjisoa 1994; Bertrand 1997).

Both mimosa wood fuel cultivation and dairy cattle production benefit from the climate of the Ambatolampy region, shaped by the altitude and the orographic effect of the Ankaratra range to the west. Precipitation is no higher than elsewhere, about 1,500 mm/yr (though it does increase to over 2,000 mm/yr in the Ankaratra range), but cooler temperatures together with winter fogs and drizzle reduce evapotranspiration. However, the urban market remains the primary force shaping this zone of no fire—for the periurban mimosa ring more or less disappears in areas of similar climate further north or south along the Ankaratra range.

Afotsara: A Variety of Fires in a Transition Zone

Afotsara sits in a zone of transition. The deep Manandona-Sahatsiho valley serves as a fluid boundary between the Vakinankaratra people to the north

33. Recent local statistics are unavailable; however, commercialized milk production in the district of Ambatolampy already averaged 60,000 liters per year in the 1960s (ANM Monographies Ambatolampy).

FIGURE 3.5. **A hillside in the Afotsara region.** One can see irrigated rice fields in the valley, *tapia* woodlands on the lower slopes, and open pastures on the upper slopes.

and the Betsileo people to the south. From east to west, the valley and Mount Ibity separate in a short space the cool, more humid eastern plateau and the warmer, sunnier lands of the Middle West.[34] While the valley itself is densely settled, the uplands to the east and the mountainous grasslands to the west are relatively open zones undergoing agricultural colonization (figure 3.5). Fire is quite common in the open grasslands and *tapia* woodlands of this region. As table 3.2 noted, over one-half of the grasslands and one-third of the woodlands burned in 1998. The woodlands, as we shall see in chapter 4, are actively managed with fire. The open *tanety* grasslands of Afotsara are managed primarily as late dry season and rainy season forage reserves for cattle, as well as zones for crop field expansion. Population density is moderate, concentrated mostly in lightly burned valley areas. An increasing number of people cultivate pines for sale as construction wood and firewood.[35] However, land pressure and economic incentives like wood fuel are only beginning to push the balance towards more fire conservancy.

34. In the classification of the Inventaire Ecologique Forestier National (DEF 1996), the valley serves as the border between the "western slopes mid-altitude" phytogeographic zone to the west and both the "eastern slopes mid-altitude" and "mountain" zones to the east.

35. A covered pickup truck (*bachette*) load of pine firewood, split and cut to lengths of 60–80 cm, sells for 35,000–50,000 MGF (1998; US$6.00–8.00) in the Afotsara area.

Behazo: No Land Left to Burn

Behazo is very densely populated and has undergone serious agricultural intensification (see chapter 1; figure 3.6). Fire plays a very small role in this landscape; as one farmer said, "There is no land to burn here."[36] Some 95 percent of village lands are cultivated; in this situation fire is too dangerous and is used only to prepare fertilizer. Farmers burn to clean up field edges or prepare some crop fields, but mainly they avoid using fire. Fire can easily escape and cause damage in such tight quarters, and unburned grass is valuable as fodder or roofing material.

Fires do burn, however, in some of the large mountaintop pastures surrounding the Behazo valley. Parts of these areas remain somewhat open, despite decades of agricultural expansion, due to their altitude, poor hydrology, or infertile soils. Other mountaintop areas are fully covered in woodlots (pine, eucalyptus, and mimosa), new rice paddies, and sweet potato fields or fallow fields. Open pastures are dominated by *A. rufescens* and *L. simplex,* with a carpet of *C. dactylon* and a local dandelion (*Centella asiatica,* local name *talapetraka*) covering fallow sweet potato fields. Herders light pasture renewal fires nearly annually in these zones.[37]

Andringitra: Between Cattle Country and a National Park

Pasture fire is omnipresent in the Andringitra-Namoly basin case study area, which is part of a cattle-exporting region (see chapter 1). Ambalavao, 50 km to the north, is the nation's second largest cattle market. Some 30,000 to 60,000 cattle are sold there each year. Sendrisoa *commune rurale,* to which the Namoly basin belongs, has the second largest cattle population in Ambalavao district (Bissakonou 1995). In Namoly, reported cattle ownership is relatively high at 0.87 head per person. Ten percent of families have over twenty head of cattle (Projet Andringitra 1996), and it is suspected that some families own herds of over 100 head at least partly hidden in the forest and in the park.

Cattle raising in the Namoly basin takes advantage of the regional diversity of ecological zones, and fire is a key management strategy. The burning of these extensive pasture zones, however, is not purely governed by pastoral needs, but also by the politics of Andringitra National Park. The interplay of

36. Interview, Behazo, 24 June 1999.

37. Some farmers from Behazo pasture their cattle in available mountain grasslands and along streams and paths from December to May, supplementing their feed with cut grasses. Others do not make the daily long walk to the mountains, satisfying their cattle with orchard undergrowth, trail-edge grasses, and cut fodder. From July to October, the base of cattle nutrition is rice straw, cut grasses, and fallow crop fields.

FIGURE 3.6. **A view in the Behazo region.** Note the dense settlement and crop fields climbing high up the valley sides, leaving little room for fire.

this mountainous region's ecological zones with the sociopolitical contexts of human settlement and project politics is best illustrated by investigating fire management along a rough transect (figures 3.7 and 3.8). This transect extends from within the Andringitra park across the Namoly basin to Mount Iaody, fifteen kilometers to the north. The transect demonstrates how the complex mountain pasture ecology interacts with the strong political influence of the project to shape fire use, in what would otherwise be a loosely settled, extensive cattle-raising zone like Kilabé. I describe the five zones of the transect from south to north.

Zone A, the Andohariana plateau, is an altimontane prairie located at 1,900 to 2,000 m.a.s.l. This prairie is a natural grassland mosaic of a dozen different vegetation communities comparable with the *paramos* and *puna* of the Andes in South America; it is shaped by grazing and periodic fires and dominated by *Panicum* spp. (Bosshard and Mermod 1996; Edmond 1996; Bosshard and Klötzli 1997; Rabetaliana 1997; Rabetaliana et al. 1999). Before the first itinerant people arrived, fires were occasionally ignited by lightning and rockfalls on the plateau. When people and cattle arrived, they began burning the plateau to fight *Erica* bush encroachment, and the plateau was frequently utilized for pasture by nearby villagers, mainly for the female cattle during the

FIGURE 3.7. **Facets of the northern periphery of Andringitra National Park.** Photos from boundary between Zones C and D. (a) View south of the Namoly basin in the foreground, and of the protection zone, the Andohariana plateau, and the Andringitra mountain range in the background. (b) View north across the Fivanonana plateau to Iaody mountain.

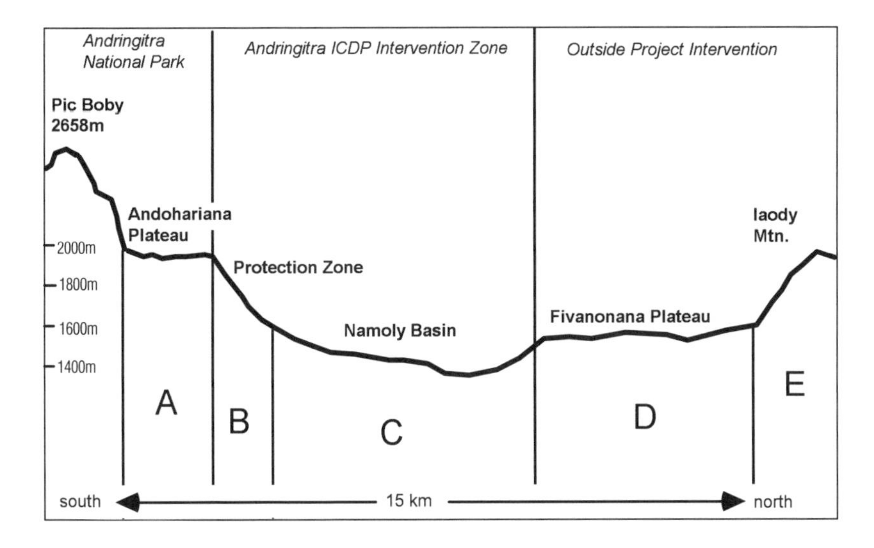

FIGURE 3.8. Schematic view of the northern periphery of Andringitra National Park.

difficult dry season when the cooler, moister plateau held more nutritious grasses. The plateau was made part of the original Strict Nature Reserve in 1927. Reserve rules strictly forbade grazing and fire, but that did not stop illegal burning, natural fires, and grazing. There were especially large fires in 1944, 1949, 1983, 1991, 1995, and 1996 (Vignal 1956a; Ratsirarson 1997), and Forest Service agents tolerated grazing within the park as a compromise with local villagers to forestall more burning (Ratsirarson 1997). The new management plan that came with the redefinition of the reserve as a national park in 1998 allows for grazing,[38] prescribed fires, and ecotourism. After catastrophic fires in 1995 and 1996,[39] however, burning was not allowed on the plateau until at least 2000 except for fire ecology research purposes. Several researchers have recommended using a rotation of controlled fires in the prairie to maintain its biodiversity (Bosshard and Mermod 1996; Bosshard and Klötzli 1997).

Zone B includes the slopes that descend from the Andohariana plateau to the Namoly basin, located at 1,600–1,900 m.a.s.l. This "protection zone" is a mosaic of grasses, ericaceous shrubs, woodlots (mimosa, eucalyptus), and, since about ten years ago, increasing numbers of crop fields. Historically, this

38. A total of 136 UBT (tropical bovine units) from nearby villages will be tolerated.

39. Probably local protests to the arrival of the WWF Project and perceived threats to resource access.

zone was managed with frequent fire for cattle pasture, and it continues to be used as pasture by neighboring villages. During the days of the reserve, this zone was burned preventively to serve as a firebreak for the reserve, and eucalypts were planted to mark the boundary. From 1965 to 1975, fires were lit here during the summer rainy season until April (Ratsirarson 1997). More recently, before the project's arrival, firebreaks were burned in this zone during the winter drizzle (June to August). Now, the protection zone has become Zone of Controlled Use #7-4, which includes four separate village-managed pastures. There is no settlement here. Fires in this zone are well monitored by the authorities, and more or less managed.[40]

Zone C is the Namoly basin. Grouped in four *fokontany,* 4,248 inhabitants are found within the basin's 50 km^2. Valley bottoms are cultivated in rice, and *tanety* hillsides are sprinkled with dryland crops, woodlots, and pastures. Elevations range from 1,400 to 1,600 m. Pastures consist primarily of *Aristida* spp. and *H. rufa,* as well as *L. simplex, I. cylindrica,* and *H. contortus.* A *chef de secteur* of the Forest Service has been stationed here since colonial days, charged with the protection of the strict nature reserve. Since 1994, this has been the primary zone for intervention for the WWF conservation and development project. The pastures scattered between the villages and crop fields are used throughout the year. While cultivation rights on hillside pastures belong to different lineage-based groups, grazing rights are held in common. Historically, fires burned annually in the dry season (July to October), as elsewhere on the island. Nowadays, this is a zone of more or less managed fire. Since colonial days, each village seeks authorization to burn from the Forest Service; authorizations are normally given for November and December (or after three days of rain), or for March and April.[41] After the large wildfires of 1995, which were punished by a collective fine of 20,000 MGF (1995; US$5, or one week's income) per taxpayer, a pact controlling fire use was entered into by the villages, the project, and the Forest Service: all fires must be agreed upon by all three parties, following a principle of fire rotations. After a three-year hiatus, fire authorizations were given again in 1999. Crop field fires do not need authorization. In 1999, I observed that large zones were burned legitimately in March and April, and at least twenty to thirty hectares burned illegally in May during a locust passage. Informants claim that some people still would prefer to burn in September and October, but they refrain, fearing punishment.

North of the Namoly basin lies the Fivanonana plateau, Zone D. A rolling

40. However, I observed unauthorized fires in this zone in October 1996 (escaped crop fire that reached the plateau), September 1998 (probably a pasture fire), and May 1999 (fire scar from March or April).

41. DEF files, Ambalavao.

plain at 1,500 to 1,700 m.a.s.l., the plateau is covered by a shortgrass prairie of *Aristida congesta, I. cylindrica,* and *H. rufa.* A few of the stream valleys are developed as rice fields. Less than 1,000 people, or one *fokontany,* live in this 20 km^2 zone. This sparsely populated plateau lies outside of project intervention. As might be imagined, the plateau is a zone of extensive pasture, with roughly 1,000 head of cattle. The large pastures are chiefly used in the rainy season from October to April; in winter, the cattle stay in the valley bottoms, in the forests, or on the mountain pastures of Zone E. Dry season burning is critical in this zone of extensive grazing. Officially, fires should be lit only after written authorization from the Forest Service. However, no application or authorization for burning in this region was found in the files of the Forest Service in Ambalavao, and the region has burned freely without control from June to October for many years. As a key informant told me, the herders here "do not worry about fires since there is no fear" of punishment;[42] the Forest Service's attention and efforts are focused on the reserve, not here.

Zone E, Iaody mountain, caps the transect to the north. This mountain massif climbs from 1,700 to 2,000 m.a.s.l. north of Fivanonana village; it is vegetated with a prairie of *horombato* (probably *Aristida*) and with small patches of natural forest. The area is far outside project intervention and contains only a scattering of small hamlets. The mountains are a zone of winter pasture for the northern half of Fivanonana village; cattle are put here once the heavy rains end in March through September or October, when they are brought down to trample the rice fields. Due to the altitude—and associated drizzle and reduced evapotranspiration—the pastures remain nutritious longer through the dry (winter) season. As in Zone D, fires burn freely here, usually late in the dry season.

The Andringitra-Iaody transect demonstrates two main patterns. First, in this lightly populated zone centered on subsistence agriculture and extensive cattle raising, people take advantage of the ecological variations inherent in mountainous terrain to establish transhumance patterns for their cattle raising. Landscape burning is integral to these extensive cattle-raising strategies. Second, the presence of a well-funded and strongly enforced project—with an enlightened position on pasture burning—has actually led to changes in fire use in the project intervention zones, A, B, and C. The contrast in enforcement of legislation and formal fire planning between these zones and Zones D and E, outside project influence, is strong. This analysis demonstrates how a strong political approach, whether by the state or a project, can influence land use and burning regimes.

42. Interview, Namoly, 2 June 1999.

CONCLUSION: FIRE'S LOGICAL PLACE

The analysis and comparison of the case studies (table 3.5) demonstrates that the primary determinant of burning regimes across the highlands is the character of regional land use, as shaped by population density and regional markets. Secondary factors include antifire enforcement and environmental contexts. Specifically, zones of low population density (e.g., Kilabé, Andringitra) have large pasture areas available without competition from agriculture. Cattle raising on these pastures is extensive, as there is insufficient labor and capital for intensification. Extensive cattle raising necessitates fire for pasture maintenance; hence these areas see frequent and widespread burning. When and where population is denser (e.g., Behazo), intensive agriculture occupies most land, squeezing out pastures and associated burning. Finally, market opportunities, such as wood fuel and dairy production (Tsimay) or orchard production (Behazo) drive changes in land use that lead to a restriction of fires.

Some regions are transition zones between oft-burned extensively used lands and dense, intensive land use. Many of the case study areas have outlying pastures, on surrounding hills, that for reasons of distance, climate, or soils remain more open and are maintained through burning (e.g., Afotsara, Behazo, or the Fivanonana plateau near Andringitra). Conversely, oft-burned study sites also have areas where intensified land use leads to localized reductions in fire (e.g., in zones of agricultural expansion east of Kilabé, the cultivated valleys of Afotsara or the Namoly basin of Andringitra).

The clear lesson is that where human density and/or intensive land use increases, the large-scale use of fire decreases. This is true both across geographic space (as the comparison of the case studies suggests) and through historical time (as the individual trajectories of the case studies show). The theme of population and market pressures leading to intensification (Boserup 1965; Coulaud 1973; Netting 1993; Turner et al. 1993; Tiffen et al. 1994; Blanc-Pamard and Rakoto Ramiarantsoa 2000), and by association leading to less fire (Dez 1966; Vogl 1974; Mainguet 1997; Pyne 1997; Mistry 1998; Kepe and Scoones 1999) is nothing new. Colonial and postcolonial administrators and technical agents recognized that intensification squeezes out fire. As a result, they repeatedly sought to replace what they saw as irrational pasture burning with intensive cattle husbandry, including field mowing, fodder cultivation, and rotational grazing.[43] Their efforts, however,

43. See, e.g., circulars from Gov. Garbit (JOM 9 Oct. 1915 and 20 Nov. 1920), circular from Gov. Berthier (JOM 9 Feb. 1929), circular from Gov. Cayla (JOM 2 July 1938), circular from Gov. Annet (JOM 6 Dec. 1941), Parrot (1924), Bosser (1954), RDM (1980), Aquaterre (1997).

rarely succeeded, for the additional labor or capital required made little sense to herders happy with extensive pastures.

A secondary factor, brought to light by the Andringitra case study, is enforcement of antifire politics. Real fire control and enforcement in Madagascar has long been minimal or sporadic at best (chapters 6 and 7). However, as the Andringitra case shows, a stronger political presence—especially with the financial help of the multi-million-dollar park project—can influence fire use.

THIS CHAPTER investigated fire in grassland zones from the perspective of rural land users. I demonstrated that fire is a logical and useful tool that has a specific place in many rural production systems. By far its most important role is in the maintenance of pastures; but burning also contributes to agricultural production, pest control, wildfire prevention, and myriad other goods. The actual character of fire use varies from region to region depending upon the economic, demographic, and political context, as demonstrated by the case studies. These variations show not only that the Malagasy peasants rely on fire as an efficient, convenient, natural, and inexpensive management tool, but also that they know when they want fire and when they do not.

This conclusion contradicts the prevailing view among Malagasy environmental policy makers and urban elite that fire is a national embarrassment and an unequivocal environmental scourge. This is not to say, however, that we know everything there is to know about pasture fires. There is still a great need for further research on the actual dynamics of fire and range management in Malagasy grasslands. In particular, knowledge is thin on the specific effects of different fire regimes—i.e., different frequencies and seasons of burns (e.g., Bassett and Koli Bi 2000; Laris 2002)—on grassland communities, never mind the interactive effects of other factors such as grazing intensity, precipitation variations, soil types, and seed availability.

The chapter also contradicts the view that Malagasy peasants burn out of ignorance, laziness, apathy, pyromania, and simple backwardness or attachment to ancestral practice (Charon 1897; Perrier 1921, 1927; RDM 1980; Neuvy 1986). Fire is a well-adapted tool, used in extensive land use across the globe, and a key part of much Malagasy land use. As a result, rural Malagasy have strongly resisted most government attempts to restrict burning, as I will show in part 3. First, however, in the next two chapters, I continue the case presented here by investigating two additional arenas in which fire plays a role in peasant livelihoods: *tapia* woodland management and slash-and-burn forest farming.

4 [WOODLAND FIRE

Fire and Rural Economy in the Tapia *Woodlands*

Mananjara samy monina ny monina any anatin'Imamo
Mielo ravin-tapia, ka na tsy vokin'ny voany aza,
Mialokaloka amin'ny raviny.
Velona mitafy arin-drano, maty mifono lambamena;
Ny vidin-dandy loharanon-karena;
Ary ny soherina tombon-dahiny ihany.

The fate of the people of Imamo:
Protect the *tapia* leaves, for even if the fruit doesn't fill you up,
You get shade under the leaves.
The living wear *lamba arin-drano,* the dead are wrapped in *lambamena;*
The sale of cocoons is a source of wealth
And the chrysalis is an advantage too.

TRADITIONAL POEM; IN RAKOTOARIVELO (1993, 58)

The fires that burn immense surfaces of *tapia* forest each year are slowly but surely causing the disappearance of these forests whose conservation to date is motivated by the economic profits of the *landibe* cocoon harvest.

FOREST SERVICE REPORT, DISTRICT OF AMBOSITRA, 9 FEB. 1935 (AOM MAD-GGM-2D19BIS#6)

FEW ENDEMIC FORESTS exist in highland Madagascar, a region dominated by vast grasslands, rice paddies, dryland crop fields, and pine or eucalyptus woodlots. One exception is the *tapia* woodlands, dominated by the *tapia* tree (*Uapaca bojeri*). These woodlands or wooded savannas grow in several zones scattered across the western highlands. In this chapter, I argue that the *tapia* woodlands are fire dependent, and that villagers manage them through fire, selective cutting, and protection due to the livelihood value of woodland products such as wild silk and fruit. The case of the *tapia* woodlands, while perhaps unique, is particularly pertinent to this book on fire, as many observers have blamed fire for the degradation of the woodlands, emphasizing that *tapia* woodlands are remnants of previously grander and more diverse forests (Gade 1996; Humbert 1949; Koechlin et al. 1974; Perrier 1921; Rakotoarivelo 1993).

The evidence I present here does not deny that presettlement forests in *tapia* regions were different, but it does make it clear that fire-caused environmental changes in the *tapia* woodlands can be seen as positive: they support a number of livelihood goals. It also shows that fire now plays a key role in the maintenance of the woodland in its present state, and that this management has not, as often asserted, reduced the extent of these woodlands.

When I began my research on fire in Madagascar, I was attracted to the *tapia* woodlands, for I thought that their mere existence in the oft-burned highlands must indicate the presence of some local fire control system not paralleled in other areas. That is, I hypothesized that, for some unknown reason, communities in a few scattered regions were motivated to regulate the ubiquitous annual fires, and that the *tapia* woodlands were testimony to this regulation.

The data turned the original hypothesis on its head. *Tapia* woodlands are not a result of fire protection; they owe their very existence to fire. Communities do not manage against fire, they manage *with* fire. Just like the grassland and farmland fires discussed in chapter 3, fire fits logically into the ecological and economic context of the *tapia* woodland regions. The *tapia* woodlands display strong adaptations to fire and play an important economic role for rural residents as a source of nontimber forest products (Koechlin et al. 1974; Gade 1985).

This chapter again counters the antifire received wisdom so entrenched in Madagascar. In this received wisdom, the *tapia* woodlands are seen as degraded remnants of the former islandwide forest, spared by their tolerance of fire and their usefulness to the population, yet slowly disappearing in the face of repetitive fire. There are three components to this received wisdom that I will address.

First, the received wisdom describes the *tapia* woodlands as degraded remnants of once larger, more diverse forests (AOM mad-ggm-2d19bis#6;

Girod-Genet 1898; Koechlin et al. 1974; Gade 1996). Before the arrival of fire-starting humans some 1,500 years ago, the highlands were more wooded than today (Burney 1996, 1997), and current *tapia* woodland areas were probably more densely forested and more diverse. Over the past millennium, degradation—in the strict ecological sense of reduced species diversity, thinning vegetal cover and undergrowth, and exploitation that changes the character of the woodland from its presettlement state—cannot be denied. In most commentary about the *tapia* woodlands, however, the term "degradation" is used as a negative normative term, implying that the changes are bad. I argue that the changes are of economic value to the locals and that the term "degradation" masks transformative, creative processes critical to the livelihoods of rural people. The term "transformation" may be more appropriate (Beinart 1996; Richard and O'Connor 1997).

Second, many authors do point out that the *tapia* woodlands are spared deforestation due to their tolerance of fire and their economic usefulness to the locals (Gade 1985; Girod-Genet 1898; Humbert 1947; Koechlin et al. 1974; Perrier 1921; Raison 1984). The woodlands host an endemic silkworm used to weave burial shrouds, and they produce large quantities of marketable fruit. However, other than Gade (1985) in his groundbreaking reassessment of the *tapia* woodlands, nobody considered fire as a natural, normal part of these woodlands, and no one shows specifically how the locals harness the trees' fire adaptation to shape the woodlands and meet the needs of their dynamic rural economy. That is the role of this chapter.

Third, multiple authors assert that the *tapia* woodlands are shrinking in the face of repeated burning (Gade 1985; Grangeon 1910; Humbert 1947; Perrier 1921; Ramamonjisoa 1995; Vignal 1963). While the trees may tolerate fire, too much fire eventually kills the trees and impedes regeneration. Based on diverse pieces of evidence, this chapter argues that there is no convincing evidence for twentieth-century *tapia* woodland decline, and in fact, in the Afotsara region, there is significant evidence for stability and even cases of increase.

THE *TAPIA* WOODLANDS

The *tapia* woodlands are an endemic, sclerophyllous, scrubby forest that loosely resembles Mediterranean oak forests or southern Africa's miombo woodlands. They are found in several widely scattered clusters, from Imamo just west of the capital, south to the Isalo massif (figure 4.1). The woodlands, found as small forest patches or as lightly wooded savanna, cover approxi-

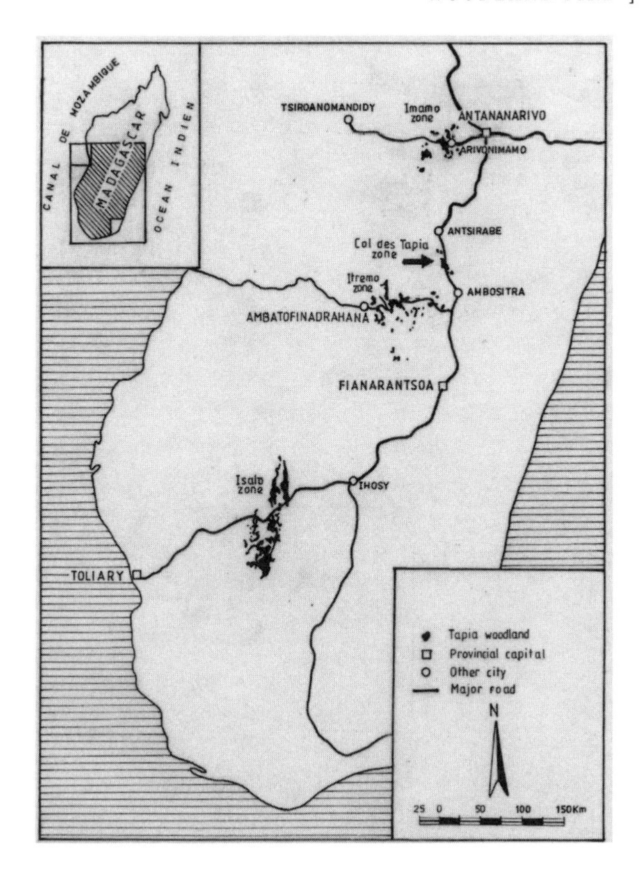

FIGURE 4.1. **Locations of *tapia* woodlands in Madagascar.** Based on FTM 1:500,000 series maps; cartography by Simon Roger.

mately 2,600 km^2 (DEF 1996). *Tapia* woodlands are the most xerophytic of the broadleaf evergreen forests of Madagascar, growing on the western ramparts of the central highlands, where the dry season lasts up to seven months. This zone, ranging from 800 to 1,600 m.a.s.l., is warmer, drier, and sunnier than the rest of the highlands; precipitation ranges from 1,000 to 1,500 mm/yr and the mean annual temperature is 17 to 22 °C. Edaphic factors likely determine *tapia* woodland distribution; most are found on nutrient-poor or rocky soils underlain by granites and gneisses.[1]

1. Some argued that the woodlands persist only in rocky areas where fires penetrate less easily (e.g., Basse 1934; Humbert 1949; Perrier 1921; Salomon 1978). While this may play some role in the ruiniform canyons of the Isalo massif, rocky areas do little to stop fires in the Itremo, Imamo, and Col des Tapia regions.

Colonial naturalist Humbert (1949) asserted that *tapia* woodlands occupied tens of thousands of square kilometers in the western slope zone before human settlement. Koechlin et al. (1974) argue instead that much of the presettlement western slope zone actually hosted a transition forest between eastern rain forests and western dry forests and that the *tapia* woodlands represented a special formation restricted in its distribution to poor soils. More recently, palynologist Burney (1997) showed that some highland areas were not forested during parts of the Holocene. *Tapia* woodlands emerged out of a presettlement mosaic of grassland, savanna, heathland, woodland, and riparian forest, shaped by a natural lightning regime, grazing megafauna, and climate fluctuations (Dewar 1984; Burney 1996).

Tapia woodlands relatively unmodified by humans are known as "mid-elevation sclerophyllous forests" or "western slopes sclerophyllous forests of the Sarcolaenaceae–*Uapaca bojeri* series" (DEF 1996; Lowry et al. 1997). Trees from eighteen families and twenty-six genera grow in these forests (Koechlin et al. 1974; DEF 1996); the *tapia* tree (*U. bojeri*) and trees from the Sarcolaenaceae family are endemic to the region. Most of the tree species found in this habitat are similarly shaped, with tortuous trunks, low branches, and dull, dark green fleshy leaves; they reach a maximum height of eight to twelve meters. Typical understory plants include immature trees, shrubs (including *Erica, Leptolaena,* and *Aphloia*), herbs, and grasses.

Most extant woodlands are significantly modified by human burning and cutting practices and are considerably less rich in species. Many western slope woodlands are open savannas dominated by a few species (especially *tapia*) with a grass understory. These forests were named *bois de tapia* (*tapia* woodlands) by Humbert and Perrier (Koechlin et al. 1974) or "open sclerophyllous forests" by a recent forest inventory (DEF 1996).

The *tapia* tree, *Uapaca bojeri* (Euphorbiaceae family) composes upwards of 80 percent of adult trees in most woodlands, whether in the Col des Tapia region (table 4.1), Imamo (Rakotoarivelo 1993), or Itremo.[2] *Tapia* is one of twelve species of the *Uapaca* genus in Madagascar.[3] It grows to heights of 8 to 10 m and reaches diameters of 20 to 40 cm. *Tapia* trees have oval ever-

2. In Itremo, remote, higher altitude woodlands (such as at the 1,600 m.a.s.l. Col d'Itremo) are less dominated by *tapia,* with significant presence of trees of the Sarcolaenaceae family (author's observations).

3. It is the only *Uapaca* species found in the sclerophyllous woodlands of the highlands; others are found in the more humid forests of the east and elsewhere, e.g., *U. densifolia* at Ambohitantely (Rakotoarisetra 1997; see Schatz 2001). Fifty additional *Uapaca* species are found in Africa, for example among the trees of the miombo woodlands (Campbell 1996).

TABLE 4.1. Tree species found in the Col des Tapia woodlands.

Genus, species, family	Local name	Dominance[a]	Uses
Uapaca bojeri (Euphorbiaceae)	tapia	88.3	edible fruit, woodfuel, silkworm fodder, medicine (stomach, heart)
Sarcolaena eriophora (Sarcolaenaceae)	voandrozana	4.9	woodfuel, charcoal, small edible fruit
Pinus patula/kesiya (Pinaceae)[b]	kesika	3.7	construction, woodfuel
Leptolaena spp. (Sarcolaenaceae)	fotona	1.4	woodfuel, charcoal, small edible fruit
Trema spp. (Celtidaceae)	andrarezina (tsivakimbaratra)	0.3	commercialized ashes
Xerochlamys bojeriana (Sarcolaenaceae)	hatsikana (katsikana)	0.3	woodfuel, roots used to flavor rum, tannins, medicine (for pigs)
Popowia boivinii (Annonaceae)	fotsiavadika	0.2	medicine (stomach)
unidentified	ndretsimora (andriatsimora)	0.2	antivenom for scorpions, medicine (diarrhea)
unidentified	fanazana	0.2	tea, medicine (stomach, paranoia)
Agarista salicifolia (Ericaceae) (formerly *Agauria*)	angavodiana	0.2	medicine (cuts/bleeding)
Schefflera spp. (Araliaceae) (formerly *Cussonia bojeri*)	tsingila	0.2	medicine (stomach, cuts/bleeding; general medicine of last resort)
Vaccinium emirnense (Ericaceae)	voaramontsina	0.2	woodfuel, fruit edible and tasty
Aphloia theaeformis (Aphloiaceae)	voafotsy	[c]	popular tea
Micronychia spp. (Anacardiaceae) (formerly *Rhus taratana*)	taratana	[c]	medicine (stomach; general medicine of last resort)
Tambourissa spp. (Monimiaceae)	ambora	[c]	medicine (teeth)

[a] Percentage of trees encountered in a dozen 500 m^2 survey plots with dbh > 5 cm belonging to the given species.
[b] Nonnative.
[c] Additional species present outside survey plots.
Source: Author's fieldwork (with help from L. A. Rakotoarivelo), local residents, and Randriamboavonjy (2000). Linnaean names from Randriamboavonjy (2000) and Ratsimamanga (1968), and updated according to Schatz (2001) when possible.

green leaves typically about 9 cm long that vary from dull olive green to yellow green. Thick, fissured bark protects the tree from fire. The species regenerates chiefly by resprouting from roots and stumps, but also by rhizomes and seed establishment (Gade 1985; Rakotoarivelo 1993; Randriamboavonjy 2000).

The *tapia* woodlands in the Afotsara case study (introduced in chapter 1) cover approximately 50 km^2 in discontinuous or loosely stranded patches (fig. 4.2). The woodlands are sprinkled on both sides of the 25 km long, 1,300 m.a.s.l. Manandona and Sahatsiho river valley, spilling over the Col des Tapia towards the Ilaka basin. Outlying patches exist west of the 2,000 m.a.s.l. Ibity massif and east of Ilaka along the Isandra River. Forests are found on slopes of all aspects up to an altitude of 1,600–1,700 m; above this the woodlands give way to pasture. In these woodlands, forest surveys recorded fifteen tree species (of at least ten families and thirteen genera) and at least thirty-four nonherbaceous understory species (tables 4.1 and 4.2), as well as a variety of herbaceous species, epiphytes, aloe, lichen, orchids, and mushrooms. All woody species, except for *Pinus* spp., were native to Madagascar.

WOODLAND ECONOMY

The *tapia* woodlands are a significant but not dominant element of the local subsistence and exchange economy. Adjacent communities gain revenue and subsistence resources from woodland products, including wild silk, marketable fruit, and firewood. These benefits lead to the management and protection of the woodlands.

Silk

Tapia leaves are the preferred fodder for *landibe,* an endemic silkworm (*Borocera madagascariensis* Lepidoptera: Lasiocampidae). *Landibe* produce cocoons twice a year, in November–December and May–June. At this time, locals painstakingly gather the cocoons from tree branches, leaves, and grass tufts (fig. 4.3). The chrysalis is removed and consumed. The empty cocoons are boiled, spun, and woven into silk fabrics (fig. 4.4). Almost all *landibe* silk is used to produce ritual burial shrouds, *lambamena,* used at both funerals and reburial ceremonies (*famadihana*) throughout the highlands.[4]

4. Burial and *famadihana,* or the turning of the dead, are key elements of highland cultural life. For more details on the silkworm, its biology, cocoon harvesting, and weaving traditions, see especially Gade (1985), Grangeon (1906, 1910), and Paulian (1953).

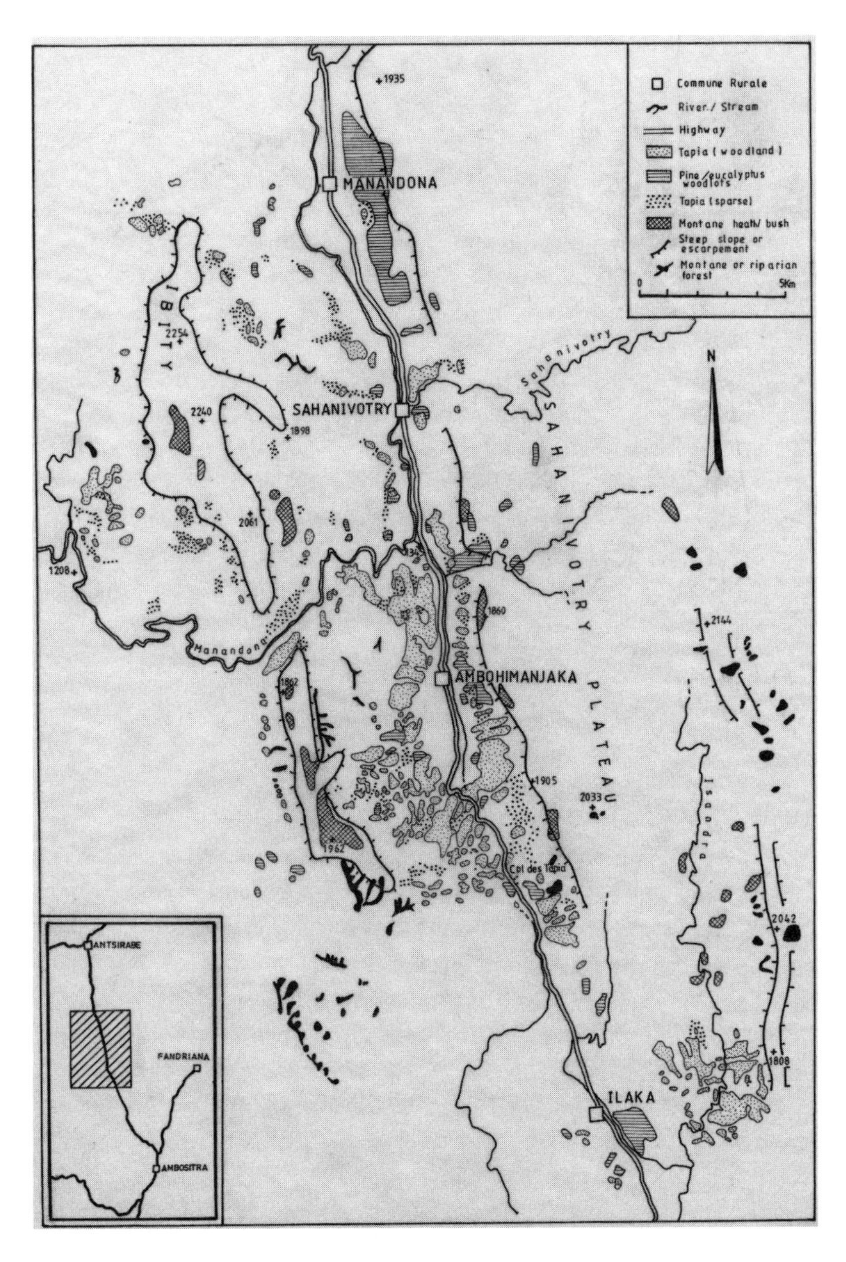

FIGURE 4.2. **1991 woodland cover in the Col des Tapia zone.** Based on 1991 air photos (source: FTM). Interpretation by author, cartography by Simon Roger.

TABLE 4.2. Understory species found in the Col des Tapia woodlands.

Genus, species, family	Local name	Uses
Aloe spp. (Liliaceae)	vahona	medicine (stomach, cuts/bleeding)
?*Asteropeia multiflora* (Theaceae)	fandambana	woodfuel
Brachylaena ramiflora (Asteraceae)	hazotokana	medicine (broad usage)
Carissa edulis (Apocynaceae)	voahangitanety	–
Cassia mimosoides (Caesalpiniaceae)	kelimanendilanitra	medicine, ceremonial (ward off hail)
Caucalis melanantha (Apiaceae)	kisetroka	medicine (headaches, cold, blurred vision)
Dombeya greveana (Sterculiaceae)	ambiaty/ombiaty	commercial ashes
?*Dracaena reflexa* (Liliaceae)	ravoanjo/ranjo/ raivoanjo	toilet training of dogs (by rubbing leaves on the ground to deter dog from using that location)
?*Embelia* spp. (Myrsinaceae)	tateradela	medicine (for children, for cuts)
Erica spp (Ericaceae) (formerly *Philippia*)	anjavidy	woodfuel, medicine (cough)
Helichrysum rusillonii (Asteraceae)	ahibalala	medicine (stomach, cough, fever, last resort)
?*Helichrysum* spp. (Asteraceae)	tsetsatsetsa	–
Maesa lanceolata (Myrsinaceae)	rafy	small sweet edible fruit, medicine (teeth, chest pain)
Plectronia spp. (Rubiaceae)	fatsikahidambo	medicine (paranoia)
Psiadia altissima (Asteraceae)	dingadingana	–
Psorospermum spp. (Clusiaceae)	tambitsy	medicine (general)
Pterocaulon decurrens (Asteraceae)	ariandro	–
Radamaea montana (Scrophulariaceae)	tambarasaha	woodfuel, cosmetic
Rubus apetalus (Rosaceae)	rohifotsy	medicine (antivenom, heart, cuts)
Senecio faugasiodes (Asteraceae)	hanidraisoa, koboi boy	medicine (cracks in feet, wounds)
Solanum spp (Solanaceae)	sevalahy	–
Tetradenia fructicosa (Labiaceae)	bororohana	medicine (paranoia)

TABLE 4.2. (*continued*)

Genus, species, family	Local name	Uses
Vernonia glutinosa (Asteraceae)	ramanjoko	medicine (fever)
Vernonia spp. (Asteraceae)	kijejalahy	medicine (STD, fever, stomach, general)
(vine)	vahy famonololo	used to tie things, medicine (paranoia)
unidentified	arivoniraviny	wood for craft-making
unidentified	fano	–
unidentified	hazomiarotena	–
unidentified	(keli)boloana	medicine (fever; general medicine of last resort)
unidentified	kiripika	–
unidentified	reniomby	medicine (for cattle illness)
unidentified	voamasonomby	medicine (constipation)
unidentified	voatainosy	woodfuel, tiny edible fruit, flavoring for rum
unidentified	voatsitakazaza	tiny edible fruit

Source: Author's fieldwork (with help from L. A. Rakotoarivelo), local residents, and Randriamboavonjy (2000); Linnaean names from Randriamboavonjy (2000) and Ratsimamanga (1968).

This silk-based economy has been important for several centuries, serving as a source of revenue for both locals and the government (through royalties and concessions). *Landibe* silk was already sold in highland markets in the early 1800s (Callet 1958). It had considerable value to adjacent villages: according to oral history accounts, locals in one village near the Col des Tapia used the silk they had harvested to buy their freedom during the Merina conquest (Raharijaona 1990). A Norwegian missionary based nearby in Ilaka noted in 1878 that the entire village departed for weeks at a time to collect cocoons—traveling as far as Ambatofinandrahana, 50 km away (Bekker 1878). Before 1896, one-tenth of the harvest in the Itremo area went to fill government coffers, and specific nobles controlled the harvest.[5]

After the conquest in 1896, the French colonial government, like the Merina monarchy before it, claimed all forested land as state domain. It allowed for traditional use rights but immediately sought to regulate the collection of wild silk by auctioning off collection rights (Grangeon 1910; JOM 7 Mar. 1899). France's silk industry was interested in the potential of new sources of

5. ANM D100s, ANM II.CC.196.

FIGURE 4.3. A collector of wild *landibe* silk cocoons.

silk; Governor Gallieni as a result wrote of "this important issue that directly affects the economic future of Madagascar."[6] Colonial administrators debated the placement of district boundaries based on anticipated silk revenues in areas of *tapia* woodlands.[7] As a lightweight but valuable product, *landibe* could potentially be a boon to the economy of isolated districts. Collection rights to the Ambositra Province woodlands in 1910 were valued at an astonishing 40,000 frs, up from 13,500 frs in 1898; collection right owners charged a royalty of 0.50 frs per basket (ANM D195s; Grangeon 1910).

6. Circular, Gov. Gallieni, 16 Nov. 1897 (ANM D100s/4).

7. See Kull (2000b), based on ANM D435.

FIGURE 4.4. **Spinning boiled *landibe* cocoons into silk threads.**

From 1921 to 1927 the colony established a *landibe* research center in Ambatofinandrahana, in the Itremo region, but it was short-lived. Over time, colonial efforts came to focus on the husbandry of Chinese silkworms (*Bombyx mori*, known locally as *landikely*) and on the cultivation of mulberry bushes for the fine silk export market.[8] Wild *landibe* silk was

8. Despite three to four million francs of government investment between 1897 and 1928, only five metric tons, worth 311,000 francs, of *Bombyx* cocoon and raw silk export were achieved. The importation of cheaper textiles in the late 1920s and the global crisis in silk prices in the 1930s contributed to a near abandonment of the effort. However, *Bombyx* silk husbandry, with mulberry cultivation, has persisted at modest levels until today (for a history of the fine silk effort in Madagascar, see ANM D100s; ANM Agr 78; Frappa 1947; Platon 1953).

simply found to be of a rougher, inferior quality, and saw favor only on the domestic market.

Despite the government focus on fine silk, *landibe* collection continued to be regulated. In the Col des Tapia area, government-auctioned collection rights were controlled in the 1930s and 1940s by colonist J.-M. Castellani.[9] During World War II, regulations tightened as administrators feared textile shortages. By a government decree, the Itremo and Col des Tapia woodlands were closed to *all* usage except for officially organized harvests, enforced by numerous guards (ANM L806; JOM 30 Sep. 1944).

New legislation in 1946 established *landibe* cooperatives, limiting silk harvesting to cooperative members with permits and controlling the transport of cocoons (JOM 18 Jan. 1947). Guards known as *fantsika* (literally, nails) were paid to control illegal harvesting. This legislation is still officially in force, but ignored. The silk cooperatives continued into the 1960s, when they were disbanded due to peasant frustrations with their corruption.[10] In 1978, woodland control was theoretically decentralized to the local level, yet this policy was never formalized and legislation remains unchanged.[11] Today, in Afotsara, the mayor's office is supposed to collect a fee from silk buyers. In 1997, annual permits were issued to only three local buyers for 5,000 MGF each; most sales are not reported to the commune.

The November–December silk harvest provides crucial cash income during the meager months before the rice harvest. In the Col des Tapia region around Afotsara, two-thirds of households collect *landibe.* One-third earn cash income from the harvest, while the rest keep cocoons for family use. One in five women is involved in spinning and weaving. In 1998, 200 female *landibe* cocoons sold for 500–750 MGF, or US$0.10–0.15, and finished 2 × 2 m *lambamena* cost between 150,000 and 300,000 MGF (US$30–60) in Antananarivo.[12]

9. Castellani, from a prominent family, also had numerous forest concessions in the Ambositra-Antoetra area. Jules Castellani (perhaps the same man) was elected to the French senate in 1951, representing conservative colonist interests (Brown 1995).

10. A poor woman, a descendant of slaves, told me of going out as a child to collect *landibe* illicitly with her parents. She accused the *fantsika* of using their positions to unfair advantage, enforcing severely yet stealing cocoons from the forest themselves (interview, Afotsara, 5 Dec. 1998). Nonmembers of silk cooperatives went out at night to collect cocoons illegally (interview, Afotsara, 8 Dec. 1998). The story of the cooperatives is based on these interviews as well as others (Forest Service official, 28 June 1999; Afotsara, 13 Apr. 1999).

11. Olson (1984), Gezon and Freed (1999); interviews with mayors in Afotsara region, 5 Aug. and 4 Nov. 1998.

12. Source: household census in Afotsara in 1998 and Randriamboavonjy (2000). Cocoons are sold by weight or by number (*isan-dandy*). Each *isan-dandy* normally refers to a pair of larger

Most marketed, unfinished silk comes from the Isalo woodlands; the Itremo and Col des Tapia woodlands also contribute but to a lesser extent, while Imamo hasn't produced *landibe* since the 1960s (Razafintsalama and Gautschi 1999). Recently, financial strains on family budgets and the availability of cheap, long-lasting synthetic cloths have put a dent into the *lambamena* economy. Meanwhile, however, a few artisanal tourist shops have begun to sell vests woven with *landibe* silk. A small development project, with funding from President Ratsiraka's office, began working with a number of villages in the Col des Tapia area to protect *tapia* woodlands, augment *landibe* populations, and teach weaving techniques; other NGO-based projects have been established in the Imamo region.

Cocoon harvests vary from year to year in terms of quantity and location. Oral traditions tell of grand silk harvests in the past, when the worms ate the trees bare of leaves. Recent years' harvests in Afotsara have been smaller, most likely due to unlimited harvesting as well as natural fluctuations in insect populations. Informants, however, ascribe the size of the harvest to *zavadolo,* literally "spirit matters," evoking the unpredictability and lack of human control in silkworm populations. People also blame poor harvests on the fact that traditional sacrificial ceremonies, called *soron-dandy,* have not been performed in years. In his day, Castellani donated a steer to the ceremony each year. The last ceremony was held around twenty years ago. In 1997, due to concern for dwindling *landibe* harvests, Afotsara elders met to write down and preserve the traditions,[13] yet in 1998, a planned *soron-dandy* was hijacked into an election propaganda session by the political party which had donated the steer.

Tapia *Fruit*

The *tapia* tree produces large quantities of a tasty small fruit, *voan'tapia;* trade in this fruit provides crucial income to woodland communities from mid-September to early December. The fruit is a small (2–3 cm diameter) round drupe, green to yellow on the tree and brown when ripe. As only fallen

female cocoons, or four smaller male cocoons. Individual female cocoons average 300–400 mg, male cocoons average 90–200 mg (Grangeon 1906); thus a mixed batch might number 5,000 cocoons/kg.

13. Rituals vary from site to site. For example, in one area, a pair of silkworms is brought from the hills to a place where people dance dressed only in ferns. As a white-headed cow is sacrificed, people sing, asking "please give us silk for we have no clothes." In another, villagers assemble at a high granite outcrop above the forest, gravesite of an important ancestor. They sacrifice a black steer with a white head, and share it together with rice cooked in milk (interviews, 1998, Afotsara).

fruit are ripe, a strong taboo (*fady*) prohibits plucking fruit from the branches. In the Col des Tapia region, nine out of ten households collect fruit; two-thirds earn cash income from the sale of the fruit to urban consumers.[14]

Tapia fruit was sold in highland markets over 200 years ago (Callet 1958); people in Afotsara remember when fruit was carried to Antsirabe by porters or oxcart. Nowadays, villagers convene roadside each morning, selling fruit to collectors; these middlemen accompany dozens of sacks of fruit to market. Fruit is sold for 400 to 1,200 MGF per *katinina* basket, about 4 kg or 7 l, equivalent to US$0.02 to 0.06/kg (1998). During the harvest, the Col des Tapia region woodlands produce about 4 kg of fruit/ha daily; this totals to an annual fruit production on the order of 600 to 1,500 tons.[15]

Fruit income comes at a crucial time, after the festive season (July and August), when families have little liquid cash. The fruit harvest provides a crucial income boost, allowing people to hire rice-transplanting labor and buy agricultural inputs. Increasing cash availability is reflected in the burgeoning number of items at markets and roadside stands by December. At this time, *tapia* prices drop due to the arrival of other fruits on the market (e.g., litchi, plum) and most people are too busy preparing rice fields to collect the remaining fruit.

Forest-Dependent Households

The *tapia* woodlands also provide many other resources. First, 92 percent of families in Afotsara rely on *tapia* forests for wood fuel. Second, during the rainy season, seven varieties of edible mushrooms are found in the woodlands, symbiotically associated with tree roots. All families collect mushrooms; one-tenth sell mushrooms locally and to regional markets. Third, as noted in tables 4.1 and 4.2, many woodland species have medicinal value. These plants are largely used locally; three-quarters of households report using medicinal plants. Fourth, three insectivorous mammals of the tenrec family, *sora* (*Echinops telfairi*), *sokina* (*Setifer setosus*), and *trandraka* (*Tenrec ecaudatus*), are occasionally hunted in and around these woodlands and savored for their meat. Fifth, woodland insects, such as *saroa* caterpillars (*Antherina suraka:* Lepidoptera: Saturniidae) and *landibe* chrysalises, are collected, consumed, and sometimes sold; they form an important part of

14. Household survey, Afotsara, 1998; Randriamboavonjy (2000).

15. Estimate based on household survey in Afotsara, where 500 kg are collected from 125 ha of forest daily. Middlemen estimate that Ambohimanjaka *commune rurale,* which includes at least half of the Col des Tapia woodlands, exports four tons of fruit daily over a period of 2.5 months.

hungry-season protein intake (Gade 1985). Finally, the berries of several trees and bushes are edible and provide snacks. Especially during the hungry period before the rice harvest, the *tapia* forest serves as an invaluable source of dietary supplements (Ramamonjisoa 1995).

Tapia woodlands provide significant income and subsistence to certain sectors of adjoining communities at a critical period of the year. A survey in Col des Tapia woodlands (Randriamboavonjy 2000) showed that woodland products supplied, on average, 6.5 percent of household cash income. However, dependence upon woodlands for cash income varied from zero to 40 percent. Five percent of families earned more than a quarter of their cash income from the woodlands, while two-thirds earned only 0 to 5 percent from the woodlands. Families with a high percentage of woodland income tended to be those with the most available labor (children) or those with the least land and overall income. In Afotsara, based on my household census, the household collection of *tapia* woodland products—fruit and silk—was significantly correlated with family size ($p = .001$), but not with land ownership and wealth. In a similar analysis of woodland-dependent households in the Imamo *tapia* zone, Razafintsalama and Gautschi (1999) determined that poor households are most dependent on the woodlands for resources and monetary income. Poor families relied on woodlands for over 500,000 MGF/yr worth of products, while rich families used only 100,000 to 165,000 MGF/yr.

An incident in June 1996 further demonstrates the value of the woodland resources for the poor. That year, the mayor of the *commune rurale* and the elected councilors were concerned with dwindling silk harvests. They decided to prohibit silk collection during the winter season to allow *landibe* populations to recover. Later, when a particularly zealous councilor arrested two poor local women for illegal silk collection, a mob assembled at the *commune* office. People blew whistles, blew horns, and rang church bells. The Lutheran pastor and village elders mediated and resolved the conflict, reopening the silk harvest to all. Several central issues animated the popular concern. One was lingering bitterness over previous colonial and cooperative control over silk harvesting; several poorer silk collectors told me they resented *commune* controls.[16] Second, the people were fed up with *commune*

16. One man—a richer man of noble descent—told me that around 1960, his grandfather was coming home from the mountain pastures. He passed through the *tapia* woodland and grabbed a half-dozen cocoons that happened to catch his eye. Upon arriving in the village, the *fantsika* (guard) arrested him for illegal harvesting. He became angry, blew the cocoons to the wind, and laid a curse on the cocoon harvest that supposedly reduced harvests for more than a

leaders, accusing them of being too harsh, of abusing their power, and of making laws that were too strict for the people; the arrests focused their anger.[17] Finally, the protesters repeatedly voiced concern for the *madinika*, the poor, for whom the silk harvest served as a crucial income source.[18] Clearly, the *tapia* woodlands play an important role in the regional economy, especially for the poor.

DOCUMENTING STABILITY

Most previous authors assert that *tapia* woodlands—like all other forests in Madagascar—are declining, albeit slowly due to their fire tolerance and economic importance. Girod-Genet (1898) calls the *tapia* woodlands a sign of previously grander forests. Perrier (1921) says the woodlands are disappearing in the face of the prairie; a 1935 Ambositra forest service report assumes that vast parts of the denuded landscape once hosted *tapia* forests (AOM mad ggm 2d19bis). Grangeon (1910), Humbert (1947), Vignal (1963), and Ramamonjisoa (1995) all cite a general decline in *tapia* woodlands. Koechlin et al. (1974) state that the *tapia* forests are degraded and disjointed remnants of previously bigger forests. Gade (1985, 1996) sees *tapia* woodlands as a stage in the long process of change from forests to grasslands, a tenuous plant-silk-people symbiosis at risk of destabilization. Recent reports on the *tapia* woodlands invariably begin with a statement about their decline (Rakotoarivelo 1993; Rambeloarisoa 1999; Randriamboavonjy 2000). However, evidence from the Col des Tapia area suggests that woodland extent and composition saw little change during the twentieth century: declines in one area were offset by advances in another. Below I support this assertion with the available evidence, including air photos, archival documents, landscape photos, and sylvicultural analyses. While some of the evidence remains tentative, triangulation among sources shows a consistent pattern, leading me to place confidence in the results.

The oldest evidence comes from archival documents. Norwegian mis-

---

decade. The poor woman mentioned in note 10 said that if the *commune* were to prohibit silk collection in Afotsara, then the next year the ancestors would be angry and there would be no more *landibe*. Interviews, Afotsara, 1998.

17. Interviews, Afotsara, 1998; minutes of meeting in *commune* office; correspondence with informant.

18. From the scanty minutes recorded in a notebook at the *commune* office: "*Henjana loatra ny dinan'ny commune. Tsy mitsinjo ny madinika*" (the commune's rule is too strict; the poor are not taken care of).

sionaries lived in Manandona and Ilaka from 1870 to 1894. A reading of the descriptions in their letters only provides sketchy details but does not indicate any radical difference from today.[19] A map of the Col des Tapia region produced by missionary Johan Smith in 1888–1889 (figure 4.5) shows, in a generalized form, the location of *tapia* woodlands. The two forest clusters on his map, when compared with figure 4.2, very tentatively suggest that the principal woodland zones have not changed in overall extent. One might argue that Smith's map indicates zones of continuous forest cover as opposed to today's patchy landscape. A tour by colonial forest service chief Girod-Genet ten years after Smith, however, confirms that the forest was patchy a hundred years ago. Except for the current proliferation of pine plantations, his description of forest extent and quality could very well have been written today:

> From Antsirabe to [Manandona] there are almost no trees or shrubs, except near villages. At [Manandona] there is evidence of former forests, due to the presence of numerous trees . . . That is where the *tapia* zone begins. At a two hours' march from the village, going south, one sees numerous trees of this species, at first isolated, then in small bosques, and then in stands that vary between 15 and 150 ha. The Manandona valley, and that of its affluent, the Sahatsiho, are forested almost purely by *tapia;* the *tapia* woods, seen from afar, remind one of olive plantations in Provence. The woodlands include only *tapia* trees; there are no other species mixed in . . . All the heights near Ambohimanjaka are covered with this precious species. It goes without saying that in several spots, *tapia* woods are very open . . . Beginning at the [Col des Tapia], a two-hour walk from Ambohimanjaka, *tapia* stands become rare, trees become isolated." (Girod-Genet 1898, 2349–2350)

A cadastral map of the region, made in 1935, shows the extent of the woodlands at the time, for they are zoned as state domain. Comparison with 1991 shows that general forest boundaries are mostly stable, with only slight losses where a few houses and crop fields have been placed inside the forest area (figure 4.6).

More recent evidence is based on photographs. Borie (1989) compared

19. For example, Franz Bekker wrote in 1876, "Right after Ambohimanjaka [heading south] the path is very hilly again, but extremely interesting because the path goes through a pretty forested area of *tapia* trees. These trees have short trunks and large crowns; they are similar to fruit trees in Norway. In addition to the trees there is a pretty undergrowth and murmuring brooks. One climbs up until Ilaka, then the path descends steeply" (NMS Boks 134/1).

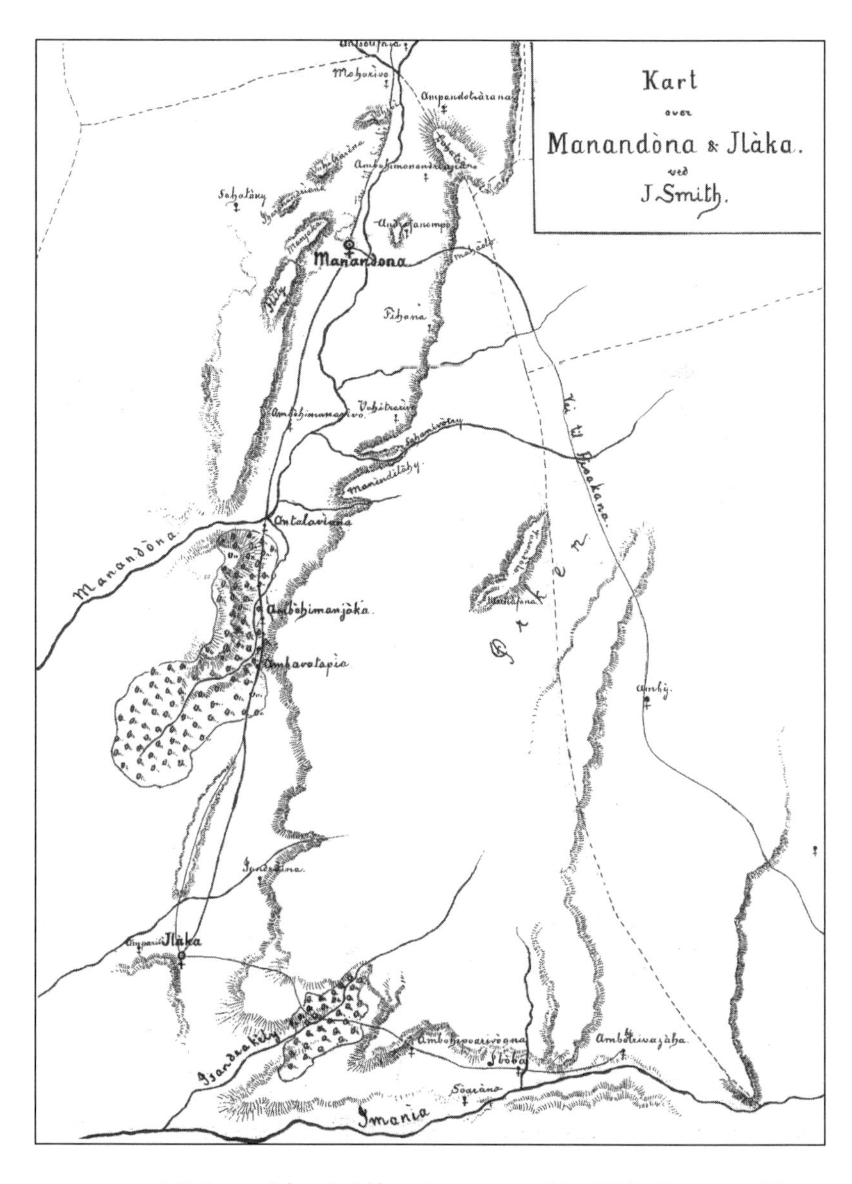

FIGURE 4.5. Missionary Johan Smith's 1888–1889 map of the Col des Tapia zone. The two patches of *tapia* woodland indicated on this map with circular symbols (around Ambohimanjaka and Ambavatapia, and along the Isandrakely River) correspond roughly with the largest patches mapped in 1991 (see fig. 4.2). Source: NMS, Kart Madag.

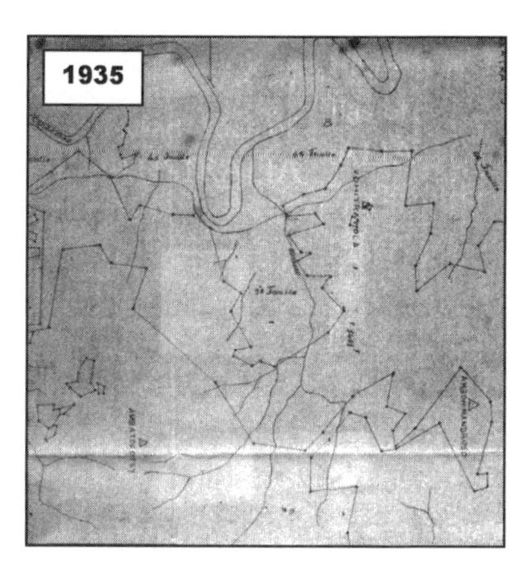

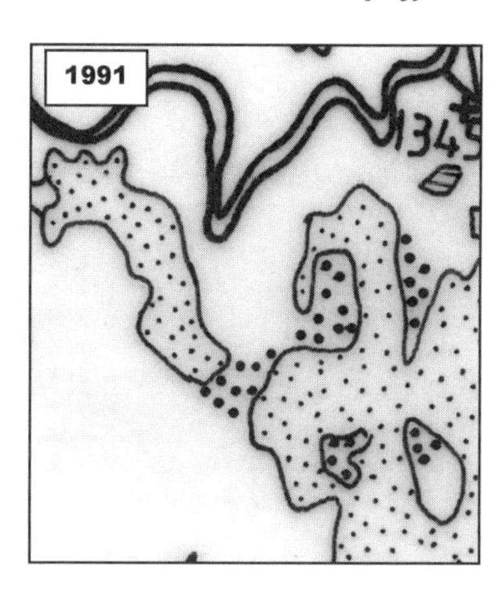

FIGURE 4.6. **Tapia woodland change evidence.** Details from 1930s cadastral map (from Service Topographique, Ambositra, feuille Ilaka B21) and from 1991 woodland cover map (fig. 4.2) for a small part of the Col des Tapia region. On the cadastral map, surveyed lines indicate the general boundary between private lands and state-claimed *tapia* woodlands. The latter category corresponds nicely with woodland areas mapped from 1991 air photos.

aerial photographs from 1965 to 1986 for the northern part of the Manandona valley, finding that the *tapia* woodlands were spatially stable. My own analysis of the smaller-scale aerial photographs from 1949 and 1991 of the entire Col des Tapia region documents overall stability. Using stereoscopes at the laboratories of the University of Antananarivo, I divided the region's forests into thirty-two sectors and compared changes in each. Of twenty-eight woodland sectors where comparison was possible, twenty showed no change, three showed woodland loss, and five showed possible instances of thickening or expansion (table 4.3). Finally, from archives and informants I found five historical landscape photographs of the region dated 1956 through 1972. When compared with today, one photo demonstrates stability, three show *increases* in woodland density or extent (figure 4.7), while one shows both areas of growth and of thinning.

Sylvicultural analyses of the Col des Tapia woodlands indicate a heavily exploited forest that nonetheless succeeds in maintaining itself (Randriamboavonjy 2000).[20] The high density of young trees, including representa-

20. An analysis of tree frequency by diameter class shows a negative exponential curve, a sign of a heavily exploited forest. The height/diameter ratio of the woodlands varies between 25 and 39, which represents a "stable" forest (a ratio over 100 indicates a rapidly growing, immature forest of tall, thin trees) (Randriamboavonjy 2000).

TABLE 4.3. Changes in the *tapia* woodlands of the Col des Tapia region from 1949 to 1991.

Woodland sector	Observed change
Tombonboanjo	same *if not bigger*
South of Kiboy	difficult interpretation
Fierenantsoa	same
Ambohiponana	same? (difficult interpretation)
Vohibongo	same
Vohitrafenana	*much less*
Fierenantsoa south	same
Ambero	*slight increase*
Andriamilarivo (north of gorge)	same
Maromanana	same
Fiakarandava	same
Farasotrina (south of gorge)	same
Sahatamiana (Ambolo south)	same? (difficult interpretation)
Manandona	difficult interpretation
Fierenantsoa-Mahaiza	*less*
Sahanivotry Itsoka	same *if not thicker*
Sahanivotry Itsoka (south of river)	same
Sahamalola	same
Faravondrona (Imaova)	same
Faravondrona (Vohitravoho)	same
Ambohitrambony	same
Marovato	same? (difficult interpretation)
Ambohinaorina	difficult interpretation
Ambohimanjaka west, north	*less* (thinner and cut by cropfields)
Ambohipo	similar; pine invasion
Ankeniheny Ampieka	same? (difficult interpretation)
Ambatobe	difficult interpretation
Ampasambazimba-Tsinjoarivo	same
Ankaranosy-Vatonrdaisoa	similar; *if not denser*
south of Somadex road, west of col	same
Fandrianjato	same
Isandra	same, *if not thicker*

Source: Air photos from 1949 and 1991 (FTM), as interpreted by author.

tives of all mature species, shows that stand replacement is assured. The existence of even-aged stands in the field site and elsewhere (Gade 1985; Rakotoarivelo 1993) suggests that regeneration may be episodic or occur in pulses, and thus may be difficult to capture in short-term studies.

Outside the Col des Tapia region, evidence for *tapia* woodland change is

FIGURE 4.7. ***Tapia* woodland change.** Ambatobe (a) circa 1970 and (b) in 1998. Note the thickening of *tapia* woodlands, partial stabilization of erosion gullies, reforestation with pines, new fruit trees near homes, and changes in house numbers and locations. Ambatobe is just north of Ambohimanjaka (S 20°11.349′, E 47°05.462′). Photos (a) by P. Ottino (with permission from NMS) and (b) by author.

inconclusive. In Imamo, two important cases of deforestation have been noted (Paulian 1953; Rakotoarivelo, personal communication), yet overall dynamics during the past century remain undocumented. The Itremo woodlands were mapped by missionary Smith in 1888 and by French officials around 1898.[21] These maps document a patchy woodland that appears roughly similar in extent to today's Itremo woodlands, based on current maps and field reconnaissance, yet a detailed comparison remains to be done. Thus, while authors complain of declining *tapia* woodlands, evidence is lacking. Hopefully, additional archival research, air photo interpretation, ecological fieldwork, and paleoecological studies will further document *tapia* woodland dynamics.

SHAPING THE WOODLANDS

We have seen that *tapia* woodland edges have been more or less stable for the past century, and how important the woodland economy is to nearby communities. These woodlands are an anthropogenic formation, modified from presettlement forests by burning and cutting. Their management is crucial to local livelihoods. In this section, I demonstrate how the *tapia* woodlands are fundamentally shaped and maintained by nearby communities through burning, cutting, and protection.

Burning

The Malagasy primarily shape the woodland by controlling the fire regime. Presettlement lightning fires were probably infrequent, yet very hot due to fuel buildup. Since settlement, the woodlands burn much more frequently but less intensely; in the most oft-burned forests, fires are restricted to the herbaceous understory. In Afotsara near the Col des Tapia, from March through December 1998, seven burning events occurred in the *tapia* woodlands; 36 percent of monitored woodlands burned (table 3.2). Woodland fires occurred in March, April, July, August, and November. According to most informants, the area of woodland burned that year was higher than average due to repeated locust invasions.

The characteristics of *tapia* trees suggest that some burning regimes favor the maintenance of *U. bojeri*–dominated open woodlands. *Tapia*'s thick, fissured bark, its fire-retardant leaves, and its ability to vigorously resprout

21. Smith's map is found at NMS (file Kart Madagascar); the French map in ANM D435, subfolder "Délimitation de la Province d'Ambositra 1898–1901."

from roots, stumps, and branches[22] make it a classic example of a pyrophytic species such as Mediterranean oaks or southern African miombo woodland species (Kuhnholtz-Lordat 1938; Campbell 1996).[23] My own experimental observations show that two-thirds of burned seedlings resprout in the following rainy season.[24] Koechlin et al. (1974) report that both germination and growth in *tapia* are favored by direct sun exposure, an environmental condition created by fire disturbance. In fact, they note that the oft-burned, more open-canopy Isalo woodlands have more abundant *tapia* seedlings than the less-burned forests southwest of Ambositra. Similarly, Rakotoarivelo (1993) shows from a comparative study in Imamo that *tapia* seedlings are much less frequent in a shadier, less-burned forest than in a frequently burned open canopy forest. Randriamboavonjy (2000) argues that *tapia* seedlings are shade tolerant but need light to develop to adult dimensions.

These characteristics suggest that *tapia* gains its competitive advantage from fire and flourishes in an anthropogenically determined fire disturbance regime. Due to their economic value, humans have used burning to favor *tapia*-dominated woodlands. All the same, overfrequent burning probably prevents seedlings from maturing into adult stages.[25] Unfortunately, little ecological research has been performed in these woodlands, let alone on *tapia* fire ecology. It remains to be determined how *tapia* woodlands react to different frequencies, intensities, and seasons of fire.[26]

Fire also plays an important role in silk and firewood production. Light understory fires during or after the rainy season (January to May) may in-

22. Randriamboavonjy (2000) demonstrated that *tapia* regeneration in the Ambohimanjaka region consisted of 61 percent resprouts, 24 percent rhizomes, and 15 percent seed establishment.

23. Other *Uapaca* species demonstrate various levels of pyrophilia. *U. densifolia* in the Ambohitantely forests of the northern highlands is pyrophytic (Rakotoarisetra 1997), while in the miombo woodlands of Africa, where one-third is burnt annually, *U. kirkiana* and *U. pilosa* are considered semitolerant of fire while *U. nitida* is fire-tolerant (Campbell 1996).

24. In July 1998, I established ten 1 × 5 m plots and noted all seedlings. Three plots burned in August 1998; I recensused plots in December 1998, April 1999, and July 1999. Mortality in burned plots was 35 percent, in unburned plots 3 percent. In the similar miombo woodlands, the majority of seedlings experience annual shoot diebacks caused by water stress or fire, yet the root often survives and produces a new shoot (Campbell 1996).

25. It is not known how many young trees must survive for adequate stand replacement; one may hypothesize that few survivors are necessary due to the longevity of *tapia* trees and their possible tendency to regenerate in pulses (indicated by the existence of even-aged stands).

26. In Zimbabwe's miombo woodlands, long-term experiments showed that a rhythm of late dry season burning every one to two years produced grass-dominated plots, while plots burned in the late dry season every four years supported a near closed-canopy woodland. Other studies argue that early dry season fires are least harmful to woodland trees (Campbell 1996; Bassett and Koli Bi 2000).

crease silk production by controlling populations of a parasitic ant.[27] Colonial foresters noted this practice with disapproval, yet some Forest Service offices issued burn authorizations within the woodlands for these reasons through about 1980.[28] According to people in Afotsara, burning the woodlands also stimulates resprouts of grass and young *tapia* leaves favored by *landibe.* Finally, burning the woodlands plays a role in the production of firewood by creating dead and downed wood one is allowed to collect.

Cutting

Villagers also shape the woodland through their cutting practices. People gathering wood fuel cut plenty of *tapia,* yet they are more aggressive in cutting other trees, like *Leptolaena* spp. and *Sarcolaena eriophora,* thus favoring *tapia* dominance. While the Forest Service prohibits the cutting of all live species (except pine), *tapia* has the strongest prohibitions—socially enforced—due to the value of its associated products. In addition, long-established and generally respected practices of wood fuel collection emphasize dead, downed, and sick branches and may have a pruning effect, removing inefficient lower branches, reducing the danger of crown fires, and perhaps increasing fruit and branch production.[29]

This is far from saying that cutting is always beneficial; indiscriminate cutting has led to forest loss. In the 1970s and 1980s, damage was done to Col des Tapia woodlands as people cut and burned entire *tapia* trees to make ashes that are used for soap and tobacco production. In 1998, a local man was charged with cutting 104 *tapia* trees for charcoal production. In the Imamo region, certain stands have been devastated for charcoal production. Such clear-cutting may have lasting consequences, since *tapia* woodland regeneration, like that of miombo woodlands (Campbell 1996), occurs primarily through coppice regrowth and root suckers and since *tapia* seeds have relatively low dispersability and no extended dormancy. The woodlands are, however, largely protected against such misuse (see below).

27. Interviews, Afotsara, 1998; see also ANM D100s; AOM mad ggm 2d19bis6; Grangeon (1906, 1910); Paulian (1953). The *landibe* harvest in Ambatofinandrahana district in 1951 was very low; the collectors blamed this upon the ban on annual burning of the *tapia* woodland (ANM Monographies 69–84).

28. Interview, Forest Service officer, 29 June 1999. This was also historically the practice in Afotsara (interview, Afotsara, 10 Dec. 1998).

29. Rambeloarisoa (1999) suggests pruning lower branches as a silvicultural treatment for *tapia* woodlands. In Spanish oak forests, pruning is used to increase acorn production; southern Africa's miombo woodlands are managed for various goods by local people through selective pruning and cutting, among other techniques (Campbell 1996).

Overharvesting of firewood can degrade woodlands. However, a calculation of firewood use versus forest growth in Afotsara suggests that firewood collection at its current level is sustainable.[30] Population growth may increase local demand for firewood, yet it may not degrade *tapia* woodlands due to protective rules (below) and increased reliance on alternatives such as introduced pine and eucalyptus (three-fifths of households in the Col des Tapia region grow pine or eucalyptus). Increased urban demand has been shown to cause massive private investment in fast-growing pine and eucalyptus woodlots in some areas (Bertrand 1999), while increased localized demand is met from a mix of sources including native trees, introduced species (pine, eucalyptus, and mimosa), and trimmings from hedges and fruit trees.

Protection

Finally, the economically valuable *tapia* woodlands benefit from protection by both local traditions and government rules. Ancestral traditions spoke for the need to protect the woodlands, and *dina* or local agreements (see chapter 8) against tree cutting existed in some areas, such as Manandona (Borie 1989). In the field site, no formal *dina* exists, yet it is commonly understood that live trees should not be cut. One informant told of her uncle chopping an entire tree at night to access hundreds of cocoons at its summit; he was caught and made to plant trees by the local leaders.[31]

In all areas, the woodlands are guarded by local traditions and lore, including restrictions on cutting live trees, on plucking fruit still on the trees, on breaking large branches to access the spiny *landibe*, or on outsiders commercially exploiting forest products. In the same vein, writing about the Imamo region, Rakotoarivelo (1993) catalogues several poems, stories, and proverbs regarding the value of the *tapia* forest. Some woodlands are even protected as sacred groves to commemorate deceased nobles.

On top of these local rules, the colonial and postcolonial Forest Service has placed strict restrictions on forest burning (see appendix 2) and cutting

30. Fuelwood use averages 7–13 kg per person weekly. For the study site population of 581, this amounts to 200–400 t/yr. Assuming that *all* fuelwood is collected from the 125 ha of *tapia* woodlands in the study site, there is a fuelwood pressure of 1.6 to 3.2 t/ha/yr. However, one-tenth of households use only pine or charcoal, and most others supplement *tapia* fuelwood with pine, eucalyptus, and aged fruit trees. Assuming that this represents a 20 percent reduction, we are left with 1.3 to 2.6 t/ha/yr. Growth rate estimates are unavailable for *tapia* woodlands; for the wet miombo woodlands of Zambia, growing in the same climate, estimates of mean annual biomass growth range from 2.15 to 3.37 t/ha/yr (Campbell 1996).

31. Interview, Afotsara, 8 Dec. 1998.

while allowing for traditional use rights. Decrees in 1900, 1913, 1930, and 1987 made it illegal to cut trees in state forests, which include *tapia* woodlands, without authorization. Fires were banned in and near all forests (without special authorization) in 1900, 1907, 1913, 1930, 1937, and 1960. Local use rights to forest products were affirmed by legislation in 1900, 1913, 1930, and 1987. Enforcement during the colonial period was stricter than it is now, and there have been episodes when indiscriminate cutting led to forest loss. However, drastic violations result in both local and Forest Service action. The man mentioned earlier who cut 104 trees near Afotsara in 1998 was brought to Forest Service authorities and prosecuted. As long as communities continue to be interested in the forests, due to their importance to people's livelihoods, they will protect the forests from destructive cutting.

Forest products, like fruit or silk cocoons, are protected loosely through a kind of common-property regime. An "ethic of access" (Peluso 1996) governs the use of each resource. Nonresidents are allowed to collect occasional forest products for personal use, but not for commercial exploitation. Fruit collection is open to all locals, first come first serve. As noted earlier, the silk harvest was historically tightly regulated, but access is now uncontrolled among locals. It is forbidden to break off large branches to access cocoons, though twigs are commonly broken to avoid touching the cocoons' spines. Fuelwood collection for household use is limited to dead or downed wood both by custom and by Forest Service regulation. It is not uncommon, however, to see live branches cut.

New legislation in 1996 opened the way to officially decentralize management of state-owned renewable natural resources to adjacent communities. In theory, this new approach would aid woodland protection by increasing stakeholder involvement in resource management. However, policy implementation has been hampered by the lack of enabling legislation, high costs, and the complexities of local governance (see chapter 8). Pilot attempts within the *tapia* forests of Imamo suggest that the process remains fraught with stumbling blocks (Razafintsalama and Gautschi 1999), but forest management in Imamo has now been transferred to nineteen local communities (*Midi Madagasikara*, 27 July 2002).

Probably the most important threat to *tapia* woodlands is neither cutting nor fire, but invasion by exotic tree species. Spontaneous colonization occurs from private and village woodlots, consisting particularly of pine (*P. kesiya* and *P. patula*) in the Col des Tapia region (see table 4.1) and *Eucalyptus* spp. in the Imamo region. According to villagers and local foresters, pines and eucalypts may damage *tapia* woodlands by shading out the heliophilic woodland species and changing soil characteristics; additional research is urgent.

CONCLUSION

Gade (1985) argues that the *tapia* woodlands are the result of a people/plant/animal symbiosis. The present book builds on Gade's groundbreaking reassessment of the *tapia* woodlands, emphasizing how fire is crucial to managing the woodlands for people's livelihoods. However, several of Gade's central tenets deserve reconsideration. First, his analysis (see especially figure 6 in Gade 1996), is based on the idea that this anthropogenic woodland is an intermediate seral stage in a Clementsian successional gradient between a diverse presettlement "natural" forest and degraded grasslands. New ecological theories, such as state-transition models (chapter 2) suggest it may be more productive to view the *tapia* forest as one of several possible "states"—discrete vegetation groupings that remain fairly stable for the duration of a similar management regime. The transition to the present state of *tapia*-dominated woodland serves important sectors of the village by improving silk and fruit harvests. Second, Gade's view of the *tapia* woodlands as a stable, homeostatic system (the "*tapia*-protein-silk association") that is threatened by a variety of outside factors is questionable. This characterization plays down the changes and conflicts that have always occurred, including colonial meddling in the silk harvest or regulation of fire regimes.

Humans have fundamentally shaped the *tapia* woodlands, in a sense "creating" them through hundreds of years of burning, cutting, and protection, for their livelihoods. The presettlement highlands were a dynamic mosaic of grasslands, savanna, heath, woodland, and riparian forests, shaped by a natural lightning fire regime, grazing megafauna, and climate fluctuations. As humans arrived, their fires, set to clear land for pasture, agriculture, and hunting, removed most woody vegetation. The *tapia* woodlands, found on certain soil types, were preserved because of their economic value and fire-ecological characteristics. Thus, the *tapia* woodlands should be seen as anthropogenic forest, shaped through a complex history to meet local subsistence and economic needs.

I am not arguing, however, that the woodlands are the result of careful planning. The evidence that I have presented shows that the processes that led to the current situation, and which will continue to shape the woodlands of the future, are complex and difficult to predict. Villagers differ in their knowledge of the woodlands and may even conflict in their goals—one person's silk-protection fire may ruin another's fruit harvest. Criminality is not unknown, as seen by illicit ash or charcoal cutting. Rising or falling prices of forest products like silk and fruit, never mind the state of the general economy, will influence forest use. Finally, government rules regulating woodland management are just as much a part of the story as local rules and enforcement. These factors must be included in the story of these woodlands, a strong example of "anthropogenic nature."

The long-term management of the *tapia* woodlands depends upon the recognition of the human role in forest shaping and of the dependence of neighboring communities on forest resources. The new policy of decentralized resource management introduced in chapter 8 can be a positive step in this direction, as long as it can gain local legitimacy, negotiate the politics of stratified local communities, and find a way around restrictive antifire laws. The future of the *tapia* woodlands, whether stability or change, depends upon varied, unpredictable, and continually evolving processes such as politics (e.g., the new decentralization policy), economics (e.g., changing demand for *lambamena*), and ecology (e.g., invasive pines, climate change). Any attempt to manage the *tapia* woodlands in this context must be adaptive and attuned to both the human and the ecological dynamics that shape the woodlands.

The antifire received wisdom sees the *tapia* woodlands as "degraded remnants" of diverse prehuman forests, reduced by incessant burning to a fire-tolerant species that coincidentally had economic value to the locals. In contrast, I have argued in this chapter that fire is a key management tool used by rural Malagasy in the management of these woodlands, specifically for their livelihood goals tied to artisanal crafts in wild silk, trade in fruit, firewood supply, and food supplements. As this and the previous chapter have shown, the Malagasy set fire to grasslands, crop fields, woodlots, and pyrophytic *tapia* woodlands in order to earn a living. They also use fire as a crucial agricultural tool in humid forest environments; I turn to these *tavy* fires in the next chapter.

5 [FOREST FIRE

Slash-and-Burn Farmers on a Forest Frontier

Tany misy afo misy olona.
Where there is fire, there are people.

MALAGASY PROVERB

THE MOST CONTENTIOUS fires in Madagascar are *tavy* fires, the fires of slash-and-burn cultivation in the forests.[1] As *tavy* is intimately linked to the ongoing loss of humid forest cover, it has long been the most vilified of the island's fire practices. For example, French geographer G. Petit wrote in 1934, "*Tavy* is not just the ruin of the forest, but this temporary and shifting cultivation also has grave social consequences" (Petit 1934, 35). The government has long tried to stop *tavy*, controlling it more strictly than other burning practices, yet it hardly succeeded, running into stubborn resistance from farmers. *Tavy*, after all, is the strategy many Malagasy families rely upon to feed themselves. This chapter looks at *tavy* from the farmers' point of view, as a way of making a living and managing natural resources. But it also looks at the consequences of *tavy* as practiced in Madagascar. I show how *tavy* fits into farmer livelihood strategies, why authorities worried so much about the effects of this practice, and how they attempted to manage the practice. This long story of conflict and negotiation will set the scene for the final part of the book.

1. *Tavy* is the widely used term in Madagascar for slash-and-burn long-fallow cultivation, whether in primary forest or secondary forest or brush (commonly referred to as *savoka*). The verb *mitavy* refers specifically to the slashing, drying, and burning of the vegetation. The word *tavy* itself also means "fat." Regional differences exist in terminology, including *tevin'ala, kapakapa, tetik'ala, tetika,* and *hatsake.*

Tavy fires occur in the forest areas of the entire island, east and west. For the purposes of this chapter, however, I will focus largely on the humid eastern rain forests, as these have long been the flashpoint in the long-standing conflict over forest management. As noted in chapter 1, the humid tropical zone of Madagascar extends in a narrow 1,200 km–long band up and down the eastern side of the island. In this zone, a coastal plain gives way to hilly relief and finally climbs abruptly towards the highlands, over a distance varying between 10 and 100 km. The natural vegetation varies from lowland rain forest to montane cloud forest reaching 2,000 m.a.s.l. These forests, practically continuous before human settlement, are now like a thin beggar's robe, full of holes and tears, no longer forming a continuous band (Olson 1984; see also figure 1.4). Colonial logging and *tavy* frontier expansion have reduced the forest cover by two-thirds since human arrival (Green and Sussman 1990; Jarosz 1996).

A number of different kinds of fire burn in the eastern humid zones. On hills and slopes long cleared of forests by repeated *tavy* cycles, annual grassland fires maintain pastures just as in the highlands (see chapter 3). In addition, large wildfires occasionally burn standing forests, especially along the dryer western frontiers of the humid forests. Vignal (1956a) describes an escaped agricultural fire that burned over 1,500 ha of primary forest in the Zafimaniry lands east of Ambositra. Likewise, 1,600 ha of primary rain forest and montane heath in Andringitra Strict Nature Reserve were incinerated in 1995. The most important fires, however, are *tavy* slash-and-burn fires. Some 200,000 to 700,000 ha of brush, *savoka* secondary vegetation, and forest are cut and fired each year for *tavy*.[2] Below, I focus in detail on the *tavy* cultivation cycle.

THE LOGIC OF *TAVY*

Global Slash-and-Burn

Tavy is the Malagasy name for a family of agricultural techniques utilized around the world, including Swedish *svedjebruk* (swidden), Mayan *milpa*, and Indonesian *sawah*, and technically known as shifting cultivation, long-, forest- or bush-fallow cultivation, or simply slash-and-burn.[3] Historically,

2. Estimate by FAO/PE2 and reported by Ramamonjisoa (2001). The methodology is not specified, but as stated in chapter 1 note 5, the figure makes intuitive sense. This statistic should not be confused with estimates of *overall forest loss*—e.g., 111,000 ha/yr (Green and Sussman 1990; table 5.2)—for most *tavy* occurs outside forests, fallow *tavy* plots can regrow woody vegetation, and *tavy* is not the only cause of forest loss.

3. In French, the most common terms are *culture sur brûlis* and *culture itinérante*. *Essartage* is also slash-and-burn cultivation, but usually reserved for historical peasant practices in Europe.

such techniques have been used by people as varied as Scottish and Irish settlers of the Appalachian highlands (Otto and Anderson 1982) and Indian farmers seeking a living outside floodplain rice fields with a practice called *jhum* (Pyne 1995). These practices are still used today in broad swaths of the world, from the Amazonian lowlands to the hills of Indochina (see Bartlett 1955, 1956; Peters and Neuenschwander 1988; Brookfield and Padoch 1994).

The essence of slash-and-burn agriculture is fertility management. While permanent agriculture relies upon fertilization, weeding, and soil manipulation to guarantee good harvests year after year, slash-and-burn relies upon time and fire. Soil fertility is maintained through fallow periods longer than cultivation periods, and the burning of the standing vegetation before planting provides a crucial input of nutrients. Slash-and-burn requires large areas of land (Peters and Neuenschwander 1988).

As chapter 2 noted, slash-and-burn was frequently vilified—together with other fire techniques—by urban elites, foresters, colonists, and others. It was declared to be primitive, backwards, and destructive to soils and forests. It was only with the work of ethnographic researchers like Conklin (1954) that people began to appreciate the possibility that slash-and-burn cultivation was a rational, often complex land management strategy that—at low population densities—did not threaten the rain forest. The new view was summed up by Peters and Neuenschwander (1988, 84), who argued, "Traditional swidden farming is an ecologically balanced system of tropical agriculture . . . there is no reason why the continued existence of shifting cultivation should be so thoroughly condemned." Researchers continue to explore current practices of slash-and-burn (e.g., Brookfield and Padoch 1994; Padoch and Peluso 1996; Coomes and Burt 1997; Panda 1999; Coomes et al. 2000; Sorrensen 2000; Messerli 2000). They document the economic and agronomic reasons, from a farmer perspective, that make swidden practices attractive. Their findings resonate closely with the Malagasy case, which I describe below.

Slash-and-Burn in Madagascar

Tavy has been practiced in Madagascar since the first settlers arrived on the island. Etienne de Flacourt, who spent seven years at the short-lived French outpost of Fort Dauphin in the mid-1600s, described *tavy:* "Having cut the trees and bushes, they wait for them to dry, and, under a strong wind, they set them on fire. Afterwards, when the rain comes, they plant their food crops [rice, peas, yams, taro]" (Flacourt 1658, 191; see also p. 129).

The site of a *tavy* clearing depends upon the land, its vegetation cover, and social factors. Clearings typically begin along the valley bottoms and proceed uphill. Farmers prefer gentle slopes, but land scarcity or soil exhaustion leads them to occasionally cultivate on slopes as steep as 40 degrees.[4] They avoid physical barriers such as rocky terrain, ravines, or escarpments and distinguish among soil types and topographical locations (Coulaud 1973; Peters and Neuenschwander 1988; Laney 1999; Gautier et al. 1999).[5]

The vegetation cover is crucial to *tavy* selection. When choosing among plots, farmers rely on indicators such as the plant species that are present, the color of plant leaves, and the length of time in fallow, all of which indicate fertility (Coulaud 1973; Dandoy 1973; Le Bourdiec 1974; Brand 1999; Laney 1999).[6] Farmers cultivate the vast majority of *tavy* fields not in primary forest but in previously cultivated land. In the Beforona region, 95 percent of the clearings took place in secondary forest or bush, known as *savoka* (Goldammer et al. 1996). There are a number of reasons. Forests are typically more remote from village sites, involving more travel, and they require more labor to clear. Cut secondary vegetation dries faster and burns more completely than primary vegetation. In addition, the remaining forests are sometimes at less productive higher altitudes. Finally, a hundred years of regulation, while imperfect (see below and chapter 7), was much stricter on *tavy* in virgin forest (Coulaud 1973; Peters and Neuenschwander 1988; Pfund 2000). Even at the forest frontier, only 15 to 20 percent of annual *tavy* cultivation is in primary forest (Brand 1998; cf. Sorrensen 2000).

Finally, *tavy* location also depends upon social factors, especially land tenure. Traditional land rights are lineage based.[7] That is, the right to cultivate a piece of land belongs to the person who first cleared the land and to his descendents.[8] Often these rights extended from the original parcel (in lower,

4. In the Manongarivo region of northern Madagascar, cultivation took place on hillsides of 11 to 17 degree slope in the 1950s, but is now common on hillsides up to 27 degree slope, including occasional fields on slopes over 40 degrees (Gautier et al. 1999).

5. For example, one remnant forest in Zafimaniry country, Manariamboa (located northeast of the village of Faliarivo), has not been cut and burned because it is full of rocks and deep holes (interview, Faliarivo, 17 Aug. 2001).

6. Farmers have a good idea of specific plant species as indicators of soil fertility. See, e.g., Dandoy (1973). Some farmers even cultivate *tavy* in secondary forests colonized by exotic reforestation species such as mimosa (personal observation, Antoetra, 2001; Borie 1989).

7. Based on Coulaud (1973), Dandoy (1973), Brand (1998, 1999), Laney (1999), Pfund (2000; personal communication, 10 Aug. 2001), and McConnell (2002b). Rights to permanent cultivated fields, orchards, and gardens are individual or family based. Some villages also have areas of common land.

8. Laney (1999, 90) calls these lineage-based groups "land management units," which often (but not always) include multiple households across two or three generations.

flatter lands along water courses) up to the hillcrest or even over much of the upstream watershed. Over time, different families claim rights to certain portions, and annual divisions of *tavy* parcels within a lineage group's lands depend upon these family ties, age, social status, and need. As Laney (1999, 96) notes, access by individuals to portions of lineage land is "underlain by a subtle system of precedent," yet "effectively, the whole group has some level of access to most of the family's parcels." Other forms of land tenure in *tavy* areas include community-held land or sacred groves. Farmers will sometimes avoid cutting and burning a *tavy* near a tomb or on *fady* (taboo) lands, where a special form of tenure called *sembontrano* exists (Althabe 1969; Coulaud 1973; Brand 1998; Pfund 2000; McConnell 2002b). In addition to land tenure, farmers may take into account the restrictions of forest agents, like the limits of state-claimed forest reserves or the officially sanctioned *périmètre de culture* (cultivation perimeter), discussed further below.

Once a site is chosen, farmers prepare the field for burning. First, they cut the standing vegetation with axes or machetes. In forest areas, cutting proceeds from the bottom to the top of the slope, allowing trees to fall downhill. This work is usually performed by the men in the family and tends to take place between August and October. Second, the mass of branches, wood, and leaves is left to dry for at least two or three weeks, depending upon the vegetation, slope, and exposure. In the humid eastern zones, the driest period typically occurs sometime between September and November. The size of the cleared plot varies from 0.25 to 1 ha (Coulaud 1973; Le Bourdiec 1974; Brand 1998, 1999; Gautier et al. 1999). As opposed to elsewhere, such as western Madagascar (Genini 1996) or Guinea (Sirois et al. 1998), no trees are left standing.

By law, a ten-meter firebreak must be cleared around authorized burn sites to prevent the fire from escaping. This involves painstakingly removing all vegetation from the ground surface. While this requirement was sometimes enforced, especially in the colonial era, it is often ignored in practice, and no firebreak is ever as wide as ten meters. Coulaud (1973), working in the Zafimaniry country, found farmers making narrow firebreaks only in areas adjacent to small remaining forests. He contrasted this with their regular use in Betsimisaraka country. Two decades later, Brand (1998, 1999) found occasional firebreak use in the Beforona area. While passing through the region in 1998 and 1999, I frequently saw narrow two-meter firebreaks on roadside plots. Pfund (personal communication, 10 Aug. 2001) explained that firebreaks are used here (and fires well controlled) when plots are adjacent not just to roads, but also to sacred forests, tombs, or other crop fields. Due to the humid nature of standing, uncut vegetation, few escaped fires travel far (Brand 1998).

Farmers set fire to *tavy* fields once the slash is sufficiently dry, typically in October or November, just before the rainy season. As observed by Coulaud (1973) in Zafimaniry country, burning takes place in the cooler, more humid evenings. Fires are set to burn slowly downhill or against the wind, assuring a good combustion of the vegetation and good fire control. A small group of men oversees the fire, preventing escaped flames. Some light the fire from the exterior of the plot towards the interior (Le Bourdiec 1974). In Betsimisaraka lands, farmers prefer to burn during midday and early afternoon, the hottest parts of the day (Pfund 2000). Observations from space confirm that burning occurs in the afternoon (Goldammer et al. 1996).

Fire accomplishes several things that benefit the farmers. Most importantly, it removes the tangle of brush and trees, transforming much of it to soil-fertilizing ashes. It also keeps crop predators away. Finally, it kills the seeds of some weed plants that would compete with the agricultural crops. It may also make the soil more friable (Peters and Neuenschwander 1988; Brady 1990; Brand and Pfund 1998).

Rice is the primary crop in *tavy* cultivation. Rice is planted soon after a plot is burned, as the rainy season begins. The women usually do the planting, by dropping the seeds into holes poked by a stick. The plot may be weeded up to three times, and as the rice develops, it must be guarded from birds and other pests. The harvest occurs towards April or May. Average yields of dry, husked rice vary between 600 and 900 kg/ha (Dandoy 1973; Le Bourdiec 1974; Ratovoson 1979; Brand 1998, 1999; Laney 1999).[9]

While rice is the chief crop, in some areas it is interplanted with maize or beans. Some farmers also grow potatoes, sugar cane, or gourds in a portion of the plot. A second year of cultivation is typical in some regions, especially in more fertile fresh *tavy* plots cut from forest.[10] During this second year, the plot may be planted with rice, or perhaps with other crops such as sweet potatoes or groundnuts. Many farmers will finish the *tavy* cycle by planting cassava in a final year, letting the cassava mature as the fallow vegetation invades (Dandoy 1973; Le Bourdiec 1974; Sakaël 1994; Brand 1998; Gautier et al. 1999; Laney 1999; Pfund 2000).

9. One kilogram of fresh paddy, once dried, weighs about 0.75 kg, and then once husked, gives about 0.5 kg of white rice.

10. *Tavy* cultivated in primary forest plots will often yield more during a second year of cultivation than the first year. Brand and Pfund (1998) confirm the value of this practice with soil analyses.

A notable exception to this pattern is the Zafimaniry region. Here, maize, not rice, is the dominant crop[11] and fields are cultivated for longer periods, for four to ten years (depending upon their fertility). Plots are reburned each year, planted with maize and beans the first few years, and with potatoes, sweet potatoes, and finally cassava in later years before being left fallow (Coulaud 1973; see also Bloch 1995).

After a year or more of cultivation, *tavy* plots are left to fallow for an average of about five years before the next round of cultivation. The time in fallow varies widely, however, depending upon the soils, the plot history, land availability, and family needs. In the Beforona case study, the mean fallow length was five years, but the average varied from three years in younger lands at the forest frontier to twelve years in soils used for longer periods (Brand and Pfund 1998; Brand 1998, 1999; Pfund 2000). Le Bourdiec (1974) and Dandoy (1973) also documented differences according to soil condition in Betsimisaraka country: recent forest clearings and zones of good soil were rested for five years, while old *tavy* plots were rested for seven or even ten years. In the north near Andapa, Laney (1999) documented a shortening of the fallow cycle due to increased population pressure, from ten-year fallows in the 1970s to three- to four-year fallows in the 1990s, a figure that is confirmed by Sakaël (1994). Gautier et al. (1999) found similar trends in the Sambirano region, where fallows decreased to five years. Based on fieldwork in the Maroantsetra region, Dark (ca. 1989) distinguished between stable communities that used six- to ten-year fallow cycles and displaced, poverty-stricken communities that recultivated fields after only three years in fallow. Even the Zafimaniry, who cultivate their *tavy* fields for four to ten years, only rest them for about five years (Coulaud 1973). While they rest and regain fertility, fallows are used for numerous purposes, including wood fuel, medicinal plants, construction wood, fibers, and pasture (Pfund 2000).

While the above description emphasizes *tavy* in the eastern humid forests, the basic process is not all that different in western dry forests. For example, in the Morondava region, the forest undergrowth is cut between June and September and then burned in October. This fire kills all the surface vegetation save large, standing baobabs. Maize is cultivated for two to three years, with a yield of perhaps 2 t/ha. Afterwards, the plot is reburned to clear

11. Maize was probably introduced to Madagascar by the Portuguese in the sixteenth or seventeenth century, and was probably adopted by the Zafimaniry in the past 200 years. Why the Zafimaniry cultivate maize is due to some extent to the climate of their land (called *tany be fanala*—land of frost, drizzle, and cold) (Coulaud 1973; Bloch 1995; interview, Antoetra, 16 Aug. 2001), but may also relate to a cultural *fady*, or taboo (Coulaud 1973, 168).

weeds and it is cultivated for another two to three years. Then the plot is left to fallow, more due to an increase in weed invasions than to decreasing soil fertility (Genini 1996; see also Réau 1996). In the Tulear region, most field preparation fires occur in September and October,[12] though I saw some in early August. Here, *hatsake* cultivation of maize is carried out by Mikea, Mahafaly, and Antandroy peoples on plots ranging from 1 to 10 ha. Yields typically decline from 4 t/ha the first year to 0.5 t/ha the fifth year. A recent boom in maize cultivation significantly threatens regional forests (Seddon et al. 2000; see also Rollin 1997).

Tavy*'s Changing Role in Rural Livelihoods and Culture*

The *tavy* process is central to the lives of many rural Malagasy farmers, both as their means of subsistence and as a source of cultural meaning (Moore 1996; Bebbington 1999). First of all, *tavy* is an important livelihood strategy. Most farmers in or near forest zones rely on *tavy* for at least some of their subsistence or income. It is an efficient way to grow rice and associated food crops in lightly settled regions, a low-input technique well–adapted to limited peasant resources.[13] The harvest contributes to self-sufficiency in rice and facilitates control over land resources. In addition, *tavy* represents a risk-averse strategy towards food security for peasants who try to avoid the vagaries of the market economy, cyclones, and other challenges (Ratovoson 1979; Brand 1999; Kistler et al. 2001; Messerli ca. 1999; Harper 2002).

However, for most farmers, *tavy* is only one livelihood strategy among others. In the Beforona case study (Messerli and Pfund 1999; Pfund 2000), people spent between 33 and 45 percent of their time on *tavy* cultivation, but the rest of their time on market gardens, on cash crops such as ginger, on irrigated rice, or on performing wage labor. *Tavy* returned less money per labor day than other activities, but was considered essential to food security and social status. In the case of Zafimaniry country (Coulaud 1973), village men typically contributed only twenty-two workdays per year to *tavy* cultivation, using the rest of their time for other lucrative activities, including woodworking and agricultural migrant labor.[14] In the Andapa region, many

12. Interview, Forest Service officer, 6 Aug. 2001.

13. As Peters and Neuenschwander (1988, 66) note in their review, "slash-and-burn yields maximum return for minimum effort." Rakatonindrina (1989) argued that while *tavy* takes 100 days of labor per hectare, irrigated rice cultivation requires 143. The use of fire in hill rice cultivation is shown to reduce labor needs, yet other approaches like mulching can give higher yields per hour worked (Ravoavy 2001).

14. The men focused their efforts on plot preparation and burning. This, of course, left the women and children with the tasks of planting, weeding, surveillance, and harvest. Dandoy (1973)

farmers supplement their *tavy* cultivation with irrigated rice, coffee plantations, vanilla cultivation, and other farm products (Sakaël 1994; Laney 1999). As land and forest become scarcer, people adapt their strategies and seek alternatives, including more intensive land use (Boserup 1965; Coulaud 1973; Messerli and Pfund 1999).

Second, as central means of assuring a livelihood, *tavy* takes on an important cultural meaning. The burning of the plot itself can be a festive occasion, not only a flamboyant spectacle but also a key step in assuring prosperity (Coulaud 1973). Rituals or blessings may accompany different stages of the slash-and-burn process, from the clearing of the plot to planting, harvesting, and other parts of the cultivation cycle. *Tavy* is intimately linked with the way of the ancestors, a crucial connection in Malagasy cosmology, and it signifies independence, tradition, and the good life (Althabe 1969; Dandoy 1973; Ratovoson 1979; Beaujard 1995; Bloch 1995; Razafiarivony 1995; Jarosz 1996; Brand 1998, 1999; Peters 1999).

This is not to say that the place and role of *tavy* have not changed over time. In fact, the history of *tavy* is a complex story of migration, frontier expansion, soil exhaustion, market incentives, political limits, and population growth. *Tavy* in Madagascar has evolved in response to numerous factors. The colonial appropriation of lands and labor for plantations pushed people further into the hills in search of liberty and free land (Desjeux 1979; Jarosz 1996). Different groups of people have migrated at different times over history, searching out new lands to settle or new forest frontiers, including the Tsimihety arriving in the Manongarivo area (Gautier et al. 1999), or the Betsileo pushing back the forest northeast of Andringitra National Park. The use of *tavy* also responds to swings in the cash crop economy and to national economic crises, as farmers seek to balance opportunity with risk. For example, declining vanilla and coffee prices led to an expansion of *tavy* in the Andapa region as farmers sought to replace lost income (Sakaël 1994).

A final, important factor shaping *tavy* livelihood strategies is population change. Population growth leads to shortened fallows or intensification of cultivation in settled areas, out-migration, or the expansion of the agricultural frontier further into the forest (Coulaud 1973; Dandoy 1973; Sussman et al. 1994). The international literature demonstrates that slash-and-burn cultivation can be a sustainable form of land use if practiced with sufficient fal-

estimates total labor per hectare of *tavy* is 244 person-days, two-thirds of which is weeding and surveillance. Pfund (2000) estimates a total of 209 person-days, with a third for weeding and nearly as much for harvesting, and relatively equal contributions (in time) by men and women.

low time in a stable region of low population density (Conklin 1954; Peters and Neuenschwander 1988; Brookfield and Padoch 1994). In Madagascar, Coulaud (1973) estimated that with a population density of up to about twenty people per square kilometer, *tavy* would be efficient and sustainable.[15] In some Malagasy regions, however, population density has exceeded this limit, leading to a mobile agricultural frontier.

THE CONSEQUENCES OF *TAVY*

From a farmer's perspective, *tavy* is a logical, efficient agricultural practice in loosely populated zones or in settlement frontiers. However, as opposed to grassland or *tapia* woodland fires, where burning is used to maintain a desired vegetation state, the fires of *tavy* in Madagascar's forest frontier are intimately associated with a transition from primary forest to secondary scrub, and eventually, to open grasslands. For this reason, the "external and scientific view of *tavy*" has often been very negative (Brand 1999, 155). As a result, *tavy* is a chief source of conflict between farmers and the state, and it has been the subject of tighter regulation than grassland or *tapia* fires.

The most complete study on the effects of slash-and-burn fires, *tavy* cultivation, and the fallow cycle on soils and vegetation in Madagascar was performed by the interdisciplinary Terre-Tany/BEMA team in the Beforona region (Brand and Pfund 1998; Brand 1998, 1999; Pfund 2000; Kistler et al. 2001). They found that the impacts of slash-and-burn cultivation differ according to time frame (after one fire or after many cultivation cycles), according to geographic scale (on a specific plot versus over an entire watershed), and according to sector (vegetation, soils, water, etc.). The initial cutting and burning of mature forest initiates the largest changes, especially in terms of vegetation species diversity and the potential to support biomass. As farmers know, soil fertility increases in the short term after the plot is burned, due to the import of ashes. After cultivation, vegetation regrowth on *tavy* plots depends upon the number of times a plot has been used, the surrounding vegetation, and topography. Over the longer term, especially on hillcrests and slopes, repeated *tavy* cycles lead to the dominance of a vegetation state characterized by grasses and ferns and the reduction of soil-based nutrients by about one-third. Burning this much-reduced biomass brings only small amounts of new nutrients into cultivation; thus such fields are

15. Rakatonindrina (1989) cites an FAO study, based on numerous tropical examples, that places the maximum sustainable population density for slash-and-burn cultivation at 25 people/km^2.

often abandoned to oft-burned pasture or used for collecting thatch grass or perhaps cultivating pineapples. Finally, the Terre-Tany/BEMA team concluded that the effects of *tavy* on hydrology and physical soil erosion are insignificant compared to the effects on vegetation and nutrients. Their results are borne out by other studies, such as the global review by Peters and Neuenschwander (1988). The details are discussed below.

Nutrients

The impact on the nutrient cycle is one of the key agroecological characteristics of slash-and-burn cultivation. Extensive *tavy*-based agriculture relies upon nutrients freed from forest or fallow biomass into the soil. In the short term, the ashes resulting from *tavy* fires provide an important input of calcium (Ca), magnesium (Mg), phosphorus (P), and potassium (K), though some is quickly lost to wind and water erosion. On the other hand, large quantities of carbon (C) and nitrogen (N) are lost to volatilization—from 95 to 98 percent of C and N in the biomass, and around 20 percent of the C and N in the topsoil (Nye and Greenland 1964; Peters and Neuenschwander 1988; Brand and Pfund 1998; Brand 1999).

In the medium term, nutrients reaccumulate as a field is left fallow. Aboveground nutrients begin to reaccumulate in the biomass immediately upon fallowing; within one year the secondary vegetation has 36 to 57 percent of the nutrients of the preburn vegetation. The topsoil first continues to lose some nutrients—like N, P, and K—and begins to reaccumulate nutrients only after four or five years of fallow (Brand and Pfund 1998; Brand 1999).

The longer-term impact of repeated *tavy* on nutrients is a slow degradation of system nutrient status, especially on hilltops and slopes (figure 5.1).[16] Nutrient losses are due to the volatilization of C and N, the erosion of *tavy* ashes, leaching through the soil, and—to a small extent—export due to harvesting (only 1 to 7 percent). In principle, a sufficiently lengthy fallow period would allow nutrients to fully recover. In Madagascar, a slow degradation results chiefly from short fallow periods. Over years of repeated *tavy* cultivation, fallow vegetation evolves from secondary forest towards grassland (see discussion below). As a result, the corresponding nutrient stocks in the biomass decrease dramatically. Grassland vegetation contains only 1 to 6.5 percent of the nutrients of the forest; in this vegetation state, most of the nutri-

16. Pfund (2000, 259) states that on lower slopes, where the soils stay more fertile and regeneration is more generous, the limiting factor for continuing *tavy* production is not nutrients but instead the increased number of adventitious weed plants.

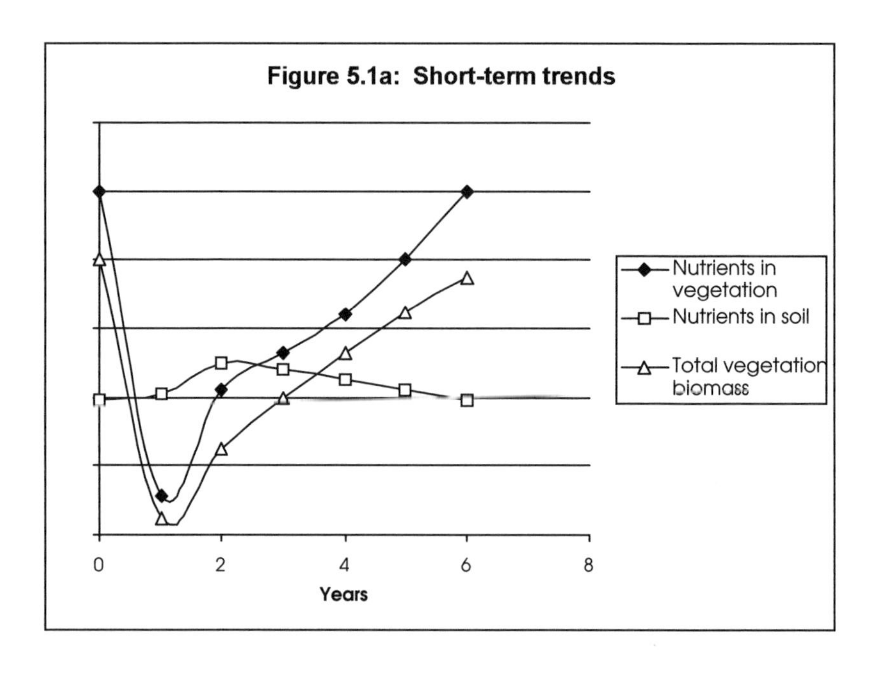

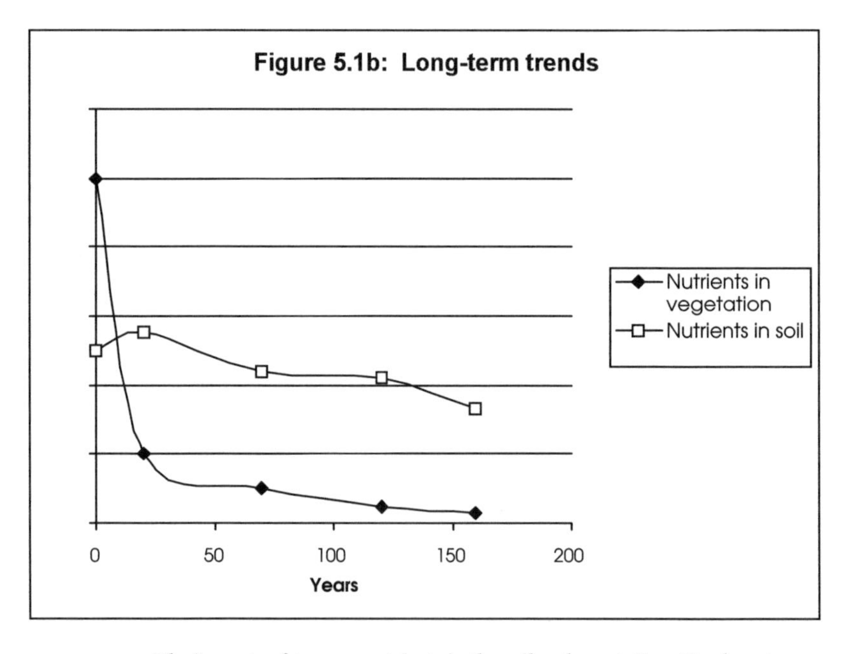

FIGURE 5.1. **The impacts of *tavy* on nutrients in the soil and vegetation.** *Tavy* burning and cultivation occur in year one. These generalized, relative figures are based upon data reported by Pfund (2000).

ents are soil based rather than biomass based. Thus, the nutrient input of *tavy* cultivation shrinks as the fallow vegetation states become less luxuriant (Brand and Pfund 1998; Brand 1999).

Soils and Hydrology

Physical soil erosion, including downstream sedimentation, is often listed as a grave consequence of *tavy* (e.g., Rakatonindrina 1989; see also Peters and Neuenschwander 1988). However, several observers consider this a less severe impact of lower priority (Coulaud 1973; Brand 1998, 1999).

Erosion obviously increases after slash-and-burn. On average, however, soil erosion rates over a five-year crop and fallow cycle are only 4.4 to 5 t/ha/yr (Brand 1999); soil loss is "minimal under sound systems of slash-and-burn agriculture" (Peters and Neuenschwander 1988, 31). Steeper slopes are most susceptible to damaging erosion, such as cyclone-induced landslides, yet rivers tested in the Beforona area are not overwhelmed with sediments (Brand 1998). Over the longer term, however, frequent *tavy* and the associated transition of vegetation from rain forest to grassland accompanies a degradation of soil quality, including reduced porosity, infiltration, and soil humidity, and increased erosion rates (Peters and Neuenschwander 1988; Brand 1999).

In terms of hydrology, cleared and burned areas have higher temperatures, higher evapotranspiration potential, and more runoff, yet the overall effects on the water cycle seem to be less than previously assumed and are not priority issues (Brand 1998, 1999).

Vegetation and Land Cover

The repeated slashing, burning, and cultivation of forestlands has a revolutionary impact upon the vegetation, in both the short and the long term. The first round of slash-and-burn on a forest plot is decisive in terms of loss of species (Fujisaka et al. 1998; Pfund 2000). While an isolated *tavy* clearing will revert to forest over time, repeated *tavy* cycles in Madagascar's humid forests tend to convert vegetation cover to fire-maintained grasslands. To understand the consequences of *tavy* on the vegetation, we must distinguish between the impacts on one plot over time and those on land cover at a regional or national scale. I begin at the scale of a plot.

Tavy destroys the standing vegetation of a forest or fallow plot. After cultivation, farmers leave *tavy* fields to fallow. The type of vegetation that grows in these fallow plots varies widely depending on previous vegetation cover, nearby seed sources, frequency of cultivation, soils, topography and climate

(Peters and Neuenschwander 1988; Brand 1999; Pfund 2000). In recent clearings or near forests, typical fallow successions begin with herbaceous species such as *dingadingana* (*Psiadia altissima,* Asteraceae family; also common are Solanaceae and Euphorbiaceae), grasses, and resprouts of woody forest species. After three to eight years, a thicket of woody pioneer species develops, up to eight meters high with *Trema* spp. (Ulmaceae) as the most common pioneer tree. After six to ten years, this bushy thicket transitions into secondary forest, developing a new canopy at ten to fifteen meters (Gautier et al. 1999; Peters and Neuenschwander 1988; Brand 1998).[17] Local names for many of these stages vary regionally. In Betsimisaraka country, typical names include *ramarasana* for the first year of fallow (sometimes including crops like cassava or melons), and *jinja, dedeka,* or *savoka* for longer fallows (McConnell 2002b; Pfund 2000). The Zafimaniry call a field where harvested corn stalks are still present *haoka,* followed by stages such as *poka, ampatrana, jinja,* and *savoka* (Coulaud 1973). Some name a fallow stage after the dominant tree, such as *trematrema* in the northwest (Gautier et al. 1999).

Fallow vegetation successional transitions change as the number of cultivation cycles increases. Over repeated three- to ten-year *tavy*-fallow cycles, the vegetation regrowth is slower, less woody, and lower in height (see table 5.1). The accumulated growth of plant biomass after four to five years of fallow varies from 40–60 t/ha in first-cycle *tavy* fallows to 15–25 t/ha in sixth-cycle *tavy* fallows to under 5 t/ha in fallows used for 100 or more years. Fallows of *tavy* cut in fresh forest are recolonized by forest species or *dingadingana.* As the number of cycles increases and as neighboring plots are also converted from forest, the fallow vegetation becomes less woody and less diverse. Brand (1998), based on work together with Pfund, Volonirainy, and others, classified a number of different types of such *savoka* vegetation formations, or secondary bush, after their visually dominant species, including *dingadingana, takoaka* (*Rubus mollucanus*), *radiaka* (*Lantana camara*), and *longozo* (*Aframomum augustifolium*). After an extended number of cycles, regrowth on the nutrient-poor ridge top and hillside soils is limited to grasses (especially *Imperata cylindrica, Aristida* spp., and *Hyparrhenia rufa*) and ferns (e.g., *Sticherus flagellaris*), these zones being managed with frequent burning as pasture.[18] Once a plot has transitioned to grassland, *tavy* or

17. Detailed lists of plant species found in fallow *tavy* fields of different ages are found in Dandoy (1973); Laney (1999); Brand (1998); and Pfund (2000).

18. These grasslands (also labeled savannas, pseudosteppes, or prairies) are used for cattle pasture and sometimes as a source of thatch or for pineapple cultivation (Pfund 2000, 201). Cattle pasturing requires frequent grassland fires (see chapter 3), especially as stocking densities are even

TABLE 5.1. **Vegetation states and management regimes in regions of *tavy*.**

Stage	Management Strategy	Vegetation States
Pioneering Clearing	extensive *tavy*	after 4–5 years fallow:
		• 40–60 t/ha phytomass
(1–2 *tavy* cycles)	(2–5 year fallows)	• 80–90 percent woody spp.
(0–10 years)		• 4–6 m height
		secondary forest or *savoka* with *dingadingana* or *trema*
Prime *Tavy* Activity	extensive *tavy*	after 4–5 years fallow:
		• 15–25 t/ha phytomass
(10–20 *tavy* cycles)	(4–7 year fallows)	• 60–90 percent woody spp.
(50–100 years)		• 3–4.5 m height
		savoka with *radiaka* or *takoaka*
Old *Tavy* Zones:		
ridge-tops and slopes	regular (semiannual) pasture fires and extensive grazing	pasture of grasses and ferns
		• 5 t/ha of phytomass
(200 years)		if 4–5 years of no fires:
		• 8–12 t/ha phytomass
		• 40 percent woody spp.
		• 1.5–2.5 m height
lower slopes and valley bottoms	intensive agriculture: irrigated rice, fruit orchards, vegetable gardens, perhaps mixed	crop fields, orchards, or, if left fallow, various kinds of *savoka* depending upon soils and seed sources.
(200 years)	with some zones of *tavy*.	

Sources: Brand (1998) and Brand and Pfund (1998).

pasture fires cause little further change in vegetation (Peters and Neuenschwander 1988; Brand and Pfund 1998; Brand 1998, 1999; Pfund 2000).

The proximate cause of this long-term vegetation transition from rain forest to grassland (at least on hills and slopes) is repeated clearing, burning, and cultivation. The mechanisms by which this transition occurs include slow decreases in soil nutrients, changes in microclimates, and losses of seed sources as forest remnants become more and more dispersed.[19]

lower than in the highlands and grass growth is faster due to the warm, humid climate. Historically, the Beforona corridor required pastures as a zone of transit for the highlands-lowlands cattle trade (Pfund 2000, 103); nowadays people in these villages raise cattle but out of sight from the road (Pfund, personal communication, 10 Aug. 2001).

19. Many authors, especially historically, also blamed regression on a paucity of pioneer for-

The transition from rain forest to grassland occurs faster or slower depending upon soils, topographic location, surrounding land use patterns, and frequency and length of cultivation. The transition can potentially be infinitely slow, at a sustainable low-impact level of *tavy* (Coulaud 1973), or as rapidly as fifteen to twenty years (Erica Styger, personal communication, 15 Nov. 2001). By tracing the history of land use in lower zones of the Beforona region, Brand (1998) and Pfund (2000) determined that grassland developed on ridges and slopes after 100 or 200 years of use, often in combination with land use as fire-maintained pasture. In the Zafimaniry region, where *tavy* plots are cultivated longer—four to ten years at a time—plots are abandoned to pasture after only seventy years (Coulaud 1973). Across the island, while ridge tops and slopes may remain as grassland for significant periods without fire, lower slopes and valley bottoms continue to support other, more diverse types of fallow vegetation or are converted to irrigated rice fields, home gardens, or orchards (table 5.1).

If we now move the scale of analysis from the level of a plot or valley up to the regional scale,[20] the clear story that emerges from the past century in Madagascar is that of net deforestation.[21] The general pattern begins with the arrival of people in previously unused areas, creating *tavy* clearings in the most accessible and productive soils. With time, a village is established and cultivation expands, following watercourses upstream. Eventually, the landscape fills in with *tavy* up to its natural limits of steep slopes, altitude, or rocky land. Over the years, as high-quality *tavy* land becomes scarcer or exhausted, fields require longer fallows to regain fertility and people begin cultivating irrigated rice in valley bottoms as well as home gardens and orchards. Hilltop and hillside *tavy* fields are abandoned to pasture. The process does not follow a clear frontier, but is better characterized by progressive forest fragmentation, where islands of *tavy* in a sea of forest are slowly replaced by islands of forest in a sea of crop fields, *savoka*, and grassland (Coulaud 1973; Kramer et al. 1997; Brand 1998; Gautier et al. 1999; Laney 1999).

---

est recolonizing species (e.g., Perrier 1936; Aubréville 1971; Koechlin 1972; Koechlin et al. 1974; Gade 1996). More recent researchers, e.g., Holland and Olson (1989), Lowry et al. (1997), Brand (1998), and Pfund (2000), have dropped this explanation.

20. See McConnell (2001) on scale issues in deforestation analysis.

21. I write *net* deforestation because there are some areas of expansion. For example, colonial resettlement of hilltop Tanala communities to valley bottom roadsides in the Ranomafana area (to facilitate tax collection and to protect state forests) led to the reexpansion of some forest areas, including the zone around what is now a chief lemur-viewing tourist trail in Ranomafana National Park (Peters 1999).

FIGURE 5.2. **Deforested land east of Antoetra in Zafimaniry country.** Charred tree stumps indicate the recent presence of forest in this zone along the trail from Antoetra to Sakaivo. Repeated *tavy* cultivation has reduced fallow vegetation cover to bracken fern and grass.

Regional analyses demonstrate this trend. The Beforona region underwent its first *tavy* clearings around 150 years ago. By 1957, only 41 percent of the region was covered with primary forest, and the last major forest areas were cleared in the 1970s and 1980s. In 1994, only 14 percent remained forested (Brand 1998), and the pace of loss quickened over time (Ravoavy 2001). In the Zafimaniry region, deforestation was already dramatic by the time of Coulaud's (1973) research. Thirty years later, my own field observations in 2001 demonstrated at least a halving of the forest cover mapped by Coulaud in the area between Antoetra, Faliarivo, and Ifasina (figure 5.2). Finally, in one 9,500 ha watershed in the Manongarivo area of the northwest, forest clearing is taking place at the rate of 121 ha/yr, or 1.3 percent a year, threatening a forest reserve that covers two-thirds of the watershed (Gautier et al. 1999).

At the aggregate national level, analysis is made difficult by widely varying estimation techniques (table 5.2). For example, four studies relied on the 1949–1957 air photo series and each came up with a different estimate of

TABLE 5.2. **Estimates of forest area and rates of deforestation for island as a whole and for eastern humid forests.** Variations in the numbers are due to deforestation and regrowth, different estimation methods, and incompatible categories. Expanded from Nelson and Horning (1993) and DEF (1996).

Year	Data type	Forest area (km²)	Forest loss (km²/yr)	Type of forest used in estimate	Source
1895	gov't estimate	200,000	–	all forests	Lavauden (1934)
1899	unspecified	120,000	–	all forests	Girod-Genet (1899)*
1921	unspecified	70,000	–	native bush and forest	Perrier de la Bâthie (1921)
1931	Forest Service map	100,000	–	all forests	Lavauden (1934)
1936	unspecified	170,000	–	"primitive" vegetation	Perrier de la Bâthie (1936)
		30,000	–	eastern forest	
1949–1957	aerial photography	166,950	–	all forests	Guichon (1960)* (see also Chauvet 1972)
		97,170	–	eastern forests and savoka	ibid.
		61,500	–	pristine eastern forests	ibid.
		191,480	–	all forests	Humbert and Cours Darne (1965)*
		107,430	–	all eastern/highland forests	ibid.
		123,780	–	all forests	DEF (1953–1974)*
		40,620	–	all eastern/highland forests	ibid.
		103,000	1,650	all closed broadleaf forests	Lanley (1986)*+
1960s	Forest Service reports	125,000	–	all forests	AGM (1969)
		67,000	–	eastern forest	ibid.
1970s	study by Delord	–	100	eastern humid forests	Chauvet (1972)
	summary of various sources	75,000	2,000–3,000	all closed forest	Grainger (1984)+
		26,000	–	closed moist broadleaf forest	ibid.
	Landsat MSS	154,844	–	all forests	Faramalala (1981) as analyzed by Mayaux et al. (2002)
		68,217	–	humid dense forests	ibid.
1984–1985	Landsat MSS	38,000	1,110	eastern rain forests	Green and Sussman (1990)

TABLE 5.2. (*continued*)

Year	Data type	Forest area (km²)	Forest loss (km²/yr)	Type of forest used in estimate	Source
1980s	recalculation of others	42,715	1,500	moist tropical forests	Sayer et al. (1992)
1990	AVHRR-LAC and Landsat MSS	58,090	–	all closed-canopy forests	Nelson and Horning (1993)
		34,167	–	eastern rainforests	ibid.
1990–1994	Landsat TM	132,600	–	all categories	DEF (1996) (see table 1.2)
		60,620	–	all eastern/highland forest	ibid.
1997	JERS 500m SAR	59,757	–	rainforest (incl. some secondary)	Matzke (2003)
1998–1999	SPOT-4	173,032	–	all forest (incl. secondary)	Mayaux et al. (2002); loss based on Faramalala (1981)
		55,328	~600	humid dense forest	ibid.

+ From Nelson and Horning (1993).
* From DEF (1996).

forest area. McConnell (2002a) describes how more recent remote sensing analyses still run into difficulty when assessing changes in forest cover, due to inconsistent categories and technological issues with pixel overlays. The most frequently cited study (Green and Sussman 1990) used Landsat data to show that 66 percent of the "original"[22] eastern rain forest has been logged or converted to agriculture and not allowed to reestablish itself, and that at 1980s rates, this rain forest would remain only on the steepest slopes by the year 2025. They also estimate, based on comparison with maps made from 1950 air photos, that about 111,000 ha of forest are lost each year.

That the forest area is decreasing, particularly in the east, is clear. Yet the actual rate of forest loss remains contested (see table 5.2). Conservation International reports a rate of 1,500 to 2,000 km² per year, but does not cite its sources (McConnell 2002a), and one must approach statistics with skepti-

22. "Original" is a troublesome term in the context of changing climates and dynamic vegetation zones. Green and Sussman (1990, 215n.) base their analysis on a "conservative" estimate of the hypothetical and undated "original" extent of the forest.

cism in a context where deforestation has routinely been exaggerated (Kull 2000a). The best-documented source remains Green and Sussman (1990). A recent analysis (Mayaux et al. 2000) claims that its findings conform to Green and Sussman, though my calculations based on their numbers (table 5.2) actually demonstrate a slower rate of loss in the dense humid forests. Deforestation is proceeding most rapidly from the east; the western forest boundary near Antsirabe and Ambositra retreated only ten kilometers in the past 150 years (Coulaud 1973; Ramamonjisoa 1995) and remained essentially unchanged from 1950 to 1985 (Green and Sussman 1990, 215n).

CAUSES OF DEFORESTATION

At this point it is important to look at the ultimate driving forces behind the repeated *tavy* burning and cultivation cycles that have led to this deforestation. Analysts have blamed a number of factors, from population growth and cultural ignorance to commercial logging or sheer desperation. For example, much recent writing asserts that the forest is being decimated by the rural Malagasy, pushed by population growth and poverty to expand and intensify *tavy* agriculture further into the forest and to decrease fallow times (e.g., World Bank 1988; WWF 1992; Allen 1995; Gallegos 1997; USAID 1997b; Myers 1999; Harrison and Pearce 2000). This analysis contains several oversimplifications and misunderstandings (see Fairhead and Leach 1998; Lambin et al. 2001; Geist and Lambin 2002). The underlying causes of *tavy* depend upon the region and the time period. Chief among the causes are a growing population of farmers seeking their livelihoods (as described above), migration to new lands (whether pulled by resource availability and accessibility or pushed by economic or political conditions), forest exploitation for markets (like timber or maize production), and political reasons (like colonial appropriation of land elsewhere).

Population growth is clearly an important factor pushing farmers to expand *tavy* further into the forest. Smaller numbers of people would require fewer *tavy* plots, allowing forests a chance to grow back. Sussman et al. (1994), for example, show that deforestation rates from 1950 to 1985 are spatially correlated with higher population density in the eastern rain forest.[23] Care should be exercised, however, when applying a population-based model of environmental change, for it is the context, not population growth *in itself*, that determines the trajectory of change. In a context of abundant

23. For a critique, see McConnell (2002a).

resources, population growth will likely cause expanding use of those resources, as it has in eastern Madagascar. However, where population growth is associated with limited resources—whether physically finite or constrained by access and tenure laws—it can, under the right conditions, facilitate landscape enhancements, like afforestation or soil conservation (Tiffen et al. 1994; Fairhead and Leach 1996; Kull 1998; Gray 1999). This is clear in the central highlands, where the island's most dense population has transformed grasslands into a landscape of crop fields, rice terraces, woodlots, and fruit trees (Rakoto Ramiarantsoa 1995a; Kull 1998; Blanc-Pamard and Rakoto Ramiarantsoa 2000; see also chapter 3). It also occurs in certain areas of the east, where population growth encounters resource limits. Laney (1999) shows how the forced closing of the agricultural frontier in the Andapa region led the population to intensify its agricultural strategies, including increased emphasis on irrigated rice and market crop production (coffee and vanilla). Laney's analysis supports Geist and Lambin's (2002) conclusion that economic and institutional factors are crucial to determining deforestation trends.

Care must also be taken in specifying the causes of population growth. Regional increases in population can be due to both high fertility and inmigration. Population growth rates are high in Madagascar: the natural rate of increase is 2.8 percent, and fecundity levels in rural areas are over 6.[24] However, a long history of internal migration (Raison 1984; Vérin 1994) also shapes forest frontier populations. Many forest farmers are relatively recent arrivals from other regions, such as the Antandroy who cultivate maize (for an export market) in western forests (Réau 1996; Seddon et al. 2000) or the Tsimihety who have settled the Andapa basin since the 1920s, lured by available irrigable land and a growing market—introduced by coastal Reunionnais plantation owners—for coffee and vanilla (Laney 1999). Both political and economic factors, like oppressive rules or market opportunities, push or pull these people to open new forest areas.

Slash-and-burn agriculture can be a logical farming strategy, at least from the perspective of a farmer seeking his or her livelihood in a complicated political and economic context. Compared to other strategies, when land is available *tavy* requires less labor input per unit yield (see note 13) and is a means to establish longer-term claims over land. This shows that contrary to common received wisdom (e.g., Jolly and Jolly 1984; World Bank 1988), poverty does not force rural Malagasy to sacrifice nature for short-

24. Based on 1993 census and 1997 Demographic Health Survey.

term needs. Instead, they are taking actions that are rational within their economic, institutional, ecological, and cultural context. Give the average Malagasy *tavy* farmer more money, and deforestation may just as well *increase* as they utilize better tools and pay for additional labor to clear more land.

The causes of deforestation vary over time and space. Many specific cases of deforestation involve not just *tavy* farmers, but also commercial interests or political factors (Olson 1984; Jarosz 1996; Gezon and Freed 1999). In fact, during the first thirty years of colonial rule, the rural population did not grow and deforestation was largely caused by logging concessions and the appropriation of land for cash crop cultivation. In this period, the colonial government appropriated the most fertile lands, redistributed them to colonists and concessionaires (French, Reunionnais, Chinese, and Malagasy, especially Merina), and tried to induce the peasants to shift to permanent fields of irrigated rice or cash crops (Coulaud 1973; Dandoy 1973; Desjeux 1979; Peters 1999). Many peasants were forced to (or chose to) escape to forest areas. At the same time, the colonial state granted vast logging concessions to companies and individuals. While the numbers remain quite uncertain, it is estimated that between one and seven million hectares of forest—out of a total of eleven million hectares—saw some logging activity.[25] Colonial logging slowed after stricter regulation in the 1930s, but forest exploitation increased again in the 1970s as government controls loosened (Peters 1999).

If one looks at the local scale, one discovers a number of locally specific causes of deforestation aside from simple *tavy* cultivation. For example, in the 1970s and early 1980s, eleven concessionaires and hundreds of porters operated in the town of Ranomafana, logging the hardwood timber *Dalbergia baroni* (Peters 1994, 1999). Gezon and Freed (1999) describe how various sectors of the Ankarana and Montagne d'Ambre forests in the far north have been impacted differently by village crop field expansion, conversion to sugar cane fields, selective logging, and cyclones. Export maize production in

25. Rapport du Service Forestier (AOM mad-ggm-d/5(18)/5); Direction des Domaines—Etat des baux et concessions forestières consentis pendant l'année 1922 (AOM mad-ggm-5(4)d13); Rapport général sur le fonctionnement du Service Forestier en 1930 (AOM mad-ggm-d/5(18)/15); Notes de l'Inspecteur Principal des Eaux et Forêts Griess (AOM mad-ggm-d/5(18)/9); Humbert (1927, 1949); Heim (1935); Guillermin (1947); Chauvet (1972); Jarosz (1996). It is likely that actual forest impact was at the lower end of the range, as some statistics were based on concessions given rather than exploited, and as much logging involved only selective harvesting of valuable species, allowing quick forest regeneration (cf. Fairhead and Leach 1998, 142). All the same, commentators of the day (citations above) decried the logging as overexploitation; indeed this was one of the impetuses for the revision of forestry laws in 1930 (see chapter 7).

recent decades has devastated some forests near Toliary (Seddon et al. 2000), as it did during World War II when 40,000 ha were felled near Morondava (Humbert 1949). Other examples abound;[26] together they suggest that while *tavy* is the proximate cause behind much deforestation, many other factors, often economic, play locally important roles. The explanation is hardly straightforward.

TAVY POLITICS

While Malagasy farmers view the ashes of a newly burned *tavy* as a sign of prosperity and nutritional security, foreigners tend to see the ash-covered slopes of charred stumps as a desolate landscape and a catastrophe (Coulaud 1973, 162). Numerous commentators have called for a halt to the deforestation that they blamed largely on *tavy* (e.g., Perrier 1921; Humbert 1927; Canaby 1933; Heim 1935; Uhart 1962; Aubréville 1971; Neuvy 1986; Rakotovao et al. 1988; Rakatonindrina 1989; see Bertrand and Sourdat 1998). Negative views of *tavy* became central to the antifire received wisdom (chapter 2), though they varied historically from racist and intolerant to a pragmatic acceptance of *tavy* as a "necessary evil." Forest Service chiefs frequently voiced the loudest complaints, disagreeing with their colleagues in the agricultural services. Few elites were sympathetic to *tavy*.

As a result, the government has always sought to restrict *tavy*. In this section, I review the government regulation of *tavy* over the past half-century. Farmers burning their *tavy* fields have had to operate within a web of regulations, but, as I show, reality on the ground often did not match government ideals.

Tavy has been more tightly regulated and controlled than grassland or pasture burning. This is because, first, the resources at stake, tropical forests, are more valuable to outsiders than grasslands, either as timber for exploitation or as reservoirs of biodiversity; and second, *tavy* is easier to control. Grassland fire may be ambiguous, easily anonymous, and multipurpose

26. Other examples include 1970s famine alleviation policies that legalized additional *tavy* in the southeast, leading to a spurt in cutting (Hardenbergh et al. 1995). During the 1947 rebellion, rebels hid and farmed in the forests at Betampona and Zahamena (Humbert 1949). Significant deforestation in the Mahafaly spiny forest occurred due to government angora goat raising and missionary successes in fighting traditional beliefs including superstitions that protected the forest (Esoavelomandroso 1986). Even back in the 1800s, there is evidence for non-*tavy*-related deforestation: at that time, the charcoal needs of the iron industry led to heavy forest cutting in the Amoronkay region, southeast of the capital (Dez 1970; AOM mad-ggm-d/5(18)/1).

(chapter 2), but *tavy* fire, tied to the clearing and cultivation of a plot, is unambiguous and not anonymous. It is easier for an agent of the state to determine the farmer responsible for a *tavy* clearing than the author of a nighttime pasture fire.[27]

Since colonial days, administrators wrote laws with the purpose of protecting the forests and encouraging permanent, intensive agriculture. The laws framed *tavy* in a negative light: it became officially banned except where authorized. Chapter 7 will discuss the overall story of fire and forest policy; here we shall look specifically at the de facto authorization system for *tavy* and how it worked. The authorization system varied over time. Sometimes, it allowed *tavy* in primary forest, sometimes just in *savoka*. Sometimes, demands for authorization had to be placed by individuals, sometimes by villages. Often administrators were called upon to delimit zones where *tavy* would be tolerated or to implement policies and infrastructure (e.g., irrigation dams) to encourage cash cropping and permanent agriculture. However, implementation and enforcement were far from uniform or thorough, especially in remote zones (Ratovoson 1979; Jarosz 1996; Laney 1999). While the goal was to stop *tavy* in its tracks, the result was far from complete, as administrators sought to balance pragmatic needs (feeding the population, avoiding revolt) with technocratic goals (stopping deforestation, encouraging intensive agriculture) (Dez 1968).

In the 1960s, the control of *tavy* reflected recent colonial practice and was based on the rules and procedures in Ordonnance 60-127 and Décret 61-079 (appendix 2).[28] Control was the responsibility of touring foresters. At that time, each canton had a forester, who would spend up to twenty days per month in the field.[29] The forester, usually a *chef de triage*, was responsible for reforesting certain areas and for arranging the building of fishponds and of small irrigation dams to encourage permanent crop fields. In July and August, the forester would visit villages to spread awareness about forest conservation and to deliver *tavy* permits. Each village was supposed to be accorded a permit that specified the number of people, the area to be burned and cultivated, and the precautions to be taken (e.g., firebreaks); it also in-

27. Interview, Forest Service officer, 7 Aug. 2001. Since enforcement was practicable, administrators could use the threat of not receiving *tavy* authorization as an way to pressure farmers into paying their taxes (Coulaud 1973, 161).

28. This description of the 1960s situation based on interview, retired Forest Service officer, 21 Aug. 2001; interviews in Zafimaniry country, 16–18 Aug 2001; and Coulaud (1973).

29. As Coulaud (1973) noted, the Forest Service was the only branch of the state one was likely to encounter on the paths of isolated rural communities.

cluded a sketch map. The forester would, in theory, visit those lands he did not already know. Foresters officially gave *tavy* authorizations only in *savoka*, not in primary forest (the only exception was to convert valley bottoms to irrigated fields). In reality, farmers often cut further into the primary forest, or burned areas without authorization, or took advantage of gaps in state vigilance.[30] A final and critical portion of the task, therefore, was enforcement, yet the overall authoritarian system was often deemed a failure.[31]

In this modernist period, the Forest Service also sought to delimit "*périmètres de culture*," or agricultural zones, with the idea that farmers would restrict their *tavy* to these zones, and would, with the help of agricultural extension services, switch to intensive permanent cropping. In some forest-based villages in Zafimaniry country, four-fifths of the land was declared state forest[32] and the rest, typically near villages and along watercourses, was declared as a *périmètre de culture. Tavy* burn authorizations were to be restricted to specific plots within these zones. These limits were established in 1946; by the late 1960s they were already out-of-date and duly ignored (Coulaud 1973; see also Ratovoson 1979).

In the 1970s, government control of *tavy*, as imperfect as it was, fell apart. Power struggles in the capital and populist concerns led to loosened regulation and enforcement, though the laws did not change. The political message of the time, a period that Madagascar saw as its "real" independence from French neocolonial control, was to "do as you please"; politicians turned a blind eye to *tavy* violations.[33] Foresters continued to tour and issued authorizations and citations, but the pace and effectiveness slowed. As Brand (1998) documents, seconded by Laney (1999) and Peters (1999), the pace of deforestation by *tavy* picked up dramatically in the 1970s.

30. Some claim that *tavy* burns in secondary vegetation (*jinja* or *savoka*) did not even need authorization (interview, Antoetra, 16 Aug. 2001). When the Forest Service was absent from the Zafimaniry region in 1970 (due to an unfilled position), villagers in Faliarivo took a gamble and cut large forest areas. The following year, the Forest Service took legal action against the village for this. Imprisonment and fines for illegal *tavy* were not uncommon (Coulaud 1973, 208; interviews, villager and assistant mayor of Antoetra, 16 Aug. 2001).

31. For example, Coulaud (1973, 326) saw it as a failure, noting that the single forest service officer in the Zafimaniry region was overwhelmed and served a conflicted role of both enforcement and extension advisor (p. 329).

32. Such state forests could officially be used by villagers only for collecting dead wood, non-timber forest products, or hunting; they were given for a fee as logging concessions to outsiders (Coulaud 1973).

33. Interview, retired Forest Service officer, 21 Aug. 2001. Laney (1999, 31) confirms relaxed anti-*tavy* politics during the 1972–1975 reign of Ramanantsoa.

By the 1980s, fiscal and economic crises dampened state efforts despite a renewed anti-*tavy* concern (e.g., RDM 1980; see also chapter 7). The number of touring foresters began to drop. In addition, the authority for issuing *tavy* authorizations was moved from the (now rarer) *chefs de triage* up to the *chef du cantonnement*. Authorizations were no longer based on the work of touring foresters but on applications signed and vetted by *fokontany* presidents (PCLS).[34] An average of 27,000 ha/yr of *tavy* were authorized by the Forest Service from 1984 to 1996, yet the actual area cut and burned annually was at least ten times as much (Ramamonjisoa 2001).

From the 1990s up to the present, conservation imperatives have led to renewed enforcement efforts in a few specific locations, while the old authorization system seems to have largely foundered elsewhere. With the influx of conservation money and interest, attention focused on protecting the last big stands of rain forest, and enforcement was jump-started. For instance, in 1994 WWF and the Forest Service began to control the limits of Marojejy and Anjanaharibe-Sud protected areas in the northeast, warning illegal settlers to leave and taking several to court or jail (figure 5.3).[35] The 1991 founding of Ranomafana National Park out of a classified forest in the southeast also resulted in tighter control, including the arrest of thirteen villagers and, unsurprisingly, conflict with local farmers (Peters 1999). *Tavy* enforcement near Fort Dauphin in the south is strongest in and around Andohahela National Park. In fact, a "punitive expedition" by the Forest Service into the park in early 2001 resulted in the arrest of twenty-two people for illegal *tavy*.[36] At the same time that enforcement has been sharpened around protected areas, there have been a number of extension efforts to encourage the development of alternative livelihood strategies, including irrigated rice, agroforestry, and improved *tavy* cropping (e.g., Uphoff and Langholz 1998; Kistler et al. 2001).

On paper, the official *tavy* permit process remains largely the same as that outlined by Décret 61-079 (appendix 2) and practiced as described earlier; each parcel must be delimited and authorized. (One small change is that demands for authorization must now bear the signature of the mayor of the *commune rurale*, due to the restructuring of local government since 1993.) In practice, however, the system functions imperfectly if at all. Foresters outside of priority conservation zones hardly have the means, the time, or the political support (see chapter 6) to go on tour; they are accused by some of being

34. Interview, retired Forest Service officer, 21 Aug. 2001.

35. See introduction to part 1. Based on field notes, July 1994; Laney (1999); Garreau et al. (2001).

36. Interviews, Forest Service officer and project staff, 7 and 9 Aug. 2001.

FIGURE 5.3. *Tavy* farmer inside Anjanaharibe-Sud Special Reserve during interrogation by Forest Service agents, 1994 (see preface).

"asphalt foresters" who never leave the pavement, or who prefer to sit in their offices, issuing logging permits for lucrative bribes.[37] Enforcement remains insufficient and paltry (Rabetaliana 2001).

In the Beforona region, for example, some communities create a collective list of all farmers who wish to burn *tavy* in village lands, and this application is brought to the Forest Service in June at the earliest. The Forest Service—based in Moramanga, over forty kilometers away over the escarpment—rarely field checks the application. They could not physically do so—in the 1980s, for example, they received around 4,000 applications for indi-

37. Interview, retired Forest Service officer, 21 Aug. 2001, and interview, anonymous project staff member, 9 Aug. 2001.

vidual *tavy* plots each year (Rasoavarimanana 1997). If the foresters grant *tavy* authorization, it often comes long after villagers have burned the intended plots. This control of *tavy* is still dreaded by villagers, yet the system hardly works now. Some villages are too remote to be bothered, others take advantage of state weaknesses, and authorizations and citations are almost unknown now (Brand 1998; Pfund 2000, and personal communication, 9 Aug. 2001).

In the Zafimaniry region, the last *chef de triage* stationed in Antoetra left around 1996, and control loosened further. At that point, *tavy* authorization applications were done at the *fokontany* level, then brought to the *commune rurale*, and from there submitted to the Forest Service in Ambositra. Authority over *tavy* has become confused since then, as some citizens interpret the new focus on decentralization (see chapter 8) as giving the *commune rurale* control, while the Forest Service sees the *commune rurale* as its *mpisoloteny*, or representative, in the field. The Ambositra Forest Service suspended the giving of all *tavy* permits—whether primary forest or secondary *savoka*—in 2001, yet I saw several freshly felled *tavy* plots along paths in August 2001.[38]

At the southern end of the Malagasy rain forest, around Fort Dauphin, *tavy* authorizations since 1990 have been given only for secondary vegetation (*savoka*) outside protected areas and state forests. Village forest and environment committees called KASTI[39] were set up by the Forest Service to vet *tavy* applications (and to promote environmental awareness). Applications for authorization are submitted at the family level. They require the signature of three KASTI members certifying that the plot is *savoka* and that appropriate precautions have been taken, plus the signature of the PCLS to ensure that the applicant has land rights to the plot. The Fort Dauphin office of the Forest Service receives perhaps thirty such requests per year, for an average of 0.5 ha per family. With such numbers, obviously, only a small minority of the many *tavy* burners actually applies for permission; these are typically the farmers located in areas of close surveillance near roads and conservation projects. *Tavy* farmers in the inaccessible Beampingaritra range fifty kilometers away fire their plots unfazed by bureaucracy. All the same, some enforcement occurs and the *cantonnement* logs perhaps forty or fifty citations (*procès-verbaux*) for forest crimes per year. These are for repeat offenders

38. Interviews and observation, 16–18 Aug. 2001 (Forest Service employee, Ambositra; assistant mayor of Antoetra commune; villagers).

39. KASTI, for Komiten'ny Ala sy ny Tontolo Iainana, or Forest and Environment Committee.

clearing illegal *tavy* in primary forest; those who illegally clear in *savoka* are only warned.[40]

In the western forests, *tavy* or *tetik'ala* follows on paper the same official permit process as in the east. In principle, communities apply for authorization, which, following a site visit, is given by the Forest Service. In Toliary *cantonnement*, the issuing of authorizations was more or less stopped in 1988 (in the 1980s, the chief of the province had used his political clout, and his need for reelection, to authorize far too many clearings; in 1988 this power was removed from his control). Since then, only two or three permits are given a year and only for special cases, like replacing croplands lost to river flooding. Enforcement has similarly dropped. The last enforcement field tour was in 1996; the number of *procès-verbaux* used to hover around twelve per year, but since then has dropped to one or two per year (cf. figures 6.2 and 6.3 in the next chapter). Like elsewhere on the island, authorizations and enforcement lag far behind the reality on the ground; significant areas of the western forests are being cleared for maize-based *tavy* cultivation.[41]

CONCLUSION

Tavy slash-and-burn cultivation is a logical, efficient agricultural practice in loosely populated zones or in settlement frontiers. It is a crucial strategy employed by rural farmers to meet their subsistence needs and livelihood goals. However, due to the dramatic loss of forest related to the expansion of *tavy*, this practice has been the flashpoint for environmental conflict on the island. In fact, the case of *tavy* serves as a microcosm for the overall argument of this book. I have demonstrated the important role of *tavy* in rural people's lives, as well as the foundations of government and conservationist anti-*tavy* ideas. The tension between the livelihood needs of the farmers and the government's repressive policies predictably led to conflict and resistance. As the preceding section shows, state repression was far from thorough, either turning a blind eye or powerless to stop the farmers from continuing to clear and burn the land. This leads to the stalemate of today.

The next part of this book—part 3—will focus more closely on the struggle over fire for all types of burning, to complete the book's argument. First, however, one last word is warranted about *tavy* fires and deforestation. In its essence, this is a conflict over the control of forests and the process of

40. Interview, Forest Service employee, 7 Aug. 2001.

41. Interview, Forest Service officer, 6 Aug. 2001; see also Genini (1996), Réau (1996), and Seddon et al. (2000).

land use change. Historically, farmers sought to protect their livelihoods and independence, while the colonial state sought to capture peasant labor into the market economy, to control exploitable forest resources, and to "rationalize" forest exploitation. Today, farmers, the government, and the global environmental community are in effect struggling over the control of these same forests—using as their tools protected-areas legislation, mapping and monitoring, project funding, and (on the part of the farmers) physical acts of cutting and burning (Ghimire 1994; Gezon 1997; Richard and O'Connor 1997; Kremen et al. 1999; Marcus 2001). Region by region, these groups are struggling to delimit which forests are to be conserved and which zones may be used for agriculture. Perhaps, as time goes on, depending on economic, political, and institutional processes, we will begin to see a "forest transition" (Rudel et al. 2002)—a transition towards reforestation of more marginal farmlands and a new conservation focus that includes such agricultural lands in its purview (Zimmerer and Young 1998).

Part Three

Fire Politics

OVER THE PAST century, strong forces within the colonial and postcolonial states have repeatedly condemned Malagasy burning practices, legislated restrictions on fires, and sought to punish burners. They did so convinced that fire was "bad," a primitive, backwards technique that threatened the island's natural resources, causing deforestation and soil degradation and blocking development. However, as we have seen in the previous part, many farmers and herders rely on fire to maintain pastures and woodlands, control wildfires, prepare crop fields, and control pests. Fire is central to many rural livelihoods as a key agropastoral environmental management tool. For these reasons, people continue setting fires; after 100 years of government antifire efforts, little has changed.

This part of the book investigates the origins of this stalemate. It focuses on the struggle between parts of the state and the rural farmers and herders over fire. My point of departure, chapter 6, is the theoretical framework of "criminalization" and "resistance." I demonstrate the mechanisms by which the state criminalized fire and how the peasants resisted this criminalization. By taking advantage of the state's distended nature and of strategic village solidarity, as well as of the biophysical characteristics of fire, rural peasants have succeeded in maintaining the stalemate over fire. Fire-setting peasants have a crafty ability to move in the interstices and ambiguities of what on the surface seems like a uniform government antifire stance, harnessing fire's own ambiguities to create and defend their livelihoods, meeting their landscape management goals. As a result, fire management is beset by frustrations, exceptions, inconsistencies, corruption, and (occasionally) protest.

In chapter 7, I chronicle this pattern of criminalization and resistance through the history of Malagasy fire politics. Distinct historical periods emerge in response to political, economic, and ideological changes. Next, chapter 8 looks at recent attempts to break the impasse between farmers and the state over fire. Since the mid-1990s, community-based natural resource management has been touted as a potential solution, yet with few results to date for fire. At the root of these difficulties, I argue, is the old antifire received wisdom we encountered in the first part of this book, as well as issues of power in local politics. Finally, chapter 9 proposes some steps to dousing the ever-simmering embers of the fire conflict without snuffing the flames of the island's fires.

6 [THE STRUGGLE OVER FIRE

Criminalization and Resistance

Ny gidro sy ny ala	The lemurs and the forest
Tsara karakara	Are well taken care of
Andaniana drala	The money is spent
Amina milliard	By the billions
Rakoto ity mijaly tsisy mpitontaly	Rakoto suffers, no one pays attention
Ny zanany tomany tsisy mpanontany	His kids cry, no one asks why
Rakoto dia espèces en voie de détresse	Rakoto is an endangered species
Rakoto efa resy tsy mba misy SOS	He's already defeated, can't call SOS
Ianareo miventy	You chant
Amin-dradio sy an-gazety	On the radio and in the paper
Milaza fa tsy mety	Say that it is not okay
Ny mandoro tanety	To burn the hills
Rakoto noana loza	Rakoto is dangerously hungry
Manao makia kitoza	So he smokes some lemur meat
Tsy lafony ny vokatra?	Cannot sell his crops?
Endasiny ny sokatra	Then he fries the turtle
Ka ny doro tanety	So the burning of the hills
No heveriny fa mety	Is what he thinks is right
Hiantso ny fanjakana	To call the government
Hijery azy any an-tanàna	To visit him in the village

MALAGASY POP STAR ROSSY, 1994[1]

1. Selected verses from song "Resa-Babakoto," album *Bal Kabosy* (1994, Production Pro).

ROSSY'S LYRICS cut right to the heart of Madagascar's fire conflict. Rakoto, a hungry farmer, struggles against a government concerned with nature conservation. Fire is a central flashpoint of this struggle—in previous chapters, we saw its importance to rural livelihoods, as well as the state's antipathy towards the practice. In this chapter, I analyze the mechanisms of the conflict and answer the questions Why has the stalemate over fire persisted so long? and What are the consequences for fire management?

The conflict over fire follows a pattern similar to many situations where states sought to assert control over natural resources managed by local populations: criminalization and resistance (Hay et al. 1975; Thompson 1975; Peluso 1992; Neumann 1998). Criminalization is the state's negative redefinition of a resource management practice, such as fire, in order to assert specific claims to resources. Resistance is the attempt—through actions and words—by resource users to forestall or fight criminalization, to protect their rights and impede interference, in order to assert alternative claims.

In Madagascar, the Forest Service and its allies, following the antifire received wisdom, criminalized fire through legislation, enforcement, and propaganda. The rural farmers and herders resisted by taking advantage of state weaknesses and the nature of fire, leading to today's stalemate and its associated frustrations and legal inconsistencies.

The key analytical categories in this chapter are the state on the one hand and the civil society—the peasants, the Malagasy farmers and herders—on the other. These categories form a dialectical couplet, or two intertwined parts of a larger category. They are, in essence, mutually constitutive parts of broader society. The state reaches deep into society—such as the farmer's brother who is a local councilor—and society reaches deep into the state—illustrated by debates and conflicts between state agencies or officials (see chapter 1).

Like the state and civil society, the concepts "criminalization" (or domination) and "resistance" also fit inside a larger category, that of the exercise of power. Practices of resistance cannot be separated from practices of domination; they are inextricably linked, mutually constitutive, and produce complex entanglements of power. Resistance is meaningless unless there is domination, and domination always causes resistance (Sivaramakrishnan 1995; Klooster 2000; Sharp et al. 2000).

Donald Moore takes the analysis of such intertwined analytical categories a step further in a way that is particularly useful to studies of natural resource conflicts. He argues that these concepts should not be seen simply as abstract two-way modalities, but should be analyzed through a "realpolitik of place and practice" (Moore 1998, 347, 351). That is, one should look beyond simple analyses of state domination and peasant resistance and pay detailed attention to how individual actors, moving within specific ecological, cultural, historical, and political landscapes, negotiate multiple fields of power. Resistance, after all, occurs in specific places by people facing their own livelihood struggles (cf. Bebbington 2000; Leach and Fairhead 2000).

These conclusions have a direct impact on my argument. While on the surface the story of fire politics in Madagascar may be one of state criminalization versus peasant resistance, it is precisely the entangled nature of these categories, and their location in a specific biophysical and historical landscape, that shapes the character of the struggle. Specifically, peasant resistance to fire criminalization is made possible by internal weaknesses and differences in the state and by the physical character of fire in wet-dry tropical climates. While I rely on the broad categories of "state and peasant" and "criminalization and resistance" (for while perhaps overly simplifying, these are heuristically useful categories), the struggle over fire is shaped by these categories in both their dichotomous, affirmative sense (state laws versus community solidarity), in their dialectical, intertwined sense (diversity and debates within the state), and also outside the categories, in relation to the realpolitik of place and practice (the biophysical nature of fire, together with local livelihood strategies).

CRIMINALIZATION AND RESISTANCE

Fire control has been an important goal of both the colonial and independent states in Madagascar. Due to the antifire received wisdom (chapter 2), most officials viewed the ubiquitous fires as primitive agropastoral tools dangerous to economic assets and natural resources. Foresters, in particular, argued strongly against burning practices. To them, fire threatened valuable timber and caused deforestation and thus soil erosion. They were joined by naturalists, who saw fire as the chief agent of environmental degradation, and by agricultural advisors, who pushed for a transition to irrigated rice, cash crops, and intensive cattle husbandry. As a result of these views, this elite portion of the state, with the support of the central leadership, succeeded in criminalizing most burning practices.

The state harnessed three broad tools to suppress fire-setting (chapter 7 will chart the specific details). First, it relied on laws and regulations (see appendix 2). The philosophy of regulating fire through rules was defended by its most extreme proponent, Forest Service chief L. Lavauden, in 1937: "To govern is to annoy. In forestry matters, where only the specialist can distin-

FIGURE 6.1. Gendarmes at the scene of a fire in the *tapia* woodlands.

guish between use and abuse, this is inevitable. It is through the quantity and violence of protest caused by forestry regulations that one can best judge its pertinence and effectiveness" (Bertrand and Sourdat 1998, 116). At the same time, due to dissenting voices, the state grudgingly made room for the continuation of some economically critical fire practices, like antilocust fires or some pasture fires.

Second, few laws are effective without enforcement, so the state made patrols, arrests, citations, courtroom trials, fines, and sentences (figure 6.1). Punishment varied from incarceration to villagewide collective fines (payable in cash or bricks) to tree-planting requirements. The state periodically attempted to improve enforcement, by giving more authority to patrolling foresters (1937, 1950), by strengthening prosecution (1930, 1972, 1976), by increasing the size of the Forest Service, by toughening prescribed punishments (1930, 1941, 1960, 1975, 1976, 1977), by giving financial incentives to foresters for citations issued (1930, 1950s, 1961–1990), or by mounting specific antifire patrols (1956, 1976–1979, 1980–1987), but its efforts can hardly be deemed a success (figure 6.2).

The third tool of the state was propaganda. States have easy access to mass media, and they make use of it to spread their message. Bureaucrats con-

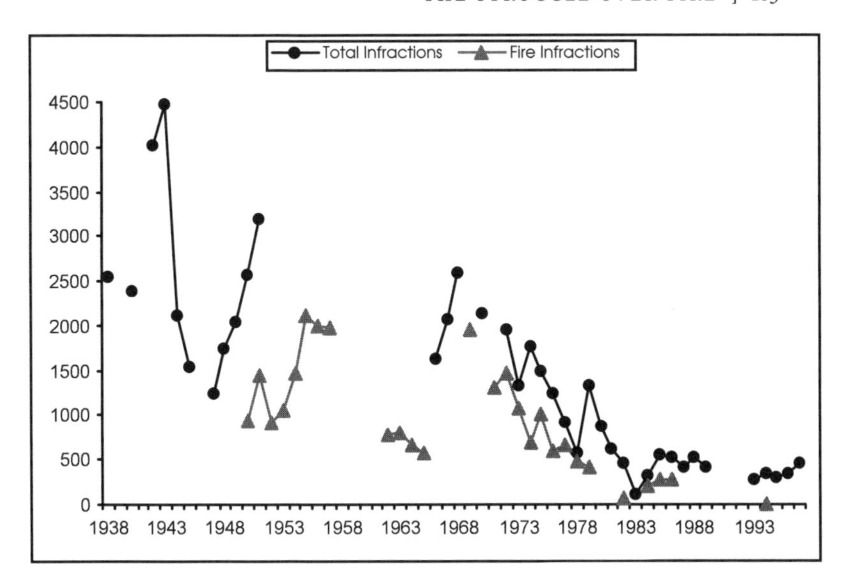

FIGURE 6.2. **Enforcement of forestry and fire regulations, 1938–1997.** For "total infractions" (total number of forestry infractions including fires), 1938–1951 are forestry, hunting, and fishing violations (source: SSG 1953), and 1966–1997 are forestry offences (source: Rapports Statistiques, Ministère de la Justice). "Fire infractions" are the number of fire-related forestry infractions based on number of *procès-verbaux* filed. Sources: 1950–1965, 1984–1986: Rakatonindrina (1989); 1969–1977, 1982, 1994: Rapports Annuels, DEF; 1978–1979.: RDM (1980).

cluded early on that peoples' attitudes must be changed, "through a slow education not only of the masses, but also of the ruling class" (Perrier 1921, 265). Approaches included public awareness campaigns, gubernatorial circulars, and political speeches. Colonial governors asked their district officers to make statements against burning in their *kabary,* speeches at local assemblies and markets.[2] Increased concern with fire since the 1970s has led to massive public awareness efforts, complete with radio advertisements, slogans, and flyers, as well as a WWF-sponsored environmental education campaign in schools. The "Year of the Fight against Bushfires" occurred in 1994–1995. Politicians continue to spread antifire messages. In September 1998, the prime minister, Tantely Andrianarivo, came to a town near Afotsara to inaugurate the build-

2. Circular No. 34-Fts from the *gouverneur du sud* to *chefs de district,* Fianarantsoa, 26 Oct. 1946 (ANM IV.D.73/1/1 folder Ambovombe, Eaux et Forêts, Incendies, 1945–1946); and letter from *gouverneur général* to *chef de province* Betroka, 29 Nov. 1911 (ANM D196, subfolder Betroka 1911–1918).

ing of a water-supply system. In his speech, he warned against burning. "The land," he stated, "is your heritage, like your wife. However, if you treat your wife poorly, she will complain or leave. If you treat the land poorly, if you burn the land, it cannot complain, it cannot leave; it will suffer."[3]

To foresters and their allies, antifire legislation, enforcement, and propaganda are transparent processes leading to the common good. Yet by prohibiting grassland burning and *tavy,* the state is criminalizing normal agricultural and pastoral practices. Most peasants perceive this as an infringement upon long-established rights and livelihood practices. They feel that they have not done anything immoral, not in the same sense as theft or murder.[4] Thus, the fire setters view state restrictions on their burning activities as a kind of unjustified "*brimade*" or hazing (Marchal 1968, 135; see also Peluso 1992; Jarosz 1996; Neumann 1998).

The criminalization of traditional resource uses has occurred throughout history. During the 1700s enclosure movement in England, "the rights and claims of the poor, [hunting, wood fuel collection, and free passage] . . . were simply redefined as crimes: poaching, wood-theft, trespass" (Thompson 1975, 241; see also Hay et al. 1975; Crummey 1986). In colonial Algeria, in the late 1800s and early 1900s, the French forest code criminalized the entire way of life of certain people by outlawing the use of certain forests as rangeland, for crop cultivation, or for the gathering of dead wood, cork, and grass (Prochaska 1986). In the postbellum southern United States, states tried to pass game and stock laws, trespass laws, and squatting laws that in essence criminalized the new free lifestyle of the ex-slaves (Hahn 1982). More recently, in cases that have parallels across the globe, Neumann (1998), Spence (1999), and Jacoby (2001) describe how the formation of national parks turned traditional villager activities like hunting, grazing, agriculture, and collection into crimes of poaching, trespass, and theft.

In Madagascar, state criminalization turned unauthorized pasture and *tavy* fires into punishable infractions. Most peasants resisted the criminalization of fire—by continuing to burn—in order to defend their right to this land management tool. Such resistance is sparked by threatened livelihoods, a sense of moral injustice, or political disagreement (Scott 1976, 1985; Isaacman 1982; Crummey 1986). The examples of criminalization presented

3. Reported by informant, Afotsara, 15 Oct. 1998.

4. The distinction is that theft, murder, and rape have always been understood as crimes. In the Afotsara region in 1998, I heard of thefts of large sacks of rice, a plow, fish, chicken, a Walkman, and money. In each case, the matter was pursued and resolved by the local elders or relevant authorities.

above all bred resistance. English peasants subjected to enclosures continued hunting and collecting wood fuel on the sly, overtly threatening the appropriators, and sometimes committing violence such as arson (Hay et al. 1975; Thompson 1975). Algerians who lost access to their woodlands continued cultivating in the forest, continued gathering wood and pasturing livestock, and eventually resorted to large-scale fire as protest (Prochaska 1986). Freed blacks and poor whites in the late nineteenth-century American south resisted stock control laws and antitrespass or antisquatting measures by tearing down fences, threatening violence, and voting (Hahn 1982). Tanzanian peasants around Arusha National Park frequently violated park laws to "defend or reclaim perceived customary land and resource rights," especially by encroaching on park lands (Neumann 1998, 13).

Resistance can be overt or covert, legal or criminal, attention-grabbing or quiet and incremental (Scott 1985). Sometimes, resistance efforts use the legal structures in place, including the ballot box, the press, lobbying, and formal protest or letters of complaint (e.g., Hay et al. 1975; Hahn 1982; Prochaska 1986). Peluso (1992, 19) argues that the character of resistance depends upon the form of domination:

> Forest peasants resist forest land control by reappropriating forest lands for cultivation; they resist species control by "counter-appropriating" species claimed by the state (or other enterprises) and by damaging mature species or sabotaging newly planted species; they resist labor control by strikes, slowdowns, or migration; and they resist ideological control by developing or maintaining cultures of resistance.

Yet resistance is not just shaped in reaction to forms of domination, but also in terms of the biophysical environment and local socioeconomic practices (Moore 1998). Resistance is shaped by a "landed moral economy" (Neumann 1998, 11), where people's expectations of subsistence rights—and thus their resistance—are based on both historically specific social relations and localized conditions of resource availability. People resist by whatever means available, whether they are found in the realm of nature itself, within the contested arena of the state, or within the sphere of images and representations.[5] Finally, it is critical to remember that the character of resistance adapts and changes as the dominating regime, community politics, and resource character themselves change (Klooster 2000).

5. Mobilizing identities and representations can be a key tool in resistance efforts (Li 1996, 1999; Moore 1996; Pulido 1996; Pile and Keith 1997; Tsing 1999).

TAKING ADVANTAGE

In Madagascar, the best way to understand forms of resistance to the criminalization of vegetation fire is through a phrase used by the peasants themselves. Multiple peasants blamed unexplained fires on the act of "taking advantage," or of seizing an opportunity (*manararaotra*).[6] In effect, peasants rooted in an effort to gain a livelihood from the land take advantage of any window of opportunity to accomplish their burning needs. They take advantage of the character of grasslands and of fire itself, of state weaknesses, and of strategic moments of village solidarity versus the state. As a result they have succeeded in fighting a century of state antifire repression, leading to the impasse, or "fire problem," of today.

Advantage taking is a subset of resistance, distinct especially from overt protest. If resistance is the attempt to forestall criminalization, to protect one's rights and one's vision of appropriate landscapes, then it can have two approaches, differentiated by goal. When the immediate goal is physical or ecological, i.e., the management of vegetation through fire for pasture or agricultural production, the approach is that of advantage taking. Protest may be implied in the action, but it is not the motive. On the other hand, when the immediate goal is to decriminalize a practice, to change state policy or ideology (with an ultimate goal of fire-based vegetation management), the approach can be through legal, political channels or through overt protest. Below, I argue that Malagasy peasants use a variety of means to resist fire criminalization by advantage taking and that all too often analysts have overemphasized "fire as overt protest" at the expense of "fire as advantage taking."

Taking Advantage of the Nature of Fire

By its very nature, fire is a powerful ally of fire-reliant peasants, at least in grassland and woodland zones. Here, fire is inevitable due to the extended dry season. The longer one waits, the hotter and further a fire burns—and burn it will, because if not lit on purpose, it will eventually be lit accidentally or by lightning. But fire is also ambiguous. Fire does not depend upon humans for ignition; it is self-propagating and can do its work in the absence of people, it is easily lit anonymously, it can accomplish multiple purposes simultaneously; the link between cause and effect is rarely straightforward or predictable (see chapter 2).

6. Multiple interviews and conversations, 1998–1999.

As a result, the peasants take advantage of the complex nature of fire in order to continue burning and escape punishment. They burn at night or out-of-sight, letting fire do its own work. They use time-delay ignition techniques. They let escaped authorized blazes run their course. They allow fires to escape "accidentally." Finally, they piggyback one fire on another, lighting additional fires ahead of a wildfire to get some burning done but avoid blame.[7] Since fire starters can easily remain anonymous, enforcement is difficult. Frustrated government officials, foresters, and tree-growing villagers all complain of the impossibility of apprehending fire starters for precisely the above reasons.[8] Fire is a trump card in the hands of those who favor burning.

The continued use of an appropriated resource or of a criminalized resource management tool is one of the most common strategies of peasant resistance. Possession is, after all, nine-tenths of the law. Fire lends itself perfectly to such uses. People often burn to fight, establish, or maintain specific land claims or resource use patterns (Prochaska 1986; Pyne 1995, 1997; Kuhlken 1999). This is why Malagasy villagers continue to burn their lands with such regularity (in the face of the antifire policies). In Afotsara, as tables 3.2 and 3.3 note, at least 155 separate fires charred a total of 696 ha during 1998, corresponding to 38 percent of village territory. Forty of the smaller fires (covering together only 4.4 ha) were legal crop-field preparation fires. The rest were all technically illegal, but most served various resource management purposes, especially pasture renewal and locust control.

Such passive resistance through continued use of normal burning practices is complemented by cases of the aggressive use of fire to assert control over resources. On the "Wild Coast" of South Africa, villagers resisted controls on wildlife hunting in a nature reserve by burning strips of grass outside the reserve to attract game (Kepe et al. 2001). In Madagascar, villagers historically burned forest remnants claimed by the colony in order to revert the land to pasture and thus village control (Parrot 1924). More recently, some farmers burned a state forest plantation in order to contest state claims to the land. A few years later, the farmers replanted the burned zone with trees in order to claim the land as their own (Rajaonson et al. 1995).

7. Multiple interviews and observations, 1994–1999; interview, Forest Service agent, 28 June 1999. For time-delay ignition techniques, see chapter 2, note 16.

8. As one informant noted, "you should not spit when nothing stinks" (interview, Kilabé, 9 July 1999). Also, interviews, Afotsara, 19, 20, and 23 Nov. and 6 Dec. 1998; interview PDFIV Ambatolampy, 26 Apr. 1999; interview, Forest Service agent, 10 July 1999; letter from *chef* of Betroka Province to the central government, 8 Feb. 1913 (ANM D196); letter from *chef* of Fianarantsoa Province to the central government, 9 Nov. 1956 (ANM D196).

Taking Advantage of Strategic Village Solidarity

Rural Malagasy communities have their rivalries, tensions, and conflicts, with respect to fire as much as everything else (cf. Gezon 1999). For example, in Kilabé, the wealthiest villagers are cattle owners who control most valley bottom rice fields and dominate local decision making. They prefer the burning of upland communal pastures and are deaf to protest from the poorer farmers who depend on those very lands for their minimal crop fields (these farmers would benefit from protecting the grassland from fire for a few years to enhance soil fertility). As one farmer said, there is no way a poor farmer can stop the cattle owners from burning—"they are naughty (*maditra*), they will burn maliciously (*manao ankasipary*)."[9] A town near Afotsara also illustrates intracommunity tension over fire. One day, the passage of locusts resulted in numerous fires. I later asked a well-off woman about one large fire that had burned acres of private pine woodlots (of which she was perhaps among the owners). She claimed that people took advantage of the locusts as an excuse to burn, in order to kill the lower branches of the pines so they could collect them for firewood (by tradition, dead wood may be collected by anyone, even in private woodlots).[10]

However, more often than not, villagers strategically use solidarity to avoid state meddling (Rakoto Ramiarantsoa 1985; Prochaska 1986; Neumann 1998). Villagers use strategic solidarity to evade responsibility and punishment, taking advantage of the easy anonymity of fire setters and fire's ambiguity of purpose. As a rule, when confronted by outside authorities, villagers blame fires on unnamed passersby (*mpandalo*), bad people (*olon-dratsy*), unknown people who burn for pleasure or out of malicious intent (*mpandorodoro, mpanao ankasokaso*), bandits and cattle rustlers (*dahalo, fahavalo*), or profiteers (*mpanararaotra*).[11] Outside of *tavy* zones, people will only rarely admit to lighting a fire, for example when a wildfire could clearly be traced back to a crop field fire that escaped.[12]

If villagers have been unsuccessful at blaming outsiders, they may choose to pass a fire off as having accidentally been lit by a young child or a very old man, in the hopes of a more lenient penalty. This was the case in the

9. Interviews, Kilabé, 9 July 1999; Forest Service agent, 10 July 1999.

10. Interview, 12 Sep. 1998.

11. Multiple interviews and conversations, 1998–1999, 2001.

12. During a large fire that burned into the Andringitra reserve in October 1996, eventually charring 100 ha, a farmer ran to the local Forest Service agent to say "help, help, my crop fire escaped" (interview, Andringitra, Nov. 1996).

story of the 21 July woodland fire in Afotsara presented in the introduction to part 1. It is also the case in the story a man in Afotsara recounted about a fire that occurred many years ago. The Forest Service arrived to enforce the law, but on account of the fire starter being (supposedly) a small boy, the punishment was only some tree planting. The historical record provides further examples. In 1919, a village in Itasy Province was fined for "bad faith" in designating two children as fire starters. In 1935, District Chief Gallaire wrote disparagingly of the "custom" of village leaders of presenting the infirm, women, or children as responsible for fires.[13]

Strategic village solidarity in the case of fire has two main causes. First, many regions have a more or less accepted norm of burning based on the regional ecology and economy. In zones of extensive pasture, grassland renewal burning serves the common good of much of the village, fighting bush encroachment, causing a green bite, and preventing wildfires, and is seen as legitimate by most of the community. Here, free-burning fire is the accepted standard of natural resource management, and this norm is reinforced by social pressures. A richer farmer who owns significant pine woodlots often threatened by fire on the outskirts of Afotsara spoke of the pressure to keep fires anonymous. If you report somebody to the authorities for burning, he said, "then his family will come and bother you—threaten you, burn your pines, kill you."[14] A Forest Service officer confirmed this situation: "People may whisper to you, yes, I saw who lit the fire that burned the hills yesterday. But there is no way that they would sign their name to a *procès-verbal* nor stand witness in a trial, because they risk retribution from *kakay* (rivals) in the village."[15]

The second cause is the desire to avoid outside intervention in village affairs by a state often seen as meddling. A strong sense of moral obligation to the community supersedes internal conflicts when dealing with external fire enforcement; community members prefer to remain silent or to address matters among themselves (Andriamampionona 1992).[16] The only situation when villagers involve outside authorities is when the accepted norms of

13. Interview, Afotsara, 6 Dec. 1998; JOM (1919, 1010); AOM mad-ds//334.

14. Interview, Afotsara, 25 Nov. 1999.

15. Interview, Forest Service agent, 10 July 1999.

16. This sometimes backfires. For example, in the case of the 13 November fire described in the introduction to part 1—a fire that damaged several hectares of young pines—the tree owner and *fokontany* president never reported the fire to the authorities. While they wanted to see justice done, they feared that the Forest Service would fine the community as a whole, for this is standard policy when there is no culprit (interview, 20 Nov. 1998).

burning are violated and local conflict resolution mechanisms do not suffice, as in cases of severe negligence, arson, or damage to property.[17]

States often respond by punishing villages as a whole, typically with a collective fine. In Algeria, collective fines were first used for illegal fire in 1877 (Prochaska 1986). In Madagascar, collective fines were used as early as 1914, formalized in 1930, strengthened in 1941, repeated in 1960, and are still used today. In fact, 40 percent of fire crimes in Antsirabe *circonscription forestière* between 1983 and mid-1997 were ascribed to villages, not individuals. Similarly, the fires in Andringitra Strict Nature Reserve in 1995 (see table 6.2 later in this chapter) resulted in a collective fine of 25,000 MGF per male taxpayer. However, collective fines in turn inspire resistance and solidarity among villagers, and according to Forest Service agents collective fines are politically difficult to apply.[18]

Taking Advantage of the State

Finally, and perhaps most crucially, fire-setting peasants in all regions take advantage of the state in order to continue burning. I highlight three aspects: the state's limited reach, its internal diversity, and moments of distraction.

First, people take advantage of the state's relative lack of power, money, and staff. In colonial days, Lavauden (1934) compared the paltry number of forestry officials, a total of sixty-six, with the national territory of fifty-nine million hectares. In 1970s and 1980s, when the Forest Service payroll numbered between 700 and 900, this still left about 70,000 ha of grasslands and forest per forester (including office staff). Forest Service staff has grown to over 1,000 in past years; however, austerity measures leave the agency with limited means (except near protected areas). Typical Forest Service offices are run-down, containing little more than a desk and a dusty cabinet with yellowing documents. There may be no functioning cars or motorbikes. In many nonforest areas, one forester historically handled three to four *cantons;* now there is often less than one agent per *fivondronana*, responsible for up to 800,000 ha. Due to desk responsibilities (giving permits to tree cutters) and financial limitations, time spent patrolling is minimal.[19]

17. Of thirteen files (dated between 1989 and 1996) of fire crimes I consulted at the Antsirabe Forest Service office in August 1998, eleven were brought to the authorities by owners of damaged property or by rivals, while one was prosecuted by passing gendarmes (the thirteenth was unclear).

18. Interviews, Forest Service agents, 28 June and 10 July 1999; 7 Aug. 2001.

19. Interviews, Forest Service agents, 12 March, 15 June, 12 Aug. 1998; 6, 7, and 16 Aug. 2001. Regarding the limited means of the Forest Service, see also Décret 97-1200 (JOM 1 Dec. 1997), Borie (1989), Gautier et al. (1999), and Henkels (1999).

As a result, fire-setting peasants know they can escape the control of the Forest Service. Foresters acknowledge this state of affairs. In interviews about forest and fire legislation, many foresters' responses were prefaced by the phrase "in principle," as in "in principle, people aren't supposed to burn this area," or "in principle, fire authorizations are given for late October, after three days of rain."[20] Of course, reality is different. Rules are most strictly enforced along roads, if at all.[21] Any gap in control is taken advantage of. For example, the one-year absence of a forester from Zafimaniry country in 1970 led to a huge burst in slash-and-burn cultivation fires (Coulaud 1973), while a colonial administrator in Midongy-Ouest complained in 1942 that herders would head off to remote areas to burn without surveillance.[22]

Second, the peasants take advantage of diversity within the state. The state, as I outlined in chapter 1, is itself a complex arena for contesting power, a varied set of institutions with differing goals, composed of individuals facing their own livelihood struggles. The antifire criminalizing part of the state, located in the capital, is an elite grouping including the once powerful Forest Service, parts of the central leadership, capital city technocrats, and their allies including logging concessionaires, influential academics, international environmental agencies, and urban elites. Meanwhile, touring foresters may turn a blind eye to recent pasture burns or allow *tavy* in secondary forests, gendarme or court judges may hesitate to enforce the laws, and district officials may argue for tolerance of fire.

Field foresters, for example, serve conflicting mandates. They work *against* villagers with fire repression and enforcement; they work *with* villagers doing extension work like irrigation canals, fishponds, or environmental awareness; and they also work for their own private gain.[23] As a result, peasants exert pressure where they can. For one, they take advantage of the fact that field foresters are far removed from headquarters. Several

20. Interview, Forest Service agent, 28 June 1999.

21. Le Bourdiec (1974). Gautier et al. (1999) determined that at the Manongarivo Special Reserve, the existence of the protected area had very little influence on where people burned *tavy* clearings, as enforcement was limited.

22. He wrote, "If we insist [on enforcing regulations], the cattle herder will emigrate to the north or the west where he won't be bothered, neither for pasture fires, nor for the requirement to always oversee cattle" (letter no. 3cf from the *chef du poste administratif*, Midongy-Ouest, to *chef de district* Ambatofinandrahana, 3 Feb. 1942, ANM IVD38/IV/11). Marchal (1968) also notes how the peasants resist government control of grazing by taking their cattle to the mountains, where they can get away with burning. The same goes for *tavy:* Coulaud (1973) and Jarosz (1996) give examples of locals simply moving to areas where control was less severe.

23. Coulaud (1973); interview, Forest Service officer, 21 Aug. 2001.

foresters reported being increasingly scared of the people, of having received threats, and thus of reduced enthusiasm for strict enforcement. As one *chef de secteur* wrote, the villagers "consider us as enemies."[24] Alternatively, villagers rely on foresters' empathy (or willingness to accept bribes); thus, from one-third to one-half of forest and fire crime citations result in comparatively lenient out-of-court settlements (figure 6.3).

There is even room for maneuver in the central part of the state. The state is of two minds about fire, and the peasants exploit this situation, testing the limits of enforcement, pressuring leaders, and so on. We can see this most clearly in the vehement complaints by government and Forest Service leaders about the lack of cooperation of other arms of the state.[25] In 1929, Governor H. Berthier berated the "indifference of the authorities in the face of the ruin of the Colony [by *tavy* and fire]," and called for stricter enforcement.[26] Soon afterwards, Forest Service chief L. Lavauden wrote, "Some civil servants, under the guise of goodwill for the natives, have not feared to openly approve these damaging practices [burning]. Every time the Forest Service has tried to stop these civil servants, it encountered complete indifference, if not a declared hostility" (1934, 953). Coudreau explained in 1937 that district administrators were human after all, and thus "hesitated to punish the poor devil who cut and burned the forest to grow the rice necessary to feed his family" (Bertrand and Sourdat 1998, 117). It was this unseen public force, a populace that said it would go hungry without fire and sympathetic

24. Personal correspondence, letter dated 2 June 1996. Also, interviews, Forest Service agents, 13 Aug. 1998, 20 May and 28 June 1999. Another forester denied the existence of tensions (21 Aug. 2001). Tensions between villagers and foresters exist over much more than fire, especially in forested zones. Here locals sometimes accuse the Forest Service of giving logging concessions while banning *tavy*, or of favoring outsiders in the granting of logging permits (interviews, Fort Dauphin, 9 Aug. 2001, and Antoetra, 16 Aug. 2001).

25. In addition to examples cited in the text, Pouperon, *chef* of Ambositra Province, defended pasture fires in 1914 as a necessity in the hilltop pastures (ANM D196, subfolder "Ambositra 1914–1924," letter no. 131). The head of the Diego Suarez region wrote to his district *chefs* in 1936 complaining of "indifference up and down the ranks" regarding fire enforcement (ANM IV.D.27, folder "feux de brousse, tavy, 1936," Circulaire no. 351 CFR, 6 Mar. 1936). Likewise, Governor A. Annet wrote in 1941 that "the ban on fires is all too often considered as a ban in principle and too many exceptions are made each year" (JOM 6 Dec. 1941). Finally, A. Joly, the *chef* of Tulear Province, wrote to the central administration in 1950 defending pasture fires—"there is no other way to renew pastures"—and advocating early season fires (ANM D196, folder "Conservation des Sols 1950"). See also Petit (1934), Dandoy (1973, 29), and the string of articles that defended some burning practices, especially in rangelands (e.g., anonymous 1904; SEM 1949; anonymous 1955; Dez 1966; Granier and Serres 1966; Gilibert et al. 1974).

26. JOM 9 Feb. 1929.

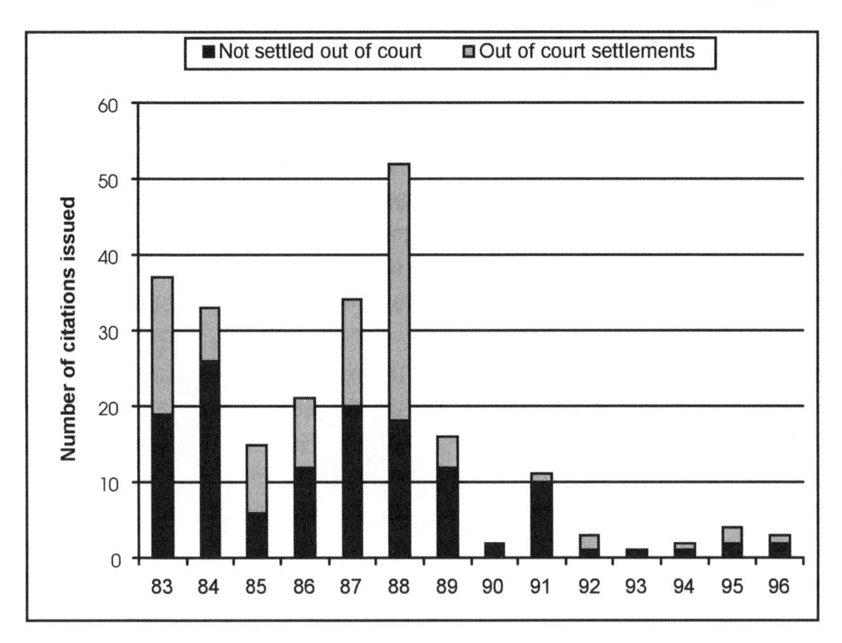

FIGURE 6.3. **Fire crimes in the *Circonscription Forestière* of Antsirabe, 1983 through 1996.** The chart shows that roughly half of all citations were settled out of court for, e.g., payment in bricks, labor, tree planting, or a small fine (as opposed to harsher official penalties of prison and fines); it also shows the reduced enforcement resulting from the phasing out of enforcement incentives (*primes*) around 1990. Source: DEF Antsirabe.

parts of the state that contributed to tempering fire legislation. As Dez (1968) noted, administrators were often caught between profire populist leaders and antifire technocrats. This is why, perhaps, the state has responded to burning infractions with "that paradoxical mix of laissez-faire and force remarked on in other conditions of conflict and change in rural Africa" (Cline-Cole 1996, 130).

Interviews expand this point. While working in the northeastern region of the island, one Forest Service agent wrote several citations for illegal slash-and-burn fires. One of the accused happened to be a relative of a National Assembly delegate, who pulled strings such that the forester was forced to drop the matter and request a transfer. Several other foresters reported that enforcement was frequently stymied by pressure from local politicians. One went as far as to claim "*tsy voahaja mihitsy ny lalàna,*" the law is not respected at all. He said he was hard-pressed to do more than environmental awareness efforts, for he knew that he would never be able to punish rich villagers due to their political influence, and punishing poor villagers was too easy and unfair.[27] Another forester working near Andohahela National Park stated that strict repression was no longer possible, for it was bad for the image of the conservation project working with the Forest Service in the area. The project, like many across the island, seeks the cooperation and goodwill of the locals, and the Forest Service cannot afford to be seen as being too strict.[28]

Politicians know that many rural villagers resent controls on burning. According to a retired high-level forester, politicians will spout official antifire rhetoric when speaking in public forums, yet at the same time they will let it be known through other channels that they will protect local burners. An elder in Afotsara recounted exactly such an event that occurred several decades ago: the mayor of Afotsara's *canton* was trying to increase his popularity, so he told the citizens to go ahead and burn (in response, the Forest Service fined the whole *canton;* each man had to plant 200 trees).[29]

Third, peasant burners take advantage of moments of state distraction. A prime moment is during locust invasions, when the state must choose between enforcing antifire laws or allowing peasants to fight the economically disastrous locusts with fire (see chapter 3). As a result, the passage of locusts serves as a convenient excuse to any potential burner: what gendarme or Forest Service officer could deny a farmer the right to protect his or her crops?[30] As fires can serve multiple purposes simultaneously, it is nearly impossible to document the difference between a locust fire and a pasture fire. This is perhaps why 38 percent of fire extent in 1998 in Afotsara was blamed in the first instance on locusts (table 3.3). As noted in chapter 2, people use the ambiguous nature of fire—that it can accomplish several tasks at once—to label pasture fires as locust fires. This could have been the case with the June fire described in the introduction to part 1.

Another moment of state distraction is the eve of the national holiday, on 25 June. At this point, people light bonfires to celebrate but sometimes also to take advantage of the moment by "accidentally" letting a bonfire escape.[31] Periods of state insecurity with banditry are also blamed for fires (see chapter 2). In the mid-1980s, people in Afotsara frequently blamed their own

27. Interviews, Forest Service agents, 28 June and 10 July 1999; 6, 7, and 21 Aug. 2001.

28. Interview, Forest Service agent, 4 Aug. 2001.

29. Interview, Forest Service agent, 21 Aug. 2001; interview, Afotsara, 9 Dec. 1998.

30. Interviews, Afotsara, 25 Apr., 19 Aug., 12 and 16 Sep., 5 and 6 Dec. 1998.

31. Interview, village near Ambositra, 23 June 1999.

fires on *dahalo* (bandits), as others have done on the periphery of Andringitra National Park.[32]

A final, crucial moment of state distraction is during elections or periods of political unrest. During such moments, there is less political will to stop fires, and people take advantage of the distracted state. There is little danger of enforcement when government employees are on strike—as they were in much of 1991—or during election campaigns when candidates seek to win votes by appealing to rural sentiments.[33] The anecdotally documented surge in fires during such periods can, of course, have two explanations: is it a reaction to state weakness, or a manifestation of protest against the state? I address this distinction below.

FIRE FOR PROTEST, OR FOR RESOURCE MANAGEMENT?

Wildland burning is frequently seen as a forceful means of rural protest (Kuhlken 1999). The argument is as follows: Fire has been one of the most frequent targets of state intervention in rural resource management (Prochaska 1986; Pyne 1990, 1995; Huntsinger and McCaffrey 1995; Sivaramakrishnan 1996), even back in the days of Genghis Khan (Ho 2000), and fire protection went hand in hand with European imperialism (Pyne 1997). Almost uniformly, governments sought to stop fire setting, which was seen as wasteful and destructive. Fire, a customary resource management tool, was criminalized by the state.

In consequence, according to this argument, fire has come to symbolize resistance and protest by rural people. In Spain, fires protest the loss of traditional common access to woodlands. In New Mexico, Hispanics burn the forest in protest against U.S. Forest Service policies. In Java, peasants light fires to protest state forest control (Peluso 1992). In colonial Tanganyika, following 1955 riots against a land-use scheme, "ribbons of smoke rose defiantly all over the Uluguru Mountains. Rules against burning were no longer enforceable, and the peasants now openly flaunted authority by firing the hills" (Young and Fosbrooke 1960, 157). In Greece, there is a strong correlation between wildfires and periods of social unrest (Pyne 1995); in Algeria fires symbolized rural resistance to colonialism (Prochaska 1986). In Ghana, peasants burned rice fields in protest at the inequalities fostered by Green Revolution

32. Interviews, Afotsara, 9 Dec. 1999, and Andringitra, 23 Sep. 1998. Government reports often mention *dahalo* as causes for fire. They probably exaggerate this purported link in order to bolster the case for repressive fire policies.

33. Interviews, Afotsara, 24 Nov. 1998 and 27 July 1999.

interventions (Goody 1980). In India, people of all classes fought the British fire suppression policy, for "without fire the land was even more worthless to them." They fought with both open arson as well as clandestinely, "a guerrilla war of biotic insurgency" (Pyne 1997, 490). Throughout, people harnessed fire's rich symbolism and iconographic appeal (Kuhlken 1999).

Fires in Madagascar are also frequently seen as symbols of protest (Dez 1968; Salomon 1981; Olson 1984; Jolly 1989; RDM 1990; Durbin 1994; Hoeltgen 1994; Rajaonson et al. 1995; Jarosz 1996; Pfund 2000). In 1971, the *New York Times* reported that "fires of anger scar Madagascar" and linked wide-ranging bushfires to peasant unrest (Mohr 1971). However, it is difficult to prove whether the fires associated with political events such as those in table 6.1 represent overt protest, advantage taking, or a combination of both. Due to a century of repression, farmers and herders can be hesitant in discussing protest burning. Interviews of farmers in Afotsara resulted in a broad range of opinions about whether fire could be overt political protest, ranging from denial to support.[34] Most farmers interviewed by Schnyder and Serretti (1997) near Ankazobe denied the use of fire for general protest.

It is much easier to link specific local fires to specific local political events than to see causation in general nationwide trends. The burning of state forest stations during the 1947 rebellion (see chapter 7), among others (see table 6.2), are plausible examples of protest fires. As Ratsirarson (1997, 15) wrote regarding Andringitra Strict Nature Reserve, "If one is too severe, the reserve is burned by unknown people." For example, Rasoavarimanana (1997) tells of Bezanozano farmers outside Moramanga burning a chunk of forest to protest a 1992 local ban on *tavy* and the brutal enforcement of that law. Another example, extremely symbolic, was the torching of the Rova, the former queen's palace, in 1995, as described in chapter 2. A final example comes from a village near Afotsara. Here, the elected village councilor—a politically well-connected rich farmer—had taken the enforcement of environmental laws to heart. In 1997, for example, he fined villagers for illegally cutting trees. Soon thereafter, a fire burned in the village woodlands. The councilor blew his whistle to summon the village to fight the fire. The very next day the rest of the woodland was ignited in protest.[35]

I believe, however, that outside of such case-specific instances, it is a mistake to overemphasize the use of fire as overt protest. In all likelihood, the number of fires assumed by the literature and by the state to be lit for overt

34. Multiple interviews, Afotsara, Nov. and Dec. 1998.
35. Interviews, Afotsara, 8 Dec. 1998, 13 and 15 Apr. 1999.

TABLE 6.1. **Linking fires to election years and periods of unrest: the anecdotal evidence.**

Date	Event	Fires	Source
1947	anticolonial rebellion	massive increase in fires and *tavy;* violence directed at Forest Service	Bertrand and Sourdat 1998; AOM mad pt//181; ANM IV.D.73/1/1
Oct. 1956	drought; anticolonial activists given amnesty and return to villages	massive fires in Fianarantsoa province	letter from Provincial Chief, 9 Nov. 1956 (ANM D196)
1969	cyclone and higher taxes cause general frustration	spectacular growth in bush fires	Desjeux 1979
1970	drought; peasant discontent with ruling party	doubling of fires	Mohr 1971; Brown 1995
6 Sept. 1970	legislative elections	significant fires on election day, especially in state forest plantations	report by E. Rabenjamina to president, 12 Jul. 1971 (ANM Vice-Pres 840)
8 Oct 1972	referendum for Second Republic	"all that hadn't burned yet"	interview in Afotsara, 24 Nov. 1998
1977, 1983, 1989, 1992	election years	more fires	Andriamampionona 1992
1991	political instability and social unrest; general strike	5-fold increase in fires in region south of capital	Rambeloarisoa 1995
		80 to 90% of all land, including almost all state forest plantations, in Miarinarivo province	Andriamampionona 1992
Sept. 1995	referendum	fires "of a political nature"	DEF annual report 1995
Oct.–Nov. 1999	communal elections	smoke and major fires across country	M. Freudenberg, P. Schachenmann, pers. comm. 1999.

TABLE 6.2. **Partial list of likely protest fires.**

Location	Date	Cause	Fires	Source
Andringitra Strict Nature Reserve/ National Park	1927	creation of the reserve	big fire	Projet Andringitra
	Oct. 1995	major ecological research within reserve: fear of increased enforcement and losing resource access	two large fires within the reserve	Projet Andringitra
	Oct.–Nov. 1996	inauguration of park lodge with important guests	blackened all around	Projet Andringitra
	Oct. 1999	inauguration of National Park, with important guests	fire within reserve	Projet Andringitra
Haute Mangoro forest plantations	1988 1990 1991	political instability (?)	10-fold increase in fires	Société Fanalamanga 1992
State forestry stations and plantations in Miarinarivo district	1991	general political unrest and frustration with the state	almost all were burned	Andriamampionona 1992
Haute Matsiatra forest plantations	1991–1992	general political unrest and frustration with the state	major fires in plantation forests	M. Freudenberger, pers. comm., 1999.
Andohahela Strict Nature Reserve	1992	total ban on fires by Forest Service	burning within reserve	Durbin 1994
Torotorofotsy (Moramanga)	mid-1990s	1992 local ban on *tavy* and brutal enforcement	big section of primary forest burned	Rasoavarimanana 1997
Ankarafantsika Strict Nature Reserve	1994–1997	criminalization of traditional activities within reserve by new ICDP project	3 consecutive years of large fires	Bloesch 1997
	1990s	expansion of cashew plantations into villagers' pastures	burning of the savanna	Bloesch 1999

protest is exaggerated. More often than not, people light fires during elections or periods of unrest not to protest the state, but to take advantage of state distraction in order to renew their pastures, reduce the fuel load, or clear brush without fear of enforcement. This is a question of intentionality. The observed logic and patterns of resource use strongly point to the conclusion that most fires are a straightforward livelihood practice (if at times politically savvy) and not overt protest. The peasants' actions implicitly protest the restrictions, but as far as motives for burning are concerned, material needs are much more relevant than political or symbolic purposes (cf. Rangan 1996; Bebbington 2000). As Leach and Fairhead (2000) note, resistance is often not targeted directly at the broader state discourse (e.g., the antifire received wisdom), but at the material effects of the discourse, such as laws, actions, and enforcement. For example, illegal forest slash-and-burn fires increased notably in several key periods of political unrest, as forest farmers used these opportunities to expand their crop fields (Olson 1984). It is highly unlikely that they did so just for the sake of protest.

In the context of a century of fire repression, all purposely lit fires are implicitly protest fires. Yet it is crucial to distinguish between moments when protest is only implicit in the presence of the fire (all purposely lit resource management fires in an antifire regulatory context) and when the explicit motive for the fire is protest (overt protest fires). While fires lit for standard agropastoral purposes may give villagers a sense of protest, of having denied the state, protest fires may be much more in the eye of the beholder. That is, a peasant may light a blaze to renew a pasture—choosing a politically expedient day to do so and avoid punishment—while the urban bureaucrat may see this fire as a symbol of peasant discontent.

CONSEQUENCES

The struggle over fire in Madagascar—the repression of fire by foresters and their allies in the state, and peasant resistance by advantage taking—has led to a stalemate, an impasse. The consequences of this stalemate are significant. On the one hand, foresters and environmentalists can report little but frustration in their hundred-year attempt to alter peasant pasture and forest management practices. Fires are still ubiquitous and deforestation is rampant. On the other hand, peasant burners, whether well-intentioned or not, still face the risk of punishment. As a result, fire practices and decisions are pushed underground, communication is stifled, and burners do not oversee their fires (aside from crop field preparation burns), letting them run semiwild.

In this context of a stalemate, the application of unpopular fire laws causes inconsistencies and corruption. For one thing, exceptions are the rule. Burn authorizations for dry season pasture fires are supposed to be given only "exceptionally,"[36] yet in some areas they have been given annually for decades. The real exception in most areas is communities that even bother to demand authorization.[37] The majority of *fokontany* do not apply for permission, either because they no longer fear enforcement or because they know that the Forest Service no longer gives authorizations in most highland areas.[38] In the *tavy*-prone areas of the east, exceptions are also the rule, as chapter 5 laid out.

In addition, many of the "exceptional" authorizations that *are* issued, at least in pasture fire zones, are unrealistic and unenforced. Authorizations are based on handwritten demands bearing the relevant information (location, purpose) and signatures.[39] Authorizations prescribe precautionary measures and burn dates. For pasture fires, the dates are usually a fifteen-day window in November or December or "after three days of rain."[40] Much of the proposed burn site has usually been torched long before the authorized window, the act of submitting a demand for authorization being considered as sufficient insurance against enforcement.[41] Some of the authorizations even specify burn periods after 15 December, when fire authorizations are legally unnecessary. Many of the precautionary conditions—daytime burn-

36. Arrêté no. 058 of 1961 sets regional dates for wet season fires (they do not require authorization); Décret 61-079 describes the procedure for "exceptional" authorizations outside of these dates (appendix 2).

37. The 1990s files of the grassland-dominated *cantonnement forestier* in Ambalavao, which covers some 150 *fokontany,* contained only about five to twelve *fokontany* fire authorizations per year. The large majority of these were for villages in the peripheral zone of Andringitra strict nature reserve, within reach of the single Forest Service agent posted in Antanifotsy (DEF Ambalavao, 1999). In Ambatofinandrahana *cantonnement forestier,* which covers vast Middle Western pastures, a maximum of eight fire authorization demands are received by the Forest Service each year (interview, 28 June 1999).

38. Historically, peasants did not apply for authorization as they feared refusal, reasoning it was better to burn illicitly with no permission than risking a definite "no" (Gov. Garbit's circular notice, JOM 23 Jan. 1915; letter from Fort Dauphin *chef de district,* 10 June 1946, ANM IV.D.73/1/1/Ambovombe). This situation has not changed.

39. As spelled out by Décret 61-079. Signatures were typically from the *fokontany* president and other local officials—vetting land ownership and precautionary measures. Recently, the signatures of the communal mayor or of the KASTI are also required.

40. In 1998, for example, all authorizations in the *cantonnement forestier* of Ambalavao were issued on 18 November for a fifteen-day burn window. Rain had fallen a few days in September and October and early November.

41. Interview, Forest Service officer, 28 June 1999.

ing, calm winds, the presence of all able-bodied men, and the cutting of a thirty-meter firebreak—are totally disregarded. As is clear from the above, the pasture fire authorization procedures as prescribed by national law are largely a farce, and exceptions have become the rule (Razafindrabe and Rakotondrainibe 1998).

The enforcement of fire crimes is similarly dysfunctional. The state's fiscal and economic crisis in the 1980s resulted in a drastic fall of fire crime prosecution, especially after the fund for enforcement bonuses (*primes*) ran out (see chapter 7; figures 6.2 and 6.3). Ever since fire was first banned, however, enforcement has been inconsistent, thwarted by the immense terrain and limited resources of the state, by administrative foot-dragging in the face of unpopular laws pushed by the Forest Service elite, and by peasant resistance. For every one illegal fire that has been enforced, there are dozens, hundreds, or thousands that are not seen or ignored, the quantity depending upon the period and region. Not one of the 150 fires in Afotsara during 1998 resulted in any enforcement. Nowadays, the enforcement of fire crimes tends to occur only when there is property damage or special interests.

Second, the impasse fosters a system deeply entrenched with corruption (Klooster 1999; Robbins 2000). Dez (1968) remarked that heavy repression inevitably leads to corruption, and revealed that in colonial days, village leaders were accustomed to taking a cut out of collective fines from their village. In the Tsiroanomandidy area, the gendarmes are referred to as *papango*, or hawks, which typically circle in the sky near a fire, ready to prey on escaping mice. The gendarmes use fire as an excuse to prey on peasant savings, demanding a bribe in exchange for not filing a *procès-verbal*, an official citation.[42] In consequence, villagers beg the Forest Service to issue annual fire authorizations like they used to—for these authorizations would protect them from the *papango*.[43]

The Forest Service itself is not above accusations of being "massagers" (*mpanotra*) of money.[44] In the Andringitra area, a farmer suggested that the Forest Service rarely issues *procès-verbaux* for illegal burning. Instead, foresters are paid off not to proceed with the investigation with sums reaching the value of one cow, 500,000 MGF.[45] In Afotsara, a zealous local official

42. Interview, Kilabé, 9 July 1999. The interviewee's father was caught with an escaped crop fire in 1988, and made to pay 20,000 MGF and a bottle of local rum to avoid prosecution.

43. Interview, Forest Service officer, 10 July 1999.

44. Ratovoson (1979, 153–154); Henkels (1999); interviews, Forest Service officers, 6 and 21 Aug. 2001; anonymous project employees, 8 and 9 Aug. 2001.

45. Interview, Andringitra, 23 Sep. 1999.

prosecuted a man for cutting 104 *tapia* trees to make charcoal (see chapter 4). This affair rose as far as the provincial Forest Service office, where again a cow was sold to settle the matter.[46] Forest Service officers defend the need for "extra income," complaining that their civil servant salaries only support their families for one week each month.[47]

CONCLUSION

Over the past century, peasants and the state struggled over fire to assert both ideological claims of how landscapes should be managed and physical claims to resource access. The chief tool of the state in implementing its vision (rooted in an antifire received wisdom) was criminalization through repression and rhetoric. Yet the state was not monolithic. Some bureaucrats and politicians spoke for moderation, while rural state agents (gendarmes, Forest Service field staff, local leaders) necessarily adapted to local contexts. The peasants who relied upon fire to manage their landscape did not perceive burning as a crime, and they resisted state control. They took advantage of internal state diversity, of strategic village solidarity vis-à-vis the state, and of the biophysical character of fire itself in order to effectively resist state control. The result is an impasse; fires continue to burn illegally and inconsistencies and corruption are the norm.

This pattern of criminalization and resistance has persisted for an entire century. This stalemate is rooted in the complex characters of both fire and the state, and in the various ambiguities that allow each side to reach its immediate goals. The multiheaded state satisfies its elite technocratic and environmental arms through the antifire received wisdom and legislation, the feeling that it is doing something to solve the fire problem, to stop environmental degradation and what it sees as senseless protest fires. Meanwhile, dominant fire-dependent peasants rely on the ambiguous biophysical character of fire to continue burning—in the face of regulation—for their livelihood needs. The fact that peasants can get away with burning then satisfies the state's populist and fire-sympathetic side. The stalemate, then, has persisted as the best compromise between multiple parts of the state and peasantry.

At a broader level, this suggests that the exercise of power in natural re-

46. Locals tell of the accused getting off by bribing the Forest Service, yet the Forest Service chief for the province stated that the result was an out-of-court settlement involving tree planting (interview, 26 May 1999). Perhaps both are true.

47. Interviews, 26–27 May 1999.

source conflicts, in struggles over appropriate landscapes and their uses, is not sufficiently captured in simple models of state criminalization and peasant resistance. These models should be expanded to include two major considerations. The first is the importance of the ambiguities and muddy middle grounds *between* dialectic categories such as domination and resistance or state and peasant. Both the broad, dialectic categories, as well as their complexities and contradictions, are crucial to shaping these conflicts. In the Malagasy case, it is precisely because of the ambiguities of fire's biophysical character and the complexities of state-peasant relations that peasants have succeeded in defending free-burning land management. Second, the struggle must be placed into its political ecological context, into a realpolitik of place and practice. The specific, contextual character of local resource ecology, of people's livelihood needs, aspirations, and possibilities, or of the regional political and environmental discourse all shape the nature of the conflict. This is demonstrated in the Malagasy case by the central role played by the characteristics of fire in different climate zones, by livelihoods based—at least in part—on fire-dependent agropastoral strategies, and by the island's particularly vehement antifire received wisdom. By incorporating these two elements—ambiguities and context—into models of domination and resistance, we can improve the analysis of the politics of resource management.

This struggle over natural resource management, over fire in Madagascar, has led to a stalemate. This stalemate was recently recognized by policy makers (e.g., Rajaonson et al. 1995), and served as the inspiration for a new community-based natural resource management policy called GELOSE. In chapter 8, I address this new twist to Malagasy fire politics. First, however, in chapter 7, I trace the ideas highlighted in the present chapter through a detailed history of fire politics in Madagascar.

7 [FIRE POLITICS

A History of State Antifire Efforts

> The French forbade grass-burning. This, namely, was of course not the custom in France.
>
> NORWEGIAN MISSIONARY J. EINREM (1912, 71)

THIS CHAPTER PRESENTS a detailed history of fire politics in Madagascar, illustrating many of the points made in the previous chapter. This story of the past century illustrates the efforts of the Forest Service, state elites, and environmentalists to criminalize fire but also highlights the dissenting voices that reflected peasant profire concerns. The story shows how the antifire received wisdom evolved and hardened, shaping fire repression in different eras. It demonstrates how that received wisdom led successive administrators to try to stop fires despite previous failures. It also shows, however, precisely those discordances within the state that the peasants used to their advantage: inconsistent enforcement, exceptions being the rule, and sympathetic administrators. Regrettably, especially in older periods, my story emphasizes the perspectives of the state, as the wealth of archival material necessarily reflects state views.

In order to best tell the story, I present it in the form of discrete periods. The periods, summarized in table 7.1, are each marked by certain distinguishing features that separate them from neighboring periods. I also address the diversity within these periods, as well as the resounding continuities between them.

TABLE 7.1. **Historical periods in Malagasy fire politics. See appendix 2 for listing of all government decrees regarding fire.**

Period	Political Context	Fire Policy Characteristics	New Ideas and Trends in Fire Policy
1896–1906: first decade as colony	1896: French colony declared; Menalamba rebellion	tentative, inconsistent policies	• first ban on pasture fires • idea of utility of fires for pastures and locust control
1907–1924: emphasis on colonial exploitation		against fire, but somewhat weak	• system of authorizations • idea of fire as a temporary solution until intensification • collective fines
1925–1947: from drafting a new forest decree to rebellion	1947: major anticolonial rebellion	heavy repression	• nature conservation • pancolonial alarm about forest disappearance
1948–1968: transition years and continued close ties to France	1960: Independence under Pres. Tsiranana's 1st Republic	repression with undertones of appeasement and realism	• fire as a "necessary evil" • counterseason fires • preventive fires
1969–1983: revolutionary period	1972: 1st Republic falls 1975: 2nd Republic under Pres. Ratsiraka	harsh words, decreasing action	• education and propaganda • village antifire committees • extreme penalties
1984–1995: economic and political crisis	1991: general strike 1993: 3rd Republic under Pres. Zafy	environmental ideals but dysfunctional state	• international environmental involvement
1996–present: decentralization	1996: Zafy impeached 1997: Ratsiraka reelected 2002: violent election dispute	decentralization of resource manage ment	• community-based management

THE 1800S: ROYAL FIRE POLICY

The precolonial nineteenth century was characterized by state building and by increasing contact with the colonial powers. This century saw the first documented government regulation of fire: King Andrianampoinimerina's edicts and the Code of 305 Articles. Many twentieth-century analysts and politicians have cited these homegrown fire regulations in order to legitimate their own restrictive policies. However, their severity and importance has probably been exaggerated.

From the arrival of the first settlers some 1,500 years ago to the early nineteenth century, there was probably no state regulation of burning. People used fire freely to expand and maintain pastures and to prepare crop fields, most likely managing fires through mutual understandings, evolving traditions, and community-based conflict resolution mechanisms. In the early nineteenth century, King Andrianampoinimerina (reigned 1797–1810), leader of a small Merina state based in the rice-irrigating plains near the current-day capital, oversaw the political unification of the central highlands. His son, Radama I (reigned 1810–1828), expanded this Merina empire over much of the island; only the Sakalava kingdoms to the west and diverse groups in the arid south were able to resist control. During the rest of the nineteenth century, European interests became important as the French sought to increase their presence and the Merina counterbalanced this with ties to the British. Missionaries became active on the island, and a sophisticated modern state, long headed by Prime Minister Rainilaiarivony (governed 1863–1896), was established to complement the monarchy (Brown 1995).

King Andrianampoinimerina's proclamations on forests and fire, recorded in the transcriptions of Merina oral history,[1] suggest an interest in forest protection. He proclaimed that his chief enemies were Hunger, Fire, and Wind and called the forest a common heritage, the means of subsistence for orphans, women, and the poor. He ordered his subjects to look after the forest and forbade the gathering of wood fuel, the cutting of trees, and the burning of the forest. Charcoal fires were allowed outside the forest only (Callet 1908; Dez 1968, Bertrand and Sourdat 1998).

Dez (1968) asserts that these measures were intended for public security

1. King Andrianampoinimerina is particularly well documented and remembered, in part due to the widely circulated transcriptions of the oral history of the time (Callet 1908, 1953–1958). These texts, the *Tantaran'ny Andriana,* have become hegemonic, eclipsing other regional histories (Larson 1995).

(so that people would not use charcoal fires to illegally make arms in the forest) and to protect supplies of construction wood; the king's measures were not focused on protecting soils or trees. In fact, Andrianampoinimerina *prohibited* the planting of trees in open land except on the twelve sacred hills of Imerina—probably to facilitate the surveillance of potential invaders. Previous leaders had encouraged the clearing of forests for exactly this purpose. In addition, it is unclear to which lands Andrianampoinimerina's prohibitions applied. Most likely, they related to patchy forests claimed by royalty in the highlands and not to the grand forest of the east. In any case, *tavy* cultivation expanded during this period of time (Dupré 1863, quoted in Coulaud 1973, 323; Dez 1968).

A second critical reference point is the Code of 305 Articles, written in 1881 under Prime Minister Rainilaiarivony and signed by his wife, Queen Ranavalona II (Julien 1900; Henkels 1999). Article 91 declared that all forests and empty lands belonged to the state; articles 101 through 106 prohibited the burning of forests, the fabrication of charcoal within the forest, the cutting of large trees, the building of houses in the forest, the creation of slash-and-burn fields in forest areas not previously cut, and the cutting of the coastal forests. Penalties ran as high as ten years in irons. In Dez's (1968) interpretation, these laws pragmatically aimed to protect sources of construction wood, to establish internal security (for easier control of the population), and to maintain external security (coastal forests were to protect the kingdom against invading foreign armies). Uncertainty exists as to the extent to which these regulations were actually applied and enforced; they certainly did not apply in remote areas out of Merina control like the Ikongo region (Coulaud 1973; Olson 1984; Bertrand and Sourdat 1998).

While Andrianampoinimerina's proclamations and the Code of 305 Articles address forest cutting and forest burning, neither regulated crop field or grassland fires.[2] In the majority of Madagascar, pastures and forests in the nineteenth century were most likely managed as common property resources by the local populations (Rakotovao Andriankova et al. 1997; Bertrand and Sourdat 1998). The highlands and savanna zones of the west were characterized by a free-burning, pasture-oriented fire regime. The landscape

2. All the same, a man in Afotsara (interview, 24 Nov. 1998) said that during the precolonial days (*fotoana gasy*), there were associations of cattle owners who would walk to Ambositra to demand permission to burn pastures from the Merina kingdom's *gouverneur*. Whether this is true or not is hard to say, yet the files of the *gouverneurs d'Ambositra* Ramonja (1870–1892) and Rainisoavahia (1894–1895) found at the National Archives include nothing about fire (ANM II.CC.180/181/196).

was dominated by endless grass-covered hills; trees grew only near hilltop villages. In eastern forest zones, *tavy* fires provided the rice upon which people depended for food.

1896–1906: AN UNSURE BEGINNING

After years of pressure, the French military attacked and conquered the Merina monarchy in 1895 and established a colony in 1896. During the first decade of French control, the colonial authorities focused on "pacifying" the island and establishing administrative structures (many of which persist today). Their approach to fire policy was marked by hesitation. The experts of the Forest Service carried with them the antifire ideology of the Métropole and their fire-fighting experiences in France and in Algeria. The colonial district officers, however, were faced with the difficult task of actually administering the island for economic profit and political stability and did not share the foresters' antifire enthusiasm.

The Service des Eaux et Forêts, created in 1896, saw its mission based upon its experience in France: protecting forests for rational economic exploitation and for stopping soil degradation and erosion (Dez 1968). Girod-Genêt, its first director, and his contemporaries were alarmed at the deforestation and destruction they saw being caused by hatchets and fires.[3] As discussed in chapter 2, their views were aided by various biases and the nascent idea of the original island-wide forest. As a result, the first colonial governor, Gallieni, wrote a circular in 1897 ordering his district officers to do their best to stop the damage done by fires and peasant cultivators to the forest (see appendix 2 for a detailed summary of all policies relevant to fire). "The peasants," he wrote, "must be taught to respect the forests, which are the property of the state . . . Therefore, while we work on the proper legislation, please take whatever measures are necessary to reduce pasture fires. As far as *tavy,* you will prohibit them completely" (JOM 15 Jan. 1898). The legislation he promised came in the Forestry Decree of 1900, which banned dry season fires within 200 m of forests, except by authorizations granted to owners of private land (which thus excluded indigenous *tavy* cultivators).

These antifire stipulations quickly met stiff opposition. Cattle raisers protested, noting the importance of pasture fires for their herds. An invasion of locusts in 1900–1903 necessitated the use of fire to combat the pests; locals argued that the prolonged invasion was due to several years of fire prohibi-

3. See chapter 2 note 2, and Girod-Genet (1899).

tion. Governor Gallieni surveyed his provincial and district officers on the subject of fires, and most wrote back in strong support of fires, which they considered necessary.[4] Fires, they wrote quite perceptively, are critical to Malagasy cattle husbandry, useful in controlling rats, locusts, and other insects and are used from the frontiers of China to South Africa. For example, the *commandant* of the east coast Marolambo district wrote:

> Pasture fires are not just necessary but indispensable. It is a question of primordial interest to the natives. There are few marshes to turn into rice fields; there are few good cultivable lands. Slash-and-burn cultivation is indispensable for rice cultivation. In these conditions, pasture fires present only rare inconveniences; one must simply regulate them. Their advantages are numerous: they destroy insects and pests, remove old grass and bushes that slow grass development, remove the yearly layer of detritus, improve the grass quality and clean the countryside . . . Just look around you: the best pastures, the greenest hills, are those that have been burned. In the regions where there are few trees and people practiced large pasture fires (e.g., Ampasimpotsy, Antanambao, on the Anosibe-to-Vatomandry road) the cattle are fatter, more beautiful, and more plentiful than around Anosibe, where trees and forests stop the fires. (anonymous 1904, 25)

As a result, Gallieni issued another circular in March 1904 authorizing free grassland burning. He stated that pasture fires were intimately related to cattle raising, one of the colony's principal assets. Mowing and haying, which would be an alternative solution, were not realistic, he admitted. As a result, he decided that pasture fires should be allowed where they did not threaten forests or houses, under the control of local authorities. He reiterated, however, a strict ban on *tavy* and other fires in primary and secondary forest. Einrem (1912, 72) rejoiced at this about-face, writing that fire was a lesser evil than dying cattle and locust invasions. He added, probably somewhat figuratively, that "after only having dark, boring September nights for many years, from now on we can enjoy the old, annual superilluminations."

1907–1924: AGAINST FIRE, BUT WEAK

The tone of the following two decades was set by the primordial economic concerns of the colony, which had to show itself to be profitable. The central

4. Many of their responses are reproduced in the *Bulletin Economique de Madagascar* (anonymous 1904).

government's position against fire, especially *tavy,* hardened, as it became concerned with protecting the lucrative hardwood logging industry as well as commercial agricultural plantations. However, district officers in pasture-dominated zones continued to defend pasture fires, and in a compromise the government decided to temporarily enter into the realm of regulating and authorizing burning until "modern" pasture management techniques (mowing, haying) could be introduced.

In early 1907, new governor Augagneur promulgated an *arrêté* that tried again to limit pasture fires, tightening the reigns from the 1904 free-burning policy. The justification, contained within the text of the law, was that while fires were useful for pasture renewal and insect destruction, they could cause significant damage to plantations, houses, and forests. The *arrêté* banned all fires not necessary for pasture renewal or for destroying locusts or other harmful insects; planned pasture renewal fires could be authorized with the proper precautions. Such authorizations, Bourdariat (1911) complained, were rarely refused, and sometimes fires escaped into colonial plantations.

The *arrêté* was followed by a circular from Augagneur in May 1909 that instructed district officers to ban all fires that could damage the forest and to severely punish forest burners. *Tavy* was to be allowed only in previously cultivated plots or in "*les broussailles,*" brushy zones of *savoka,* and was to be progressively phased out in favor of irrigated rice cultivation. While Augagneur's replacement in 1910, Picquié, repeated these instructions, the situation on the ground hardly met their expectations. In 1913, Picquié asked his provincial officers for evidence that the 1909 circular was being enforced. The results are tabulated in table 7.2. While some district officials claimed a total success, others documented large numbers of *tavy* fires. It was impossible to control such a vast country, and where control was tight, farmers moved elsewhere or took advantage of the unclear distinctions between different kinds of *savoka*—from herbaceous brush to secondary forest—to get authorizations from more sympathetic officials (Coulaud 1973, 324–325).

In 1913, a new forestry decree signed in Paris replaced the 1900 forestry decree, which was deemed too weak (Lavauden 1934). This decree, which explicitly referred to experiences in France and Algeria, pronounced a general ban on vegetation fires, both pasture fires and *tavy.* However, until improved pastures and modern cattle husbandry could be introduced, the decree temporarily allowed pasture fires at least two kilometers from forests and plantations (Dez 1968). In the same year, Governor Picquié also issued an *arrêté* that simply banned all fires for *tavy* and cultivation. In an accompanying circular, he stated that it was time to finish the progressive phaseout of *tavy*

TABLE 7.2. **Infractions for forest burning and illegal *tavy*, 1909–1912.**

Province	1909	1910	1911	1912	notes
Analalava	36	113	35	59	
Betroka			30		total for 4 years; individuals fined 15–200 frs
Diego Suarez	0	0	0	9	
Fort Dauphin	6	49	27	145	one in ten sent to prison
Mananjary					report for two districts only
Antsenabolo	5	17	1	0	
Nosy Varika	0	2	0	48	
Morondava			1		total for 4 years
Tamatave	2	0	0	11	Fenerive district only; fines of 15–16 frs
Tulear	0	0	10	4	fines of 20–100 frs
Vakinankaratra	0	0	0	3	collective fines 0.25 fr per capita
Vatomandry	0	0	0	27	

Note: Table lists only those provinces for which statistics were found.
Source: ANM D196 and D81s.

planned by the 1909 circular. All the same, he allowed that farmers could continue to cultivate hill rice on former *tavy* plots, provided they cut the brush by hand and burned it in a pile in the middle of the field.

The 1907 and 1913 laws stipulated that each province establish its own regulations regarding fire authorizations and specifying fire timing, location, and precautionary measures. As a result, province chiefs created a flurry of *décisions locales* between 1907 and 1919.[5] These provincial regulations all followed a similar model. Pasture fires would be allowed upon authorization by provincial authorities (levels varied from *chef de province* to *gouverneur madinika*, depending upon the province). The season for burning, always in the dry season, was defined by province, e.g., 1 August to 15 October in Fianarantsoa Province or 1 June to 1 November in Itasy Province. Some provinces in the west and south had longer burn periods (e.g., Morondava and Betroka legalized fires from May through November), while eastern and northern regions adjusted to their climatic zones by authorizing later fires (e.g., Vatomandry and Diego Suarez permitted fires until 30 November, or Sainte Marie until 31 December). In all cases, exceptional fires would be allowed outside of this time frame in case of locust

5. Each is printed in the *Journal Officiel de Madagascar;* correspondence regarding these *décisions locales* is found in the National Archives (ANM D196 and D81s).

invasions. The *décisions locales* forbade burning within a specified distance from forests and trees (between 200 and 5,000 m), villages and houses (between 100 and 2,000 m), crop fields (between 300 to 2,000 m), and public utilities like bridges and telegraph poles (between 50 and 500 m). Previous to burning, a firebreak of specified width (2 to 20 m) had to be cleared of vegetation. Finally, the regulations sometimes stipulated the time of day (before noon or in the evening) and other precautionary measures.

In their correspondence with the central government over the drafting and revision of these local regulations, we can see the different ideas and forces pulling at the district administrators. A 1911 letter from the mayor of the east coast island of Sainte Marie attached a handwritten petition signed by thirty-seven citizens asking to be able to burn their crop fields in preparation for planting. However, the mayor noted that a colonist's coconut plantation had almost been incinerated the previous year by a wildfire and that repeated fires destroyed the topsoil and led to the erosion and sterilization of the soil. Thus he recommended allowing only fires for the preparation of crucial food crops.[6] The Betroka Province leader wrote in 1914 that while pasture fires were bad, for they quickly led to gully erosion and soil degradation, they should not be restricted too severely for the time being, because this would disrupt the cattle industry, the primary resource of his region.[7] The *chef* of Tulear Province took a different slant, arguing in 1911 that pasture fires were not a serious danger for forests, that damages were rare, and that the fires served well against locust invasions.[8]

In a letter dated 6 May 1913, the *chef de province* of Itasy, Georges Piermé, wrote the governor to say that there was neither forest nor *tavy* (slash-and-burn forest cultivation) in his province and thus no bad consequence to fear from fires. He asked that the locals be allowed to burn in order to prepare crop fields, to destroy locusts and other insects, and to renew their pastures, as prescribed in the local decision of 1911. If they were not allowed to burn, he wrote, they would not be able to feed their cattle, and these would die. Handwritten pencil notes in the margin indicated the opinion of the letter's reader: yes to destroying locusts and renewing pastures, but no to preparing crop fields, probably due to its association with "wasteful" *tavy*.[9]

6. ANM D81s, subfolder Sainte Marie 1911–1913.

7. ANM D196, subfolder Betroka 1911–1918.

8. ANM D81s, subfolder Tulear 1913.

9. ANM D196, subfolder Itasy 1911–1915. The identity of the person making the comments is unknown. The *chef du service de colonisation* forwarded Piermé's letter to the *directeur des affaires civiles* on 20 May, saying that crop field fires should be outlawed only if they threaten forests.

Piermé's letter, among others, probably inspired Governor H. Garbit's circular (JOM 14 Nov. 1914) that remarked that some officers had confused *tavy* and pasture fires. He explained that *tavy* was strictly banned, while pasture fires were allowed upon authorization and appropriate precautions. Garbit postulated that peasants burned illicitly because they feared that confused authorities would not give authorizations for pasture fires. In a later circular (JOM 23 Jan. 1915), he also worried that the complicated process of getting formal burning authorizations from French officers discouraged peasants from seeking authorizations at all; they just burned without permission. Both cases, Garbit noted, led authorities to fine the communities, which was a regrettable approach if it became regular. In response to the latter circular, the new Itasy *chef de province*, P. Orsini, wrote the governor to admit that the requirement in the 1911 local decision that fire starters seek permission from the district chief was never practical nor observed. He submitted a local decision to decentralize authority to the indigenous *gouverneur madinika* and in cases of extreme distance or urgency to the *mpiadidy*, traditional village leaders.[10]

Pouperon, *chef de province* of Ambositra, justified his province's *décision locale* to the governor in August 1914.

> I have determined, after studying the matter, that the pasture fires are very often a real necessity in the high and denuded parts of the province. This is because the renewal of pastures—the major economic resource of these bare, dry, and lightly populated areas—can hardly be done otherwise. Also, there is no other way to destroy locusts.[11]

A year later, in yet another circular (JOM 9 Oct. 1915), Governor Garbit asked district officers to consider ways for replacing pasture fires with improved, irrigated, or seeded pastures. He argued that while pasture burning should be tolerated in the short term because it responded to an immediate need, it should be phased out quickly, for it impoverished the pastures and destroyed the more palatable grasses. Likewise, Garbit pushed for his district officers to replace all *tavy* agriculture with irrigated rice, saying, "Please act quickly, with force and perseverance." Only in the most severe cases, where villages had absolutely no bottomlands for rice irrigation, would some *tavy* be tolerated, and never in virgin forest.[12] Garbit declared that once the colony had

10. Ibid.

11. ANM D196, subfolder Ambositra 1914–1924.

12. In Beforona, with its sharp relief and limited irrigable bottomlands, authorities allowed *tavy* on *savoka* (secondary vegetation) lands, while in Vatomandry, where irrigable bottomland

left this transition period to intensive pastures and irrigated rice, burning would be unnecessary except to prepare for the passage of the plow or to fight against locusts. Garbit repeated this argument in another circular five years later, but also declared that *tavy* was prohibited everywhere, with no exceptions (JOM 20 Nov. 1920). The government increased its efforts to enforce the new regulations. Quite frequently, colonial authorities had to resort to collective fines, *sazim-pokonolona.* For the period 1914–1920, 143 collective fines were reported for fire-related crimes,[13] including illegal burning, fire damage to forests or villages, neglecting to or refusing to fight fires, or refusing to give the name of (or to help find) the author of the fire. Some fires burned into forests, tree plantations, or colonists' lands; others destroyed houses or entire villages. One, in Diego Suarez Province in 1914, destroyed a small military facility. These fines were applied to one or several villages, and the amount levied averaged three to five francs per able-bodied man.

Despite this legacy of collective fines and the record of forestry infraction enforcement,[14] many commentators complained about the ineffectiveness of the laws. It is hard to imagine that *tavy* was totally stopped or that all the minutely detailed prescriptions and authorization procedures of the local pasture fire regulations were followed. Enforcement was uneven (Jarosz 1996). Pouperon, the chief of Ambositra Province, wrote in 1914 that "despite the severe penalties contained in the 1913 decree, fires obstinately continue to appear, especially in August, September, and October." He concluded that the Malagasy were not doing this to spite the colonial government; it was more that they either had not heard of the rules or that they just misinterpreted them and believed it was permissible to light the grass far from the forest.[15] The 1913 Forestry Decree remained largely nonfunctional, due to the number of exceptions, the weakness of the Forest Service (which did not have the power to prosecute), and the lack of will on the part of some authorities.[16]

---

was widely available, *tavy* was completely banned. Peasants responded by moving to Beforona (Jarosz 1996).

13. From 1914 through 1920 (but not other years), collective fines for all crimes (including fire, hiding lepers, stealing telegraph wire, and not helping to catch cattle thieves) were published in the *Journal Officiel de Madagascar.*

14. A collection of forest crime cases for this period is found in ANM files D82s–D85s. Only a few involved fire; most relate to illegal cutting.

15. ANM D196, subfolder Ambositra 1914–1924.

16. Berthier, *chef* of Tamatave Province, wrote in 1913 that the forestry laws remained "dead letters" in many forest regions (ANM D81s, subfolder Tamatave 1917). Bensch, *chef* of Vakinankaratra Province, complained in 1918 that the administration tolerated too many fires (ANM D76s/13, letter 2312). The colonial governor wrote to the president of the French Republic in 1924

This is what probably prompted Governor Garbit to write that he would "not hesitate to take severe action against those [authorities] who, by negligence or in bad faith, tolerated infractions" of the fire rules (JOM 20 Nov. 1920). It probably also explains the creation by Governor Berthier in 1926 of two mobile "forest brigades" to enhance the repression of *tavy,* forest fires, and illegal logging. By 1930, frustration with the existing legislation and poor enforcement, combined with the new input of influential naturalists, had led to toughened rules.

1924–1947: TIGHTENING THE SCREWS

The years from the mid-1920s through 1947 were a crucial period in which the tone of state-led fire repression stiffened and in which peasant resistance became entrenched. The period began with debates over the strengthening of forest and fire legislation, and culminated in the anticolonial rebellion of 1947, which, not surprisingly, hit foresters rather hard. The principal trend that gives this period its character was the maturation of the antifire received wisdom. As foresters in Madagascar and across the colonies complained about the insufficiency of forest protection, naturalists concerned with the loss of forests gained an increasingly prominent role in policy making. The number of district officials defending local fire practices shrunk significantly (through not completely). As a result, repression increased, as did resistance.

Due to foresters' and naturalists' alarm at rates of deforestation, the Forest Service began working on a revision of the extant forest and fire laws in 1919. The process moved slowly for over a decade, culminating in a new forest decree in 1930.[17] From 1922 to 1923, a governmental commission worked to draft a proposed law.[18] This text was circulated to provincial *chambres de commerce* and *chambres consultatives,* producing numerous comments and objections.[19] Logging industry representatives, of course, wanted a strict

---

that certain parts of the 1913 decree were never applied, and others were too strict (ANM D64s/12). See also Lavauden (1934) and Bertrand and Sourdat (1998).

17. This section describing the revision of the 1913 Forest Decree between 1922 and 1929 is based on documents found at the National Archives, files Régime forestier 1918–1936 (ANM D63s, D64s, and D65s), unless otherwise noted.

18. The Commission for the Revision of the Forest Decree consisted of representatives from the Forest and Agricultural Services, the Land Tenure Office, Economic Services, Chamber of Commerce, and the Economic Office.

19. Most comments and objections related to logging and the forestry industry. Regarding fire, the chamber of commerce of Nosy Be noted that the "energetic repression" of *tavy* by the *chef de province* led to a decrease of fires and demanded that the new legislation be strictly applied to get

control of *tavy* to protect "their" forests but worked to avoid excessive controls over their own exploitations. A second commission—with additional private-sector logging representatives—worked from February to October 1924 revising the proposal, which then went to the technical services for comment. The administration submitted the proposal to the colonial government in Paris in 1926, where additional suggestions were made by the Service des Bois Coloniaux (the Colonial Wood Service, with ties to the logging industry) and approved by the Service Economique. The proposal languished in the Malagasy colonial administration from 1926 until late 1928, and was finally signed in Paris in January 1930.

Two main forces pushed the 1930 decree. The first was Forest Service concern with the pace of extractive logging.[20] Concern was widespread at the time about the abusive effects of not just peasant *tavy* agriculture, but also unrestricted commercial logging (Jarosz 1996). A broad alarm of forest disappearance due to both colonial loggers and local cultivators was ringing across Africa (Bergeret 1993). In Madagascar, the annual reports of the Forest Service in 1922 and 1930 included passionate pleas against the free exploitation of forests by concessionaires. Colonial forestry inspector Griess reviewed the Forest Service in the mid-1920s, suggesting keeping better track of logging companies because of the terrible rate of forest clearing.[21] Humbert (1927) in his important tract also complained of the effects of greedy forest concessionaires. In justifying the final revision of the new decree, the minister of colonies, F. Pietri, wrote in his report to President Doumergue that the 1913 decree had long been insufficient to reach the goals of conserving (and increasing) the forests while facilitating rational exploitation (JOM 22 Nov. 1930).

The second force was the voices of scientists who espoused strong and alarmist antifire views (see chapter 2). Notable among the 1922–1923 forest decree revision commission members was Henri Perrier de la Bâthie, one of the most prominent naturalists, botanists, and ecologists working on the island (he served on the commission as representative of the Comité de l'Office Economique). Perrier arrived in Madagascar in 1896 and explored

---

the same results (ANM D64s/5). The chamber in Mananjary also suggested taking a hard line (ANM D63s/12). Meanwhile, the *chef de province* of Ambositra noted concern for the food security of the forest-dwelling Tanala people if the new decree were to absolutely ban *tavy*—he suggested allowing *tavy* in secondary brush (ANM 64s/7).

20. The voice of the Forest Service was increasingly important, especially as it was made independent of the agricultural and colonization services in 1926 (Lavauden 1934).

21. AOM mad-ggm-D/5(18)/5, 9, and 15.

the vegetation of the island for thirty-five years (Rauh 1995; Middleton 1999; see especially Perrier 1921, 1936). He was one of the strongest advocates for the theory that Madagascar was once completely forested, and that fires were thus the ruin of Madagascar. Perrier believed that the Malagasy burn "out of simple habit and without any reason" (1921, 4). He stated that the Malagasy peasants are "of little intelligence, very passive, and lacking all forethought . . . [and] tainted by this strange and contagious mania that pushes all Malagasy to burn the dry grass" (1921, 265). He concluded that "The conquest of the island would be purposeless if we only came here to continue an aimless destruction, without a care for the future, by imitating the Malagasy and their childish actions" (1921, 265–266). He noted that the prairies of the central highlands were a result of periodic fires, and that they were mediocre pastures continually degraded by burning. He suggested that all grassland fires be halted (1921, 9). Working for the commission, however, Perrier was slightly more realistic, recommending the creation of a ten-kilometer wide protective zone around all forests and reforestations. Within this zone, he suggested that fires be banned completely, or at least that the "natives" be forced to closely circumscribe their fires. To Perrier, however, the only real solution would be to privatize all lands, at which point property owners would have an interest in protecting the value of their lands and fires would cease (ANM D63s/3).

Perrier was not alone. His colleague, Henri Humbert of the Paris Natural History Museum, made ten expeditions to the island between 1912 and 1960 and is perhaps the island's most famous botanist (Rauh 1995). Humbert's (1927) monograph, subtitled "The destruction of an island flora by fire," echoes the ideas of Perrier. His monograph probably influenced the revision of the forest decree; a copy is stored among the discussions of that decree in the National Archives (ANM D65s). Humbert's strongly argued case against deforestation and fire (see also Humbert 1953) directly led to the creation of Madagascar's first nature reserves by decree in 1927.

In addition to the views of the Forest Service and eminent naturalists, the Forest Decree Revision Commissions also encountered more pragmatic views. In its January 1923 minutes, the commission recognized that it was currently impossible to stop the practice of pasture fires, and thus it recommended developing more rigorous requirements for burning authorizations. The commission wrote, "Even though the subcommission considers the custom of bushfires a calamity for Madagascar, it decided to keep the articles [regarding authorized pasture burns] since it sees no way to stop the practice. All regulation of the fires, given the availability of personnel, is unrealistic . . . A so-called absolute ban, followed by a general exemption, is perhaps the worst."[22] Their first proposal in 1923 thus allowed pasture fires upon authorization by a local decision, no closer than two kilometers from the forest. Not everybody agreed, of course. One commentator stated that allowing fires two kilometers from the forest, on an island covered with dense dry grass, was to authorize the burning of forest on three-quarters of the island. This commentator proposed focusing efforts on protecting the forests, with the creation of a four-meter-wide well-maintained firebreak around all forests (ANM D63s/12). An opposing commentator suggested that the two-kilometer distance requirement was ludicrous and that topography, streams, and other natural firebreaks could be used to control fires much closer to the forest (ANM D64s/2).

As far as *tavy* fires in the forest were concerned, the commission initially recommended a complete ban (ANM D63s/23). By 1926, this had been revised to include the provision that when *tavy* was the only available means of food cultivation, it could exceptionally be authorized in special zones, following a four-year rotation of crop fields (ANM D64s/13 and D65s).

In 1928, while the Forest Decree was still undergoing revisions, L. Lavauden became chief of the Forest Service. Lavauden's views took those of Perrier and Humbert to their extreme. He proposed not only that forests once covered Madagascar coast to coast, but also that these forests were uniform across the island and largely evergreen, like the current rain forests of the east. He blamed the natives and their incendiary practices for the disappearance of the forest and the resulting climatic desiccation and degradation. Lavauden championed the missionary-like role of foresters and their unpopular jobs of saving the island from ruin: "Foresters are the natural guards of precious interests the conservation of which is a discomfort for many" (1934, 959). He justified extremely repressive policies and ridiculed the compromising colonial administration:[23]

> Foresters are the only ones on Earth who care what will happen in one or two centuries. But what does an administrator, a governor, or even a governor general care? The best-intentioned are not aware of the whole forestry thing. They say to themselves, perhaps all [the warnings of foresters about environmental change] are not true, and they rock themselves to sleep with the illusion that nothing in the world has changed or will change . . . The less well-intentioned

22. ANM D63s/11, Projet de décret forestier, Titre VII, Section II.
23. See also chapter 6.

know quite well that they will not be there anymore when the bad effects become catastrophic. (Bertrand and Sourdat 1998, 33)

Lavauden's complaints notwithstanding, the colonial central administration's views were, by 1929, convincingly antifire (even if field officers were more pragmatic). At this time, Governor H. Berthier issued a circular (JOM 9 Feb. 1929). He stated that the goal of colonial agrarian policy was to move people out of the hills and mountains, stop shifting cultivation, and to institute a "normal" agricultural regime, e.g., modern cropping techniques, fertilizer, and mowed and seeded pastures. *Tavy* slash-and-burn fires, he declared, were a nefarious practice that would destroy the forest and ruin the country. They had always been banned, including by the 1913 decree, and Berthier encouraged his district officers to enforce this ban even if "the texts have, in some circumstances, been interpreted in a very tolerant spirit." As far as the use of fires for crop fields and pasture renewal were concerned, he stressed that these should still be subject to his approval. He declared that fires "should never be tolerated on mountain slopes or valley sides, even if they were only covered with grass," for this inevitably led to torrential erosion.

The new forest decree, which appeared in January 1930, significantly tightened regulations and procedures for forest exploitation. *Tavy* was simply not an option for managing land classified as forest; therefore it was illegal. In article 36, the decree also banned fires within 500 m of any forest, no matter how small the forest. The decree prohibited bushfires and prairie fires in nonforest lands, except when authorized by the governor's delegated authorities. In such cases, a firebreak of fifteen meters' width was required. Articles 58 and 59 spelled out specific measures for the repression of fire infractions, with the toughest sentences reserved for burning forest: fines of 500 to 5,000 frs and/or one to five years' incarceration. In either case, native villages (*les collectivités indigènes*) were legally responsible for infractions within their territory if they could not prove that an outsider caused it.

According to the minister of the colonies, the Forest Decree of 1930 legislated a more severe repression of forestry and fire infractions, increased the powers of forest agents, and reinforced punitive measures against the locals in order to effectively fight the prairie fires that were gnawing at forest edges and the *tavy* fires that were destroying forest value.[24]

The 1930 decree set the tone for the following three decades (Rakatonin-

24. Paraphrased from report to the president of the French Republic by the minister of the colonies, 25 Jan. 1930 (JOM 22 Nov. 1930).

drina 1989). However, the repressive legislative record masks a debate that continued throughout the 1930s. Geoffroy (1931), a cattle industry expert, published an article in the *Bulletin Economique de Madagascar* presenting research from South Africa on pasture fires. This research highlighted the use of pasture fires by South African ranchers to improve forage and fight ticks and investigated the best season and timing for fires. Clearly, the cattle industry—which included several large colonial ranching concessions in the western highlands—did not support fire suppression. The 1935 annual report by the Forest Service in Ambositra district concurred, noting that in most cases, pasture fires did not cause any important damages, especially considering their importance to the cattle-based economy of the region. This report continued by chastising the administration for reacting so violently when prairie fires and bushfires burned around cities or along major roads, but ignoring the more important fires that burned forest.[25]

Such dissent was probably aided by the fact that the Malagasy were not the only ones to use fires. The colonial ranches in the Middle West, such as the immense Rochefortaise concession near Tsiroanomandidy, burned their lands for pasture renewal.[26] A poor colonist named Goissaud, farming near Vohipeno on the east coast, was fined in 1933 for burning the pasture he used for his cattle.[27] Another colonist, M. Chavanne of Ambilobe, was fined for an authorized pasture fire that escaped his control in a high wind, burning 100 ha.[28] In Diego Suarez Province in the 1930s, the *chef de province* received demands for authorizations of fires not only from the Malagasy locals (for crop field fires and pasture renewal) but also from the French military (to clear the plain of Morvan of tall grasses) and from colonial planters (to clear vegetation from a perfume plant concession, or to clear land and cultivate crops).[29]

The 1930 decree, in particular article 36, was soon deemed insufficient, ambiguous, and too complex by foresters and administrators.[30] In 1933, the

25. AOM mad-ggm-2/D/19bis/6, p. xiv.

26. Jean-Pierre Raison, personal communication, 1 Apr. 1999; interviews in Kilabé, August 1999.

27. ANM D83s, subfolder "délits forestiers 1933, Farafangana."

28. Demandes de transaction avant jugement en matière forestière, Conseil d'Administration, 7 Jan. 1942 (ANM ID 351).

29. AOM mad-ds//334.

30. "It is difficult to arrive at a clear understanding of article 36 of the 1930 Forest Decree. It does not distinguish between forest-clearing bushfires and prairie fires, which it confuses" (letter no. 1459, 10 Nov. 1936, from *chef de district* Ambovombe to *administrateur superieur régional* Fort Dauphin, ANM IV.D.73/1/2). See also AOM mad-ggm 2/d/19bis/6, p. xvi., and letter from Procurer General L. Rouvin to Gouvernement Général, 31 May 1936 (ANM D65s/2).

governor delegated his authority to regulate pasture and bushfires to regional administrators, prescribing additional precautionary measures.[31] District officers noted that fires burned each year with the same intensity as the last, despite the regulations,[32] and Lavauden (1934) complained that the 1930 decree did not solve the fire problem due to illegal softening of the laws by parts of the administration. Heim (1935) echoed the severe interpretations of Lavauden, and called for more power for the Forest Service.[33] The government even applied emergency regulations—as specified in article 36—to problem regions. Approved for the eastern parts of Tananarive Province (JOM 21 Nov. 1936) and certain areas in Tulear Province,[34] these regulations stipulated that all fires had to be announced fifteen days in advance by written notice to the *chef de district.*

Due to the decree's insufficiencies, and bolstered by a hardening antifire received wisdom reflecting economic and ecological concern for the forest, the government significantly strengthened the forest decree in 1937, tightening rules about forest exploitation, enforcement procedures, and fires. Article 36 was totally revised to begin with the unequivocal statement that "the burning, destruction, or cutting of forests, and bushfires for preparing crop fields or for pastures, are prohibited on all the lands of the colony." None of the previous decrees or *arrêtés* had stated outright that *all* fires were banned; the revised article 36 signaled a much more severe approach. All the same, reflecting dissenting voices, the decree still allowed *chefs de district* to develop rules to authorize pasture fires, as long as they were well controlled and at least two kilometers from forests. The new decree also strengthened repressive measures for all forest crimes, and authorized Forest Service agents to write *procès-verbaux* (citations) without consulting the police.[35]

In a 1938 circular, Governor Léon Cayla introduced the 1937 decree to his district officers.[36] He explained that the new decree was meant to end all forest fires; it totally banned all *tavy.* At the same time, he allowed cultivation in *savoka* secondary forest vegetation (which in practice meant allowing *tavy*), and he encouraged the *chefs de régions* to delimit agricultural zones for forest

31. AOM mad-ggm 2/d/19bis/6, p. xiv.

32. ANM IV.D.73/1/2.

33. Heim (1935, 424) wrote that current repression was insufficient and that the colony needed "new methods, more rigorous and even brutal, I dare even say revolutionary."

34. Approved 10 Aug. 1937 by the conseil d'administration (ANM ID 298/7-J), but probably never published as the stricter 1937 Forest Decree was approved one month later.

35. JOM 13 July 1937.

36. JOM 2 July 1938.

villages. People exploited these loopholes to get away with some *tavy* cultivation, at least until enforcement tightened in 1939 (Coulaud 1973). As far as pasture fires were concerned, Cayla stated that this disastrous practice should be stopped, for it ruined pastures by removing the good forage species. However, like Governors Gallieni and Garbit before him, he suggested that Madagascar was (still) in a transition to modern pasture maintenance (mowing, seeding), and thus that pasture fires had to be tolerated for the benefit of the cattle economy.

In 1941, a second revision sought to address another weakness in the 1930 Forest Decree: the difficulty of enforcing anonymous forest and fire crimes.[37] The revision modified articles 58 and 59, making native villages legally responsible for fires whether or not they caused them. It also named the village leader as personally legally responsible and reinforced the possibilities for collective fines.

Accompanying this legislation, Governor A. Annet issued a circular that urged administrators to take a much harder line against burning, for fires were increasing and the rules ignored by all—including some sympathetic district officers.[38] Thus Annet reminded his district officers of the law and told them to be stricter. He said that fire was banned on all land that could be forested—land not given over to agriculture or covered by pasture grass. He wrote forcefully that

> *Tavy* is absolutely prohibited. To allow the smallest exception to this absolute rule is to encourage laziness among the natives. Nowhere in the world does one earn one's daily bread without effort. It is inadmissible that in a country of 63 million ha, with only 4 million inhabitants, and where nine-tenths of the forest has been destroyed, that it be necessary to allow the natives to destroy the little remaining forest for subsistence. Therefore, you will ensure that this law is enforced.

Annet also proposed that all villages found existing inside the forest massifs be moved outside of the forest as soon as possible.

While he was unequivocal in his fight against *tavy*, Annet grudgingly accepted the "temporary" necessity of pasture fires for the cattle economy. He

37. Projet de décret modifiant les articles 58 et 59 du décret de 1930 (AOM fn-sg-mad 353/948). The decree itself is published in JOM (20 Dec. 1941). As a decree tainted by association with Vichy France, it was revalidated by the new Gaullist government in 1944 (JOM 29 Apr. 1944).

38. JOM 6 Dec. 1941.

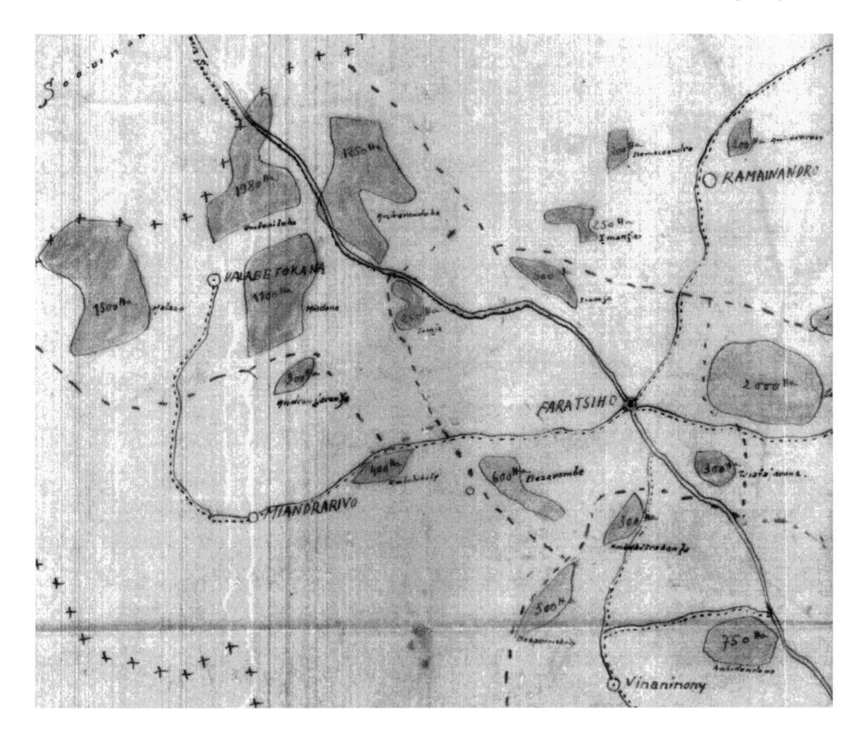

FIGURE 7.1. **Designated fire zones in Ambatolampy district, 1942.** This is a detail of a map accompanying the Ambatolampy district Décision Locale #79 regarding the authorization of pasture fires in certain zones. Source: ANM D81sup/1.

proposed the creation of irrigated pastures to get cattle through the dry season and to allow only the burning of limited zones in areas where such pastures had not yet been established. For the purpose of managing such fires, Annet asked each district to specify its pasture fire regulations. This resulted in another flurry of *décisions locales* regarding fires, echoing those of 1907–1919.[39] Ambatolampy district, for example, listed 122 pastures that were designated for pasture renewal fires (figure 7.1). These pastures ranged in size from 14 to 3,500 ha and covered a total of 50,000 ha, or about 5 percent of the district. In essence, one or two pastures were delimited per village. Authorizations would be given only to villages, not to individuals, for the very late period of 15 November to 15 February, provided the proper precautions were

39. Local decisions for pasture fires are documented for the districts of Ambatolampy, Fianarantsoa, Ambatofinandrahana, Morondava, Tulear, Ambovombe, and Diego Suarez (ANM D196; D81s; IVD38; IVD73).

taken.[40] For its part, Ambatofinandrahana district, a vast, lightly populated cattle-raising zone, authorized 300,000 ha of fire zones between July and September each year.[41]

These colonial-era fire authorizations are remembered in the field sites. While time has blurred people's memories of the exact details, they insist that the authorization system worked well, at least in comparison to today, and that the French used to be quite strict in enforcing fire infractions. In Afotsara and Kilabé, fires were authorized for the dry season, August and September; on a specific day all the men would climb to the designated pasture areas and assist in burning and controlling the fires.[42]

The 1930 Forest Decree, along with its 1937 and 1941 modifications, may have threatened severe repression, but experience showed it to be overly ambitious. Coudreau (1937), a Forest Service officer, noted that the severe measures of the decree led to an impasse as the administration, especially its lower echelons, could not realistically enforce the legislation. They were forced to compromise, and enforcement became inconsistent (see chapter 6). The vast terrain of the colony, combined with the limited numbers of Forest Service agents, conspired to make enforcement as difficult as it always had been.[43]

During the Second World War, four out of five Forest Service officers were mobilized to the war effort and Britain wrestled the island, through military action, from the Vichy French (under Governor Annet). After some months of uneasy rule, they turned the island over to the Free French. As a result of this turmoil, and subsequent wartime burdens like rice requisitions or forced labor, French authority and prestige declined (Razoharinoro-Randriamboavonjy 1971; Brown 1995; Bertrand and Sourdat 1998). The weakned colonial state was paralyzed by political instability, and thus restrictions slackened, enforcement loosened, and the Malagasy took advantage of this situation.[44] Officers in Fort Dauphin district admitted, for example, to the *chef de district* in 1946 that none or few applications for pasture fires had been received, yet that vast areas had burned.[45]

40. The local decision required firebreaks except where a natural obstacle served the same effect—e.g., a ditch or nonflammable crops. Proper precautions also included lack of wind, surveillance by community members, and two kilometers' distance from forest.

41. *Décision Locale* no. 57, District d'Ambatofinandrahana, 26 Oct. 1938 (ANM IVD38/7).

42. Interviews, Afotsara and Kilabé, 1998–1999.

43. Ministère des Colonies, Inspection des Colonies, Madagascar, no. 44: Rapport par M. Pruvost concernant District d'Ambanja, 6 Mar. 1938, p. 7 (AOM mad 330/855).

44. Letter from Governor de Coppet to *chefs des régions*, 12 Oct. 1946 (ANM IV.D.73/1/1); Bertrand and Sourdat (1998).

45. ANM IV.D.73/1/1.

Rising nationalist sentiment and anticolonial resentment resulted in a major rebellion in 1947. Precipitated by French political maneuvering that delayed hopes of independence, increasingly vocal nationalist parties, and the belated return of Malagasy soldiers from the war, this anticolonial rebellion broke out simultaneously in many towns on the night of 29 March. The rebellion spread over the following months, and it was brutally quashed by the French, causing the deaths of tens of thousands of Malagasy (Brown 1995; Dahl 1998). Not surprisingly, foresters and forest plantations were a prominent target of violence. The prohibition of *tavy* figured directly in the rebel-led massacres of Anosibe and Moramanga (Vérin 1954, quoted in Jarosz 1996). Many forestry posts were destroyed, forest plantations were burned, and a massive increase in fires and *tavy* was noted (table 6.1; Humbert 1949; Deschamps 1965; Olson 1984; Bertrand and Sourdat 1998).[46] The pot had boiled over, not just in terms of colonial domination, but also specifically in terms of the colonial appropriation of resources (like forests) and repression of natural resource management techniques (like fire)—which for many farmers and herders was their key interaction with the colonial state.

1948–1969: FIRE AS A NECESSARY EVIL

The French quickly reestablished control after the 1947 rebellion through harsh political measures. In the 1950s, the administration strongly suppressed political activity, and concentrated on economic development (Brown 1995). However, the legacy of the rebellion (the fact that French rule was more tenuous and sensitive to local unrest), together with new advocates of fire tolerance, led to a slight softening of the repressive regime against fires. While the 1930 Forest Decree continued to set overall policy directions, the dissenting profire voices of some range managers led leaders to look for pragmatic solutions.

This historical period, when fire was often seen as a "necessary evil" spans the transition years—the final decade of colonization and the first decade after Independence, achieved in 1960 (France continued to play a major role in advising and funding Madagascar after Independence). Throughout this period, burning was considered evil, a threat to natural resources. Yet, within this frame, some administrators recognized the necessity of fire to rural farmers and herders. As a result, technical solutions flourished, includ-

46. See also AOM mad-pt//181, which is a folder entitled "Rébellion 1947—destruction de postes forestiers et de forêts."

ing the idea of early-season or counterseason burning, as did regulations for authorized prescribed burns. Legislation made room for these changes, while maintaining an overall antifire frame and strongly enforcing infractions.

Enforcement of forest and fire legislation just after the 1947 rebellion was a paradoxical mix of heavy-handed force and politically pragmatic tolerance.[47] After the rebellion, the administration focused on regaining control of the island. In some forest areas, like around today's Ranomafana National Park, they forcefully relocated villages from hilltops inside the forest to valley bottoms near roads and irrigated rice fields (Peters 1999). This was to facilitate administrative control, but also justified in terms of forest conservation and the intensification of agriculture. In fact, the forced relocation out of the forest of up to 500,000 people was discussed several times in the 1940s and 1950s, in what Coulaud (1973, 320) likened to the Nazi "final solution," but it was never politically feasible at a large scale.

It took three years after the rebellion for the situation to return to normal (Coulaud 1973); by the 1950s enforcement became retrenched and somewhat more effective. The number of infractions that were enforced during this decade was relatively high, at least compared to later years (Rakatonindrina 1989; see figure 6.2). Foresters received bonuses for successful *procès-verbaux*.[48] In Majunga Province, Opération Bozaka was launched to control the amount of fires: three ten-person military patrols roamed the countryside in Mandritsara district during September 1956, in order to convince people that they had to follow the laws. According to the *chef de province*, the operation was conducted with tact and the collaboration of local officials, provoking no "incidents."[49] Interviews in the field sites found that people remembered the strict enforcement of the late colonial period (at least in comparison with later years), where forest officers might come to the burned area, take pictures of the damages, and write reports. They remembered sometimes getting permits to burn the hilltop pastures or to make *tavy* plots in *savoka;* they were scared of enforcement at the time.[50] In fact, the Fia-

47. Coulaud (1973). In 1948, the Tulear Forest Service chief complained that due to the political situation (and a lack of personnel), his agency was powerless to repress fire (ANM IVD73, letter 1746, 20 Oct. 1948).

48. Additional authority was given to the Forest Service by gubernatorial circular in 1950 (Bertrand and Sourdat 1998).

49. Letter 1776 from *chef de province* Majunga to Haut-Commissaire de Madagascar, 22 Oct. 1956 (ANM D196/subfolder Feux de Brousse 1956–1957).

50. Multiple interviews, Afotsara, Kilabé, and Behazo, 1998–1999.

narantsoa branch of the Forest Service supported its entire budget in the early 1950s through out-of-court settlements with fines (Dez 1968).

Collective fines, a central means of fire policy enforcement since the 1910s, became more difficult after the 1947 rebellion—they were seen as a colonialist measure and regional courts hesitated to apply them. However, the Forest Service usually successfully appealed these decisions to ensure enforcement (Dez 1968). A case in Fianarantsoa Province, where seven *collectivités rurales* were each fined 45,000 frs, led to a discussion on whether it was moral to imprison village chiefs who refused to pay the collective fines. The procurer general encouraged a hard-line approach, but also warned that officers should seek out individual culprits whenever possible, as the overuse of collective fines would make it difficult to recruit village chiefs. Following a complaint about collective fines by Malagasy representative Louis Rakotomalala at the Assemblée Représentative in 1956, the colonial administration defended the laws as legally, morally, and economically necessary in order to defend the nation's soils and forests.[51]

A new tool for enforcement appeared in 1950: Arrêté 1884-SE/SF, which prohibited all grazing in pastures burned without authorization and all cultivation (e.g., *tavy*) in cut or burned state forests. This policy sought to suppress the incentives to burn illegally and gave the Forest Service a better means to enforce the regulations against illegal burns. It was repealed in 1957,[52] but reappeared in the new legislation of 1960.

Fire policies were actively discussed in Madagascar and elsewhere in the 1950s. On the one side, hard-liners continued to develop the antifire received wisdom. They included botanist Humbert (1949), who called for the Forest Service to assert more control, and Bégué, chief of the Forest Service in the early 1950s, who called for additional forest officers and argued that active and consistent repression gets results (Bertrand and Sourdat 1998). Rangeland ecologist Bosser (1954) also supported strict fire suppression, arguing (based on a range retrogression model; see chapter 2) that healthy prairie was retreating in the face of degraded, burned, and overgrazed prairie.

51. These discussions of collective fines found in subfolder "Feux de Brousse 1956–1957," ANM D196.

52. In a five-page note (no. 115 CF.AP.2, 25 Jan. 1957) by Marcel Villepreux of the Direction des Affaires Politiques, responding to questions placed by Louis Rakotomalala at the Assemblée Représentative (Voeu no. 53-26, 13 Dec. 1956), the government defended the prohibition of pasturing on burned lands as the only way to reduce the dangerous and ruinous practice of fires. "By the way," Villepreux wrote, "the same prescription exists in French law, even more strictly" (ANM D196/subfolder Feux de Brousse 1956–1957).

These hard-line opinions about fire developed in the context of experience across Africa and elsewhere, where foresters and naturalists continued to strengthen their views against fire. For example, a 1948 Conference on African Soils solemnly condemned fires as a leading force of degradation, even as it recognized fire's sometimes indispensable role in primitive agropastoral systems (Schmitz 1996). A. Aubréville, the influential head of the French colonial forest service, wrote of mainland Africa in 1947:

> Each year at the beginning of the dry season this country is swept by fires . . . It seems as if things had always been so, but that is not true. Actually we are witnessing the death struggle of a plant world, slow stages in the drying up and degeneration of tropical Africa. (Aubréville 1947, quoted in Bartlett 1955, II 39)

He argued that most savannas would revert to forest if it were not for the effects of fires, but he did admit that climatic desiccation may also have played a role in forest disappearance.

Aubréville made his clearest statement on vegetation fires in a 1952 review of a profire article by a scientist named Jeffreys. This stinging critique lambasted Jeffreys for ignoring the scientific evidence on fire's damaging effects on vegetation, soils, and erosion. Several times he used Madagascar as a case in point. Aubréville's ideology as a forester shone through; his argument was predicated on the assumption that forests were better than grasslands, and thus that fires should be stopped to allow forests to regrow. At the same time, he recognized that in grasslands, fire was a useful tool for native pastoralists. Most tellingly, Aubréville worried that articles like Jeffreys's would damage the tenuous gains by antifire foresters in convincing a hesitant administration and a contrary population that fires must be stopped (Aubréville 1952; see also Bergeret 1993).

Despite these hard-line views, several fire-tolerance advocates surfaced in the 1950s, especially among administrators and range specialists (e.g., SEM 1949; Metzger 1951; anonymous 1955). Administrators heard the complaints of the rural people and did not enjoy enforcing laws that they did not always believe in themselves.[53] Together with range specialists, they were concerned about the economic future of the island, a future in which the cattle economy played a significant role. Elsewhere, researchers such as Bartlett (1955, 1956) and Conklin (1954) raised the appreciation of both pasture fire and *tavy* as widespread management tools. As a result, several ideas emerged.

53. Dez (1968, 112n) notes that the colonial administration, which bent to the protests of cattle raisers, was far from being as tyrannical as it was sometimes depicted.

The first was that of delimiting areas where fires could be authorized. For pastures, a circular in 1950 reemphasized the ideas of 1941, proposing the creation of specific pasture areas for burning, the size of which depended upon the number of cattle. In forest zones, agricultural agents and foresters struggled over the creation of "cultivation perimeters," areas of secondary forest and fallow within which farmers could cultivate their crops, even with *tavy* fires, when authorized. According to Dez (1968), by 1955 some villages had not yet had a zone delimited, while others had zones that were hopelessly out-of-date.

The second and most significant new idea of the period was that of counterseason fires or early fires. An article by Georges Metzger (1951) was forwarded to all provincial and district administrators in Madagascar.[54] He studied fire and extensive grazing and argued that fires would remain a necessary evil for a long time. He criticized those who believed that the Malagasy burned ignorantly for simple pleasure, without worrying about the consequences. Fire, he wrote, is extremely useful in removing old, dry grass and undeniably effective against ticks.[55] Metzger described the flowering ecology of pasture grasses, and suggested that burning in the mid rainy season, i.e., 15 January to 15 February, with a three-year rotation, would be an excellent way to renew pastures without causing ecological damage. Research at the state Kianjasoa agricultural research station, in the Middle West, also showed in the 1950s that early fires, in February to April, were possible and reduced the ecological consequences (Dez 1968).

In Fianarantsoa Province, district authorities took up Metzger's idea of early fires. They legalized burning from 15 January to 15 February, without written authorization, and banned all other fires completely. This would make it easier for cattle raisers to comply—no lengthy bureaucracy—and for the government to enforce (anonymous 1955). By 1955 to 1957, the idea of early dry season fires was widespread across Madagascar, and fires burned throughout the year, legally and illegally.[56]

It is important to note that even if administrators and scientists began to support some legal fires, they still did so within an overall rhetorical frame denouncing fires. Fires are bad, they argued, for they destroy the forests, degrade the land, and are generally a primitive technique; however, they are regrettably necessary for indigenous agropastoral use, until modern techniques can be installed. Thus the technical solution of wet season or early fires, which reduced

54. ANM IV.D.73/1/3.

55. Though he said that chemical means to treat for ticks are more effective.

56. Dez (1968) worried that the total number of fires actually increased.

environmental consequences yet renewed pastures, was very attractive. Unfortunately for its champions, early burns were never a perfect solution, as they were often very difficult to ignite in the rainy season, and their effects on pasture were not the same as those of dry season fires (Dez 1968).

A third new idea from the 1950s related to *tavy.* While the government had always pushed farmers to engage in irrigated rice cultivation, it had never helped them do so. In the middle of the decade, leaders realized that they needed to help the farmers to switch to irrigated rice. This is why, for the next three to four decades, Forest Service agents were often involved in building small irrigation dams and canals and in instructing farmers on different techniques for soil conservation and intensive cultivation (Dez 1968; see also chapter 3).

A final idea from the 1950s, though one that never took hold in Madagascar, was the use of preemptive controlled fires to reduce wildfire danger. In 1955, Paris issued Décret 55-582 on forest protection in the African territories. In addition to authorizing controlled fires for crop field clearance and pasture renewal, this decree stated "in regions where wildfires are dangerous, the burning of all savannas at the beginning of the dry season may be authorized." In addition, it authorized forest services to use early fires in and around state forests in order to protect them from wildfires.[57] Clearly this legislation disturbed the antifire hard-liners in Madagascar, especially within the Forest Service; the decree was not published in the *Journal Officiel* for two years.

Arrêté 25-SE/FOR/CG of 1957 finally detailed the application of the 1955 colonial decree to Madagascar and represented a meeting of the old ideas with the new. Bound by the enlightened frame of the 1955 decree, the authors sought to tighten it with regards to fire where possible. The *arrêté* repealed article 36 of the 1930 decree and replaced it with twenty-seven new articles. It stated that authorizations were necessary for *all* fires, specified the precautions to take, and created the possibility for provincial officers to ban all fires if necessary. Finally, the new law specified that *collectivités* could be held collectively responsible only for infractions within their legal "territory," not an indefinite "neighborhood" as before.

These new rules about fires had little chance to be used, for Madagascar regained its independence three years later, on 26 June 1960, and within four months the new country had written its own laws regarding fire.[58] Ordon-

57. JOM 23 Feb. 1957.

58. Colonial-era legislation, e.g., the 1930 Forest Decree and the 1957 *arrêté,* has never been officially repealed. The 1960 *ordonnances* state specifically that their purpose is chiefly to "codify" the previous texts. Thus multiple layers of legislation govern fire (see appendix 2).

nance 60-127 on forest clearing and vegetation fires differed from colonial laws in two significant ways. First, it contained no general ban on fires, even as it prohibited or restricted many specific types of fire use. Second, the new law was the first to introduce clear categories of fire. It distinguished among four types of fire and specified regulations for each:

1. *Défrichements* (forest clearings or *tavy*), defined as the cutting of woody vegetation to cultivate crops, whether followed by fire or not. Illegal in all state forests and elsewhere had to be authorized by a forest agent, only on flat lands or on the lower thirds of hills.

2. *Feux de culture et de nettoiement* (crop and cleaning fires), which included burning fallow permanent crop fields before planting and burning field edges or other places to "clean" them. These fires were allowed at all times outside of forests.

3. *Feux de pâturage* (pasture fires), which are used to renew grasses in pasture zones officially delimited by the government. These fires were allowed during the rainy season (dates specified in Arrêté 058 of 1961) and could exceptionally be authorized at other times.

4. *Feux sauvages* (wildfires), a completely new category defined as fires burning without control or limits, in any type of vegetation, without any economic utility. These fires, including escaped fires of the above categories, were always illegal.

The new law repeated some of the ideas of the 1955 French decree—i.e., that the Forest Service could use preventive fires around forests to protect them from wildfires, and that backfires could be used in fighting fires. As far as enforcement, the law raised the ante on prescribed sentences—infractions would be punished by a fine of 15,000 to 300,000 frs and/or six months to three years in prison—and it maintained the possibility of collective fines.

The law was accompanied by Ordonnance 60-128, which grouped in one text the procedures for forest agents to deal with all forestry and fire infractions, and sought to speed up this process. Soon thereafter, in early 1961, Décrets 61-078 and 61-079 fixed the precise procedures getting pasture fire and *tavy* authorizations and for enforcing the legislation. Finally, Décret 61-261 instituted bonuses for Forest Service agents based upon infractions enforced.

The fire legislation of independent Madagascar, created at the beginning of the First Republic under P. Tsiranana, was considerably more profire and realistic than colonial legislation, especially those of 1930 and 1937. As Bertrand and Sourdat (1998, 22) note, the new legislation had more of a "persuasive" than "repressive" connotation. However, for the rural farmer or herder, little had changed. The legislation was still framed within an official antifire ideology, even if it allowed certain agropastoral fires. For example, the preamble to Ordonnance 60-127 emphasized the environmental degradation resulting from fires, including soil erosion, forest loss, and catastrophic floods. For the farmer, the requirement to get authorizations for burning—often a complicated and time-consuming procedure, and not always assured of a positive response—was prohibitive. This impression was doubled by widespread official antifire rhetoric, often in relation to the national tree-planting effort, participation in which was mandatory.[59]

The effect of the legislation is hard to determine. Pasture fire and *tavy* authorizations were given in certain areas throughout this period, but the actual number of fires probably far outstripped the authorizations, as they likely had in every period. Forest Service policy was to allow one hectare of burning per head of cattle in pasture areas (Randrianarijaona 1983), which is potentially over 10 percent of the island's territory, yet most burners probably never sought permission. Nevertheless, in the field sites, the period of the First Republic (1960–1972) is remembered as a time when a system for authorizations existed, though was not necessarily strictly followed.[60] Fire enforcement (figure 6.2) continued, though at a slightly slower pace than in the late colonial period (Rakatonindrina 1989). Compared to today, however, enforcement in the 1960s was strict; foresters actively policed fire.[61] Many *procès-verbaux* were written—in both Tamatave and Fianarantsoa Province, there were 150 to 600 citations for illegal *tavy* per year, involving 400 to 1,000 ha each year (Ratovoson 1979).

59. Concern with trees and erosion was central to Tsiranana's government. In 1962, by Ordonnance 62-093, all males had to plant 100 seedlings a year or pay a tax, and the Forest Service was elevated to a Ministry of Forests and National Reforestation. Obligatory tree-planting continued through the end of the First Republic in 1972 (JOM 22 June 1963; DEF 1982; Olson 1984; Gade and Perkins-Belgram 1986).

60. Multiple interviews, Afotsara, Kilabé, and Tsimay, 1998–1999. In Afotsara, however, two people claimed that fire authorizations ended in 1960. Independence probably did have a significant effect on the relative power of the Forest Service. As Althabe (1969) notes, the removal of French colonial power destroyed, to some extent, the credibility of Malagasy rural officials, like foresters, whose job it was to perpetuate colonial regulations after Independence.

61. Interview, retiring forester, 11 Aug. 1998.

The Forest Service maintained a repressive antifire attitude throughout the period, and this attitude would soon reemerge into the public and administrative conscience. In fact, by 1966, Dez (1966, 1212) would refer to the "officially declared war against vegetation fires." However, the more moderate view—a continuation of the earlier "necessary evil" idea—also persisted, exemplified by the writings of Jacques Dez (1966, 1968), historical sociologist at the University of Madagascar.[62] Dez sternly criticized pasture fires for their environmental consequences. Yet he argued that repression was a problematic approach that could lead to resistance, revolt, and corruption. Instead, he stressed that one should understand and address the underlying logic of why the rural Malagasy light fires. They are not burning, he stressed, just out of ignorance or for fun. He supported a moderate solution, one that tolerated certain pasture fires (e.g., early season fires) and that allowed for some *tavy* while working with forest farmers to encourage a shift to intensive cropping.

1969–1983: HARSH WORDS AND POLITICAL CHAOS

Fire politics—in tandem with national politics—heated up in the following decade. The period from 1969 through the mid-1980s was characterized by the radical restrengthening of antifire politics within a turbulent political and economic context. The postcolonial First Republic fell apart in 1972, replaced by the Marxist-socialist Second Republic in 1975. By 1980, economic crises began to paralyze the nation. During these tumultuous years, the fire problem regained attention (table 1.1), numerous high-level commissions met to discuss the issue, laws were toughened, and multiple awareness campaigns were launched to stop the fires. Yet the political chaos of the time, together with unrealistically severe policies, made enforcement difficult.

From 1969 to 1971, fires were at the top of the national agenda and repressive fire politics were resparked.[63] President Tsiranana had always championed massive tree-planting efforts, out of a concern for soils and forests, requiring peasants to plant a certain number of trees each year. Drought years in 1967 and 1968 led to a particularly incendiary fire season in 1968, and Tsiranana became increasingly concerned. In October 1969, he flew over a

62. See also the work of rangeland specialists Granier and Serres (1966) and Gilibert et al. (1974), who recognized the important role of fire in extensive grazing regimes and wrote of a difficult compromise between the economic necessity for fire and its environmental consequences.

63. Unless otherwise noted, my discussion of the period 1969 to 1971 is based on ANM Vice-Presidence folders 831, 840, 844, 850, and 851.

region north of the capital and was distraught at the extent of fires. He is quoted as saying, "It is a great shame that all the living vegetation from Anjozorobe to Mahitsy is black, charred, destroyed." He wondered publicly why anybody would burn the forests and hills—the wealth of the nation.[64]

A week later, a high-level Interministerial Commission for the Fight against Bushfires met to propose antifire actions.[65] The problem was considered to be very severe, and initial reactions were severe in turn: the *conseil de cabinet* decided on 6 November 1969 to ban *all* fires including wet season and early dry season fires. The Interministerial Commission was more reasoned, discussing potential modifications to Ordonnance 60-127, such as requiring permission for all fires at any time and specifying the prohibition of fires in drought years.

In the same period, Madagascar became increasingly concerned with questions of nature conservation. In 1969 and 1970, the government prepared to host an international conference on natural resource conservation, ratifying the Algiers Convention for African Nature Conservation (Loi 70-004). The conference, which took place from 7 to 11 October 1970, was cosponsored by IUCN (the World Conservation Union) with help from WWF, UNESCO, FAO, ORSTOM, and the Paris Museum of Natural History. It emphasized the urgency of the island's environmental problems. Calvin Tsiebo, then vice president of the republic, stated in his opening remarks, "Unfortunately, our incomparable natural heritage, this unique natural capital, is gravely endangered. According to the specialists, few areas of the world suffer from such grand and rapid degradation [as Madagascar]" (IUCN 1972, 29). The conference was a milestone for Malagasy conservation; it put the issues on the front page of the newspaper. In the conference proceedings, scientist after scientist condemned the degradation caused by *tavy* and pasture fires. The conference resolutions blamed *tavy* and pasture fires for the degradation of the island's natural resources and recommended that the Malagasy government seek to stop these practices (IUCN 1972; Kull 1996).

Three new themes came out of the discussions of 1969–1971: the principle of public awareness, the idea of community-level fire management, and technical, machine-based fire fighting. First, the primary approach that was taken—after legislative changes were abandoned—was public awareness. Perhaps inspired by the ideas of Dez (1966), an antifire propaganda

64. "Fentriben'ny firenena ny dorotanety," *Vaovao*, 31 Oct. 1969, no. 672, p. 3.

65. The attendees at the commission's meetings included the president's counselor, a colonel from the army, the ministers of the interior, communications, justice, finance, and agriculture, the head of the Gendarmerie, and the head of the Forest Service.

campaign sought to appeal to peasant logic. This campaign, costing over 2,400,000 MGF in 1970, included television and radio spots, posters, slogans,[66] brochures, and newspaper articles. Materials were delivered free of charge to thousands of *communes rurales* by the national air, rail, and road transport companies. The annual "week of the tree" was replaced in 1970 by the "week of the fight against bushfires."

Second, the discussions brought to the forefront the idea of community-level responsibility for fire and forest management. The Forest Service had already, since the mid-1960s, supported the creation of local-level antifire committees (comités de défense contre les feux). In December 1969, the National Assembly proposed to the government that *fokonolona*-level conventions, called *dina*, be used to fight fires. *Dina* are traditional agreements and rules used by the *fokonolona*, or traditional community groups.[67] The government rejected this proposition as unnecessary, for the legislation of 1960 and 1961 already made possible the use of *dina* for fire management. In 1971, however, the government again proposed to legislate the creation of local antifire committees and *dina* against fires, but this decree was never approved. Nonetheless, Forest Service statistics show that—at least on paper—over 5,000 "antifire committees" existed in 1971.[68] In a similar vein, in 1978 the government proposed to decentralize forest control to local communities.[69]

Third, this epoch inaugurated the first mention of Western-style fire fighting, based on heavy machinery. The Interministerial Commission discussed using the machinery of the highway department to create firebreaks. People in Afotsara, in fact, remember this era, when bulldozers from a nearby

66. Posters with these slogans still hang in some Forest Service offices. Slogans echo the era of Smokey Bear in the U.S., and included "Mampidy-doza ny afo" (fire is dangerous), "Izay mandoro tanety mandoro tanindrazana" (those who burn the hills burn the ancestral homeland), "Ho simba sy ho hoalo ny vohitra tia doro doro" (the village that loves to burn will be broken), "Ry mpitsangatsangana, ry mpianenitra, ny tsy fitandremanao mety hiteraka dorotanety mambotry firenena" (dear tourist, dear traveler, if you are not careful you may cause a hill fire stunting the growth of the nation), "Aza fohy fisainana toy ilaikamo ka doro-kijana main-tany no iopi-an'omby" (don't be short-minded like those lazy ones who use pasture fires to raise cattle). RDM (1980) adds, among others, "Asa ratsin'ny kamo ny doro-tanety" (hill-fires are the bad work of the lazy) and "izay tia firenena tsy mandoro tanety" (those who love the country don't burn the hills).

67. See chapter 1 for further discussion of *fokonolona*, and chapter 8 for further discussion of *dina*.

68. DEF Rapport Annuel, 1971.

69. Interviews, Afotsara, 5 Aug. and 8 Nov. 1998; Olson (1984); Gezon and Freed (1999). While many refer to this policy, it does not appear in the legislative record, and may never have been formalized.

quarry would scratch firebreaks across the hills. By the late 1970s, the Forest Service owned several fire trucks, pumps, and hoses (RDM 1980).

The antifire environmental fervor of 1969–1971 ran headlong into the revolutionary politics of the following years. In the years 1972 to 1975, power struggles, riots, and a presidential assassination occurred. The postcolonial government of Philibert Tsiranana, which never really let go of its close ties with France, was quickly ousted, and at the end of three turbulent years under Gabriel Ramanantsoa, Didier Ratsiraka emerged as the new leader. Thus began the Second Republic, committed to Marxist-socialist policies and isolationism. Westerners were no longer welcome, and the international conservation initiatives spurred by the 1970 conference stalled.

Despite the regime changes, the antifire received wisdom persisted strongly, even intensified—perhaps as a means for the increasingly dictatorial state to assert its control. The 1970s were marked by increasingly harsh approaches to fire enforcement.[70] Two *ordonnances* passed in 1972 sought to strengthen fire prosecution: number 72-023 removed the possibility of the use of attenuating circumstances in trial, and number 72-039 sped up the process of enforcement "in order to ensure rapid and exemplary repression." Yet these new rules made no difference; Ratsiraka's new government claimed that the number of fires was still increasing. It then got even stricter and took additional repressive measures in order to stop the fires that "endanger the national forest heritage and threaten the economic infrastructure of the State": Ordonnance 75-028 raised the prison sentence for illegal fires to five to ten years. A year later, Ratsiraka's increasingly heavy-handed regime promulgated Ordonnance 76-019, which created the Tribunal Economique Spécial in order to rapidly prosecute economic infractions, including illegal fires. Soon thereafter, Ordonnance 76-030 put in place further measures to punish illegal fires, including requiring the active participation of local communities in capturing suspects. Sentences were again raised, to ten to twenty years or even lifetime forced labor, in the case of malicious fires. The final step in the increasingly repressive fire regime was Ordonnance 77-068, which placed illegal fires under the jurisdiction of the Tribunal Criminel Spécial, which could theoretically give the death sentence. These brutal rules hardly helped, as cattle raisers and forest farmers felt attacked by the unfair legislation (Cori and Trama 1979, 58), and as the administration largely failed to enforce them (Fanony 1989).[71]

70. Discussion of the policies from 1972 to 1977 is based on the texts found in the *Journal Officiel* (see appendix 2) and on Andriamampionona (1992).

71. According to a retired forester (interview, 21 Aug. 2001) after 1975 politicians turned a blind eye to fire regulation, and even encouraged an expansion of *tavy.*

Antifire programs and reports also proliferated in the Second Republic. From 1976 to 1979, the government spent 40 to 90 million MGF per year on an antifire program, Opération Lutte Contre les Feux de Brousse. This program created antifire brigades at important forestry stations, including fire trucks and pumps. It sponsored a new barrage of antifire propaganda and awareness efforts, printing antifire brochures and posters. Finally, it sponsored field tours of joint military, gendarme, and Forest Service teams in order to protect state forest plantations. This program clearly served largely to protect Forest Service–run plantations and forestry stations, and ignored the fire question elsewhere (RDM 1980).

In a speech welcoming the New Year of 1980, President Ratsiraka spoke of the pressing need to fight without fail for the complete eradication of *feux de brousse* (RDM 1980, 3). As a result, another commission was convened on the subject: the Interministerial Fire Study Group. Citing the disastrous environmental effects of fire, their report (RDM 1980) suggested that enforcement efforts had been insufficient. It criticized the apathy, laziness, and pyromania of the rural populations, but grudgingly recognized the reasons behind the purposeful uses of fire. The report referred to pasture fires as a "necessary evil" (RDM 1980, 108) and recommended the use of counterseason fires as well as intensive pasture management (mowing, seeding, rotation, and haying) to reduce the need for burning. In order to stop *tavy* fires, the report suggested public awareness campaigns, building small dams to encourage irrigated agriculture, and stricter repression. For all kinds of fire, the report recommended more antifire propaganda and awareness, enhanced repression and enforcement, tree planting, and community-based fire prevention (antifire committees, *dina* against fire, and designated fire watchers).

Many of the study group's recommendations were to be implemented by Opération Danga, which began in 1980. Its goals were to teach rural residents about the dangers of fires, help create antifire committees in each *fokontany*, launch the creation of intensive pastures, repress illegal fires, and enforce antifire laws. The operation, which lasted until 1987, involved the army and the gendarmes, as well as the Forest Service, and thus served to give the rural population the impression of intimidation and antifire repression. As a retired Forest Service agent said, "The people were scared of the gendarmes, so in those regions where they passed there was less fire."[72] Despite the grand goals, however, the actual activities of Opération Danga were minimal.

72. Interview, 20 May 1999.

Also arising out of the study group's recommendations was Décret 82-313, which legislated the establishment of fire rotation plans for official pasture zones in each *fivondronana.* It unrealistically required that a twenty-meter firebreak be cut around each fire zone and that each burned zone be seeded with high-value fodder species. This decree was rarely applied, both because of these requirements and because of the national economic crisis.

The idealistic socialism of the Second Republic soon encountered political and financial obstacles that compromised many of the antifire efforts on the ground. In 1977, Ratsiraka set into motion an aggressive program of investment, based on foreign borrowing and increased centralization and nationalization of the economy. As a result, by 1980 the economy had collapsed, debt mounted to over US$1 billion, and Ratsiraka sought relief from the International Monetary Fund. As the 1980s advanced, the economy stagnated, communications and transport systems degraded, and by 1985–1986, the country was in a deep economic crisis with a sharp reduction in living standards. Ratsiraka's popularity also declined, and political violence increased (Brown 1995).

As a result of the political and economic turmoil, the attention of most state agents turned away from fire. The state had other, conflicting priorities: beef and rice for the politically important urban market was crucial (Olson 1984), so rice-producing *tavy* and beef-producing pasture fires may have been tolerated. Thus, despite the ultrarepressive laws and the antifire programs, fire enforcement became more and more inconsistent. The enforcement of illegal fires dropped noticeably (Rakatonindrina 1989; see also figure 6.2). From an average of over 1,000 infractions per year between 1969 and 1975, the mid-1980s only saw 200 to 300 infractions per year. At the same time, district Forest Service offices continued to authorize pasture fires as demanded by cattle raisers. As table 7.3 shows, authorizations were given for 150,000 to 860,000 ha of fire each year, or between 0.3 and 1.5 percent of national territory, in the 1970s, and this system persisted to some extent into the 1980s. Politicians bent to the wishes of their constituencies (chapter 6), and repression was in name only.

1984–1995: BOOMING ENVIRONMENTALISM

After a period of inward-looking isolationism, crisis-ridden Madagascar reopened its doors to the outside world in the mid-1980s. This marked a new period in fire politics, lasting a decade, shaped strongly by two overriding factors: a boom in environmental concerns and activities, and a financially

TABLE 7.3. Pasture fires authorized by the Forest Service, 1971–1977, in hectares.

	Province						
Year	Antsiranana	Fianarantsoa	Mahajanga	Toamasina	Antananarivo	Toliary	TOTAL
1971	847	520,932	179	11,197	5,591	na	>538,746
1972	21	592,006	14	30,978	957	234,117	858,093
1973	4,297	222,783	na	9,203	2,617	114,244	>353,144
1974	5,140	138,675	na	8,873	2,037	na	>154,725
1975	4,564	246,041	na	7,636	1,489	44,495	>304,225
1976	na	209,234	na	6,625	na	196,283	>412,142
1977	2,220	112,964	na	13,514	8,888	146,238	>283,824

na = data not available.
Source: DEF annual reports.

and politically paralyzed government. The conservation boom resulted from a surge in global environmentalism coinciding with Madagascar's rapprochement with the world; it was also shaped by the conditionalities attached to foreign aid (Kull 1996). International environmental actors gained prominent political influence, and environmental funding ballooned. Fires were denounced with renewed vigor, while government capabilities declined.

In the mid-1980s, the momentum for conservation that had been lost since the 1970 conference resumed. In 1984, the country adopted a National Strategy for Conservation and Development, signed by every government minister. Madagascar was the first major country in tropical Africa to adopt such a strategy, called for by IUCN/UNEP/WWF (1980). A year later, a second international conference on the conservation of natural resources was held in Antananarivo. This important event discussed the implementation of the strategy and showcased the Malagasy environment. It is remembered by many as the moment when Prince Philip, the international president of WWF, confronted President Ratsiraka with the statement, "Your nation is committing environmental suicide." Several programs were initiated as a result of the conference and funded by the World Bank, bilateral donors (especially the U.S. and Switzerland), WWF, and UNESCO: soil conservation and forest management programs, biodiversity conservation, and environmental awareness, including a massive school-based environmental education program (Kull 1996).

In 1988, the government convened yet another interministerial meeting on the subject of *feux de brousse*. Neither as ambitious nor as creative as its previous incarnations, its report largely repeats the recommendations of 1980. As if unaware of the century of previous efforts, the report states that "it is time that the country think seriously about the [fire] problem" (RDM 1988, 3).

At the same time, the government was busy working with the World Bank, conservation organizations, and bilateral donors to establish an Environmental Action Plan. This fifteen- to twenty-year plan included a wide variety of programs designed to protect biodiversity, stop environmental degradation, and establish the institutional structures necessary for environmental monitoring and management (World Bank 1988). Funded by over $100 million for its initial years, the plan was put into law in 1990 as Loi 90-033, the Charte de l'Environnement (RDM 1990).

The situation on the ground in this period was a strong contrast to the grand ideas, intentions, and programs launched at the national and international level. The economic crisis of the mid-1980s had paralyzed the nation, especially as International Monetary Fund austerity measures reduced government services. Political unrest in 1991 and 1992, including a six-month general strike, led to the election of opposition leader Albert Zafy as president and the beginning of a hopeful, but poorly executed, reformist Third Republic. A stable system of law no longer existed, and the government lacked the means and finances to execute its policies.

As far as fires were concerned, the rules had not changed, yet fires burned more or less uncontrolled and enforcement lagged. As unpaid or striking government agents became less motivated to do their work, the rural population largely ignored the requirements and stopped seeking authorizations. For those who did seek authorizations, the system became quite loose or corrupted: sometimes authorizations were given after pastures had already been burned (without the knowledge of the Forest Service, which did not have the means to visit all regions). In most cases, date restrictions and the firebreak requirement were not enforced. Some villages used one authorization to cover several fires, or took advantage of authorized fires in neighboring villages.[73] The budget of the Forest Service declined at the same time

73. Interviews, Forest Service officers, 2 July and 12 Aug. 1998 and 28 June 1999.

as the Malagasy environment became a pressing national and international concern. As a result, the Forest Service began to refuse to give fire authorizations,[74] trying to show the National Assembly that it had done something about the fire problem.[75]

Austerity measures and government crisis also affected fire enforcement. Around 1990, the funds for fire enforcement bonuses to Forest Service agents ran dry,[76] and this is marked directly by a drop in enforcement statistics. In the Circonscription Forestière of Antsirabe, *procès-verbaux* for fire dropped from an average of thirty per year in the 1980s to under five; nationally, the number of forestry infractions dropped from 400 to 500 per year to 200 to 300 per year (figures 6.2 and 6.3).[77] Even when the Forest Service attempted to prosecute fire infractions, it was met with resistance at the level of the courts. Forest Service recommendations regarding forest crimes were often not upheld by the courts, which routinely reduced the penalty or dismissed cases. As a result, the Forest Service often settled cases out of court, sometimes for bribes, resulting in monetary fines or tree-planting requirements.[78] Similarly, when a fire case involved the gendarmes, they often dropped it in favor of more important crimes (as in the story in the introduction to part 1; see also Andriamampionona 1992).

Due to the mounting environmental concerns in Madagascar, 1994–1995 was declared "Year of the Fight against Bushfires"; it marked the end of the first decade of the conservation boom, a period characterized by the contradiction of a national-level proenvironment ideals, with political disorganization, anarchy, and uncontrolled burning. The Year of the Fight against Bushfires led to renewed antifire propaganda reaching the rural populations;

74. The fire authorization system had de facto ended in most of the field sites between the end of the First Republic and the early Second Republic, due to government paralysis and loosening enforcement, thereby reducing the incentive for villagers to seek permission. Authorizations for pasture fire and *tavy* were no longer *formally* given around 1990 in many areas (for example, Antsirabe in 1985; Ambositra in 1989–1990; Tsiroanomandidy in 1992; Andohahela in 1992; Toliary in 1990). Fires were allowed only in the most wide-open, vast pasture areas, e.g., west of the Itremo mountains, or in the southwest around Ihosy or Betroka (multiple interviews, Afotsara, Behazo, Kilabé, 1998–1999, and Forest Service officers, 1998–2001; see also Schnyder and Serretti 1997; Durbin 1994).

75. Interview, academic forester, 8 Apr. 1999.

76. Interview, Forest Service officer, 11 Aug. 1998.

77. The Cantonnement Forestier of Toliary had perhaps a dozen *procès-verbaux* per year until 1992; since then, its staff has stopped going on enforcement tours, due to lack of money and transport, and there are perhaps only one or two cases per year (interview, 6 Aug. 2001).

78. In 1993 there were seventy-seven forestry infractions in Madagascar; seventy-six were settled out of court (DEF Annual Report, 1993).

reports discussed jump-starting the old village-level antifire committees.[79] After 1995, however, the government and foreign donors opened a new chapter in the history of fire politics, seeking to reconcile environmental goals and poorly functioning regulations through the new approach of community-based resource management. We will turn to this current situation in chapter 8.

79. E.g., DEF *Rapport Annuel*, 1994.

8 [EMPOWERING RURAL FIRE SETTERS

Towards Community-Based Fire Management

Persons caught intentionally burning the hills, with lawful proof, will be fined 10,000 Ariary and will be brought to the authorities.

DECISION SIGNED BY MAYOR AND COUNCILORS,
COMMUNE RURALE OF AFOTSARA, 1996

A communal decision about fire? No, *tsa vao reko,* I haven't heard of any.

SEVEN DIFFERENT VILLAGERS, INCLUDING
FOKONTANY PRESIDENT, AFOTSARA, 1998

Yes, there is a communal decision about fire. But the people don't bother with it; because it wasn't done publicly, it isn't really legal.

VILLAGER, AFOTSARA, 25 NOVEMBER 1998

IN THE MID-1990S, fire politics in Madagascar took a new turn. In a series of workshops, reports, and new laws, the government, together with a host of international environmental and development organizations, pushed for a new policy of community-based renewable natural resource management. To be applied to forests, pastures, wildlife, and water, thus by extension to fire, this policy spoke of a new nonrepressive era. The policy was to promote better resource management through local-level control, leading to better environmental stewardship and thus vastly reduced fires. However, the policy remains abstract paperwork for much of Madagascar. Fires burn with their regular wildness across the island, on some days causing teary eyes and sneezing in the cities and provoking the usual denunciations of fire as the ruin of the country.

The new policy, known as GELOSE (for *gestion locale securisée,* or secure local management), is the Malagasy manifestation of the recent global trend of enfranchising local communities in matters of environment and development. Malagasy interest in community-based natural resource management (CBNRM) rose out of the growing international consensus on the importance of local and community participation in development and conservation (e.g., Adams and McShane 1992; Ghai and Vivian 1992; Wells and Brandon 1993; Western and Wright 1994). This consensus was based on ideas of equity and efficiency. CBNRM sought to undo the mistakes of past top-down conservation efforts out of a sense of justice and out of political necessity. It also represented a realist strategy to move towards more effective and sustainable conservation, and beyond failed coercive conservation measures that led to noncompliance, resistance, and unsustainable programs. This new consensus was deeply felt in Madagascar, informing conservation projects across the island (Durbin and Ralambo 1994; Peters 1998; Uphoff and Langholz 1998; Bertrand 1999; Gezon and Freed 1999; Hackel 1999; Marcus and Kull 1999; Rabetaliana and Schachenmann 1999; Wollenberg et al. 2000; Richard and Dewar 2001). Environment and development agencies saw in CBNRM the key strategy—if not the "only possibility" (Rajaonson et al. 1995, 6)—to solve environmental problems in an effective and equitable manner.

Not all forms of participatory conservation and CBNRM are equal. Different communities receive varying powers, rights, and responsibilities. Also, these powers, rights, and responsibilities are sometimes held by village elites, sometimes by the whole community, and sometimes by a faction within the community. Ribot (1999) proposes that successful decentralized management requires the *enfranchising* of local populations (see also Robbins 1998). To be enfranchised, real decision-making power must be vested in legitimate community institutions. Real decision-making power comes through *empowerment,* which occurs when the decentralization of resource management gives not just responsibilities, but also rights, to local communities. The *legitimacy* of community institutions is rooted in the notion of popular acceptance—that the community members accept that leaders be granted certain rights and responsibilities, accept that rules are fair and just, and accept that certain institutions are the proper venues for activities such as management or dispute settlement. This acceptance is more likely if leaders and institutions are accountable to local people (Agrawal and Ribot 1999; Ribot 1999). Without empowerment and legitimacy, CBNRM would serve only to expand top-down power and state control, or to give power to some at the expense of others, risking ineffectiveness due to noncompliance and resistance.

This chapter brings Madagascar's fire politics up to date by using the above ideas. In particular, it demonstrates how two factors challenge the enfranchising of rural resource users. First, dominant ideas about environmental change and management, such as the antifire received wisdom, can affect the implementation of CBNRM by creating obstacles to the empowerment of locals and by threatening the locally perceived legitimacy of policies. Second, attaining legitimate CBNRM policies and institutions is complicated by the limited accountability of local government institutions and by the tensions and negotiations inherent in the application of the term "community." In the Malagasy case, these factors deeply influence the success of GELOSE for fire management. As a result of the persistent antifire received wisdom, the state requires locals to comply with antifire legislation that carries little legitimacy and hesitates to empower peasants in the sector of fire management. In addition, weak local governance institutions and problems with establishing legitimate community institutions plague efforts to implement the new policy.

Planned interventions in development, conservation, and governance rarely go exactly according to plan; they do not have direct, predictable effects upon society. Yet they still "do something" in the sense of Ferguson (1994, xv). For example, the planned colonial appropriation of forests in Bengal was not a straightforward process. Because it encountered regional ecological diversity, regional political and historical contrasts, and tensions within the colonial state over conflicting purposes, the result was not a uniform system of state-controlled forests, but a "rich and systematic variation in the regime of restrictions" (Sivaramakrishnan 1997, 89). Scott (1998) proposes that the near-inevitability of the derailment of large-scale, modernistic planning is a result of the simplification implicit in the process, which does away with the local, situated knowledges and practices that, it turns out, are key to sustainable and efficient functioning. Powerful plans and ideas are constantly being subverted by situated practices; humans are creative and constantly carving new complexities into grand simplifying ideas. In this chapter, we will see that this is exactly what is happening to GELOSE.

THE MID-1990S: FROM IMPASSE TO PROPOSED SOLUTION

The Impasse

Fire politics in the early 1990s was at an impasse. As chapter 7 notes, the conservation boom beginning in the mid-1980s refocused attention yet again on fires and led to increasingly vocal calls for better controls on fires. The state continued and even intensified its efforts to criminalize fires with numerous pronouncements and an end to fire authorizations over much of the island. Leaders and policy makers figured that stricter rules would have the straightforward result of reducing fires. But the rules did not succeed. The peasants, rooted in their livelihood-based need to manage their lands, continued to resist, taking advantage of state weaknesses and the special nature of fire. As Klooster (2000, 284) observes, "When rural society develops a culture of resistance, actively protecting members who commit 'illegal' acts, the resource control system needs even more coercion to meet its goals. It becomes increasingly oppressive and ineffective." In the 1980s and 1990s, the debt crisis, the severe economic crisis (especially in 1985–1986), the political crisis leading to the collapse of Ratsiraka's regime in 1992, and the strict fiscal austerity imposed by the International Monetary Fund together weakened the reach of the state. Not every arm of the diverse state considered fire repression a priority or even an appropriate goal, and the system broke down. This contradiction—a weak, divided state pushing strong antifire policy—together with the biophysical character of fire, gave the peasants the opening to continue burning. As chapter 6 noted, this perpetuated the "fire problem," the impasse over fire, to this day.

The Solution: GELOSE

In the mid-1990s, analysts working to improve natural resource conservation on the island recognized the impasse over fire, noting that a century of repressive politics had failed.[1] In addition, they recognized the growing international consensus on the importance of local participation in conservation and of the decentralization of state power. As a result, GELOSE emerged. This policy was to bypass the problem of state weakness and austerity by mobilizing local resources to accomplish the same tasks. It was to bypass the alarmist and antifire views of the conservation boom by adopting a more tolerant stance towards some burning practices. Finally, it was to eliminate the resistance inherent in repressive antifire regimes by making the peasants complicit in resource management, by empowering them to implement, enforce, and benefit from management rules (Bertrand 1994; Rajaon-

1. Key actors included the late Mamy Razafindrabe and Alain Bertrand, two sociologists steeped in the traditions of common property research, as well as conservationist Bienvenu Rajaonson (see Bertrand 1994, 1999; Leisz et al. 1994; Razafindrabe and Thomson 1994; Rajaonson et al. 1995; Razafindrabe et al. 1996; Rakotovao Andriankova et al. 1997; Razafindrabe and Rakotondrainibe 1998; Henkels 1999).

son et al. 1995). It was a fundamentally new proposal, for while the government had repeatedly pushed for communities to become more responsible in supporting antifire laws,[2] GELOSE instead revolved around the idea that community-based regulation could replace top-down control.

GELOSE grew out of a masterful 1995 proposal for the decentralization of fire and renewable resource management. Written by an interdisciplinary team guided by rural sociologist Alain Bertrand, the proposal (Rajaonson et al. 1995) self-consciously placed itself in opposition to a century of failed policies, presenting a nuanced argument for the devolution of fire management. It accepted some burning as a legitimate resource management tool, discussed the legal and political logic and logistics of decentralization, and laid out a detailed program for its application. Key elements of the proposal were quickly transformed into legislation, becoming Loi 96-025 on 10 September 1996. During 1997, GELOSE was incorporated into the new national forestry policy (Loi 97-017 and Décret 97-1200) and the second phase of the Malagasy Environmental Action Plan (PE2), a major coordinated effort by the Malagasy government, foreign donors, and international NGOs.

The GELOSE law of 1996 proposes the creation of three-way negotiated contracts between the state (represented by its technical branches such as the Forest Service), the *commune rurale,* and a group of residents or resource users known as a *communauté locale de base* (CLB, basic local community group). Each contract would govern the management of one or more renewable natural resources, including forests, rangelands, or lakes. Under these contracts, community groups would regulate resource use through *dina,* locally sanctioned rules.[3] The CLB's legal rights to the resource in question, formerly state owned, would be secured through a provisional property rights registration process (*sécurisation foncière relative,* SFR). The contract negotiations would be coordinated by an environmental mediator, a person hired out of a state-certified pool of specially trained professionals. The process established for creating each GELOSE contract was quite long, with twenty-two

2. In particular in the 1960s and 1970s—see chapter 7 for details. Even earlier, Parrot (1924) floated the idea of putting highland forest islands into the control of the local *fokonolona,* positing that it would lead to better management.

3. *Dina* are rules and decisions at the local level defining acceptable behaviors and stipulating fines (Henkels 1999; Marcus 2000; Rakotovao Andriankova et al. 1997). Traditionally, *dina* are local socially guaranteed agreements; since 1960, *dina* have also had formal legal status. *Dina* have taken on particular significance in recent years, as conservation and development actors seize on the concept—perhaps somewhat romanticizing *dina* as timeless and traditional (Tsing 1999)—in order to implement participatory aspects of their programs.

steps.[4] The plan was to implement contracts in 400 rural communes during the five years of the PE2, and to all 1,100 communes by the tenth year.

It was only in the year 2000 that the GELOSE law received the first two installments of its long-overdue enabling legislation (*décrets d'application*). Despite the delay, GELOSE contracts (or contracts in the style of GELOSE) have been implemented in a few cases, often in conjunction with a foreign-sponsored project. Gauthier and Ravoavy (1997) surveyed twelve unofficial GELOSE contracts in 1997; Maldidier (2001) counted a dozen official GELOSE contracts by the end of 2000, including near the Andapa and Beforona case studies, and the process gathered some momentum. By 2003, there were over 200 official GELOSE contracts. Some of the troubles in implementing the plan, as well as the fact that most of the implementation occurred in forest management cases (not in fire management), are due to questions of empowerment, ideology, legitimacy, and community, discussed below.

EMPOWERMENT AND IDEOLOGY

The decentralization of resource management involves the transfer of rights and responsibilities to local communities. The character, and perhaps success, of CBNRM depends upon what mix of rights and responsibilities is decentralized. Such empowerment, of course, is profoundly political, and, as Cruikshank (1999, 2) has argued, is neither good nor bad but "contains the twin possibilities of domination and freedom."

On the one hand, if local levels only gain responsibilities, and few if any rights, then state power grows. The state penetrates further into society, and the decentralization process, often in spite of its best intentions, becomes an

4. The steps of the GELOSE process, according to an information sheet from the Office National de l'Environnement, 1998: (1) informational campaign in rural communities, (2) official request by the CLB at the communal office, (3) legal registration of the CLB as an association or NGO, (4) submission of the request to the commune, (5) posting and announcement of the request by the commune, (6) verification of the legality of the CLB, of its proximity to the resource, and of the legality of the proposed resource transfer, (7) communal decision by the end of the announcement period, (8) choice of the mediator, (9) contacting stakeholders (including government agencies), (10) creating a work plan, (11) mapping of the resource and the land surface, (12) establishment of requirements for technical studies, (13) preliminary technical studies, (14) creation of long-term goals, construction of scenarios, and definition of access and use rights, (15) evaluation of economic management tools and management structure, (16) additional technical studies, (17) performance of the SFR operation (a multistep process to establish resource tenure security for the CLB), (18) writing of first draft of contract and management plans, (19) creation of community *dina*, (20) verification by mediator of the termination of the preparatory process, (21) signatures, and (22) speech and ritualization of the contract (e.g., ceremony with cattle sacrifice).

expansion of bureaucratic power (Ferguson 1994). The state co-opts local labor and time to its purposes. Neumann (2001) describes such a process in Tanzanian wildlife management, whereby village "self-surveillance" is harnessed to accomplish central state goals. Ribot (1999) documents a similar process in the context of Sahelian forestry policies, whereby much of the decentralization process served to transfer only work, not power, to local levels, affording the state more microcontrol. Likewise, Gauld (2000) describes a case in the Philippines, where forest-based CBNRM policies are shaped by bureaucratic efforts to maintain centralized control. In such cases, decentralization will most likely be subverted, encountering resistance or spawning inconsistencies and corruption.

On the other hand, if local communities gain rights commensurate with their responsibilities, they are empowered. Such rights, or powers, can include adjudication, decision making, access to resources, control over finances, and the ability to tax (Ribot 1999). This expansion of real local power is the theoretical aim of CBNRM. It would enable more equitable, efficient, and institutionally sustainable management of renewable natural resources, depending how the local issues of power are navigated (see next section).

In the case of GELOSE—at least for the management of vegetation fires—not enough rights are transferred. As in many CBNRM efforts elsewhere (e.g., Gauld 2000; Sundar 2000), the state still has the deciding vote. This is because Loi 96-025 requires that GELOSE contracts and resource management *dina* conform to existing legislation and rules. Thus, a community's decision to burn a pasture must still fall within the stipulations of restrictive 1960s vegetation fire laws, and must (if during the dry season) still receive an "exceptional" authorization from the Forest Service. A *tavy* clearing would still require formal Forest Service authorization. In this case, no rights have been transferred. For GELOSE to seriously change fire policy and empower communities, the entire suite of fire and forestry laws must be reformed as well.

A crucial factor that may block the empowerment of local communities is the persistence of strongly held ideas that cast doubt upon local natural resource management techniques. As previous chapters noted, a strong antifire received wisdom persists in Madagascar, especially in the popular urban imagination,[5] among associated policy makers, and in external agencies. In recent years, the received wisdom has been reinforced by the strong environ-

5. The antifire received wisdom is ubiquitous in the news (table 1.1, figure 2.2), in speeches, even in music. Each time I explained my research subject to fellow passengers on city-to-city *taxi-brousses*, to taxi drivers, or to urban residents, the reaction was of the importance of the problem of fire and the disaster it was for the country. Nobody spoke of fire's usefulness to the peasants.

mental interests of foreign donors.[6] As a result, when the architects behind GELOSE proposed CBNRM for fire and other renewable natural resources, they found their original intentions co-opted by the antifire received wisdom in three ways.

First, the received wisdom has, over the past century, created a legacy of rules tightly restricting the use of fire. As noted above, GELOSE contracts must conform to previously existing legislation. These rules—which restrict free pasture burning to the wet season and to villages gaining "exceptional" authorization during the dry season, and which restrict *tavy* fires to secondary forests pending Forest Service authorization—would conflict with real community-based management, setting what to the peasants are unacceptable limits on burning. The Forest Service would neither physically nor ideologically be prepared to authorize the tens or hundreds of thousands of small, opportunistically ignited, and not always strictly controlled fires that the peasants would desire.

Second, the received wisdom frames the ideological and discursive space within which GELOSE develops. An illustration comes from discussions I had in 1999 with two Forest Service officials regarding the use of fire as a tool to manage pine woodlots.[7] As chapter 3 notes, fire is a useful tool for managing certain species of pine to facilitate regeneration or to clear undergrowth. The foresters knew of this use of fire, they had read about it in a forestry journal, and they recognized that the peasants harnessed these techniques. Yet they said that the Forest Service could not authorize such fires, for it would be setting a bad precedent. Thus, the received wisdom limits the range of possibilities.

Another example derives from when the Interior Ministry asked the *communes rurales* in 1996 to create *dina* governing fires, inspired by the ongoing discussions about CBNRM.[8] Mayors around the island complied; many of the study site communes had such *dina* dating to 1996.[9] In the case of Afotsara, the 1996 *dina* on fire states:

6. Strong foreign support of the antifire received wisdom is supplemented by local material interests that prop up the received wisdom: a government eager to secure foreign funds, or (at a smaller scale) the historical awarding of bonuses to foresters for enforcement, or the current use of antifire laws to extort bribes.

7. Interviews, 28 June and 2 July 1999.

8. A *note de service* reputedly came to the communes from the Interior Ministry (interview, Tsimay, 12 May 1999).

9. Such *dina* were created—among other places—in the Afotsara, Tsimay, and Behazo study sites, as well as around Arivonimamo (Razafintsalama and Gautschi 1999) and Ambatofinandrahana (interview, Forest Service officer, 28 June 1999). They came after a rash of workshops on the implementation of the GELOSE ideas.

- All adults who see a fire must extinguish it.
- It is the responsibility of the *Quartier Mobile* to blow the bull's horn to call the *fokonolona* to fight a fire.
- Not fighting fires is punishable by a fine of 5,000 Ariary.
- If this is not paid within ten days to the commune, the delinquent must contribute twenty workdays.
- If this requirement is not met, the delinquent will be reported to higher authorities.
- Persons proven to have voluntarily burnt the hills will be fined 10,000 Ariary and will be brought to the authorities; if the fire was accidental, the fine will be 5,000 Ariary.[10]

This *dina* was not subject to public discussion; most peasants—including the new *fokontany* president—claimed not to know of it. Almost none of these 1996 *dina* included viable enforcement mechanisms, and thus enforcement by communal officials was rare. They were stricter than the national legislation, made no concessions to peasant agropastoral needs, and were accordingly ignored by the peasants. Because of the power, dominance, and framing effect of the received wisdom, however, they could hardly have been otherwise.

Yet another example of the ideological framing effect of the received wisdom is the list of specific resources included in the GELOSE mandate. As proposed by Rajaonson et al. (1995), GELOSE was centered on vegetation fires (largely pasture fires and *tavy*). However, the text of the proposed applicatory decree describing the renewable natural resources to which GELOSE could be applied includes forests, forest products, phytogenetic resources, reforestations, watersheds and their biologic resources, marine resources, lakes, marshes, tidal areas, and springs, *but not pastures and grasslands.*[11] Perhaps, pressed by the received wisdom, lawmakers shied away from allowing managed grassland fires. As a result, in practice, GELOSE was slow to approach pasture management. By 2003, only a few recently-inaugurated GELOSE contracts focused on pasture management, in particular at Ambo-

10. From the communal offices of Afotsara; rules copied from the notebook titled *Fitanana antsoratra ny fanapahan-kevitry ny filankevitra kaominaly* (Notebook for communal decisions). Rules created 2 Feb. 1996 and amended 18 Sep. 1996. 5000A equals 25,000 MGF, or approximately US$4 (1998).

11. Projet d'Arrêté relatif aux resources naturelles renouvelables pouvant faire l'objet de transfert, July 1998. This proposed decree, however, reaffirms article 2 of GELOSE (Loi 96-025), which does specify rangelands.

hiby near Tsiroanomandidy. This may now change, as the proposed third phase of the Environmental Action Plan (PE3) includes a goal of transferring the management of 500,000 ha of pastures to local communities (RDM 2002).[12]

Third, the antifire received wisdom stifles communication. Due to the received wisdom, the state has long criminalized and denounced vegetation burning, showing little understanding of its usefulness and sometimes even confusing pasture fires with slash-and-burn forest cultivation. After 100 years of repression and antifire rhetoric, everyone knows that the government does not like burning; everyone can recite the destructive effects of fire, even in the most remote corners of the countryside. In Afotsara, for example, farmers frequently told me the disadvantages of fire, reciting them like a list off a government radio announcement: fires cause erosion, degrade the pastures, destroy the forest, and dry up springs.[13] The introduction to part 1 told of how children lit fires while mockingly stating, "Fires destroy the environment." They knew that outsiders like me stood for fire repression, but they also wanted to burn (and the latter factor overrode their hesitations). As a result of the antifire received wisdom, peasants lose confidence in the state, and communication stops (Mainguet 1997; Kull 1999). When confronted by the state over fire, farmers would rather politely blame passersby, repeat antifire slogans, and nod in agreement than explain in detail why they break the law. As Fairhead and Leach (1996, 116–117) note,

> The interface between villagers and external agencies, rooted in the colonial encounter and developed over antagonistic circumstances, has rendered the expression of local environmental experiences highly problematic. Faced by direct questions about deforestation couched in the environmental services' terms, villagers tend to confirm outsiders' opinions which are, by now, long familiar to them. Agreeing (or not denying) visitors' views can be a polite way of coping with extractive or repressive encounters, or to maintain good relations with authoritative outsiders who may bring as yet unknown benefits; a school, road or advantageous recognition to the village, for example.

In this way, received wisdoms and legacies of repressive policies can generate what Klooster (2000, 284) has called a "baggage of resistance" including noncommunication and noncompliance. Local points of view only slowly filter

12. Yet the PE3 proposal does not mention how community pasture management will address fire.

13. Not just in rural areas, but even more so in the cities. See note 5.

up to national decision makers. Top-down communication is hampered as well: in Madagascar, most peasants believe that the government forbids *all* fires except those in existing crop fields; few know that the legislation allows them to burn pastures freely from December to April or that dry season pasture fires could officially be authorized, at least according to the legislation. As a result, the success of GELOSE—which must accept some burning in order to succeed, never mind to foster dialogue between the contractual parties—is threatened. While peasants may sign a GELOSE contract in the hopes of securing outside funding, they will continue to burn as before, out of sight, "accidentally," and anonymously.

LEGITIMACY AND "COMMUNITY"

CBNRM programs engage with extant local institutions (power structures, forms of governance, and rules) and are likely to establish new institutions. The success of CBNRM depends upon the legitimacy of these institutions to the actors involved. Yet legitimacy is not simple, as institutional formation, empowerment, and legitimation occur within the complex arenas of regional social relations and culture. The CBNRM process can never be a surgically perfect installation of new systems. Instead, it will be a constantly negotiated and subverted process that leads to transformations in power and social relations that may or may not be intended or useful, e.g., the capture of CBNRM benefits by village elites, or the shifting dominance of local social groupings. Intracommunity relations are still relations of power, and as such are continually renegotiated (Cruikshank 1999, 18). Below, I highlight the problems of achieving legitimacy in both rules and institutions, giving special attention to the problems of "community" institutions.

The key to legitimacy, according to Ribot (1999), is locally accountable representation. Leaders, rules, and institutions must be accountable to the people. In Madagascar, what is legal is often not legitimate; what is legitimate is often not legal (Rakotovao Andriankova et al. 1997). The restrictive national fire rules, for example, are legal but far from legitimate or just in the minds of the peasants. Likewise, legitimate local fire management systems—e.g., informal, opportunistic burning strategies negotiated by the co-users of a pasture (see case studies below)—are highly illegal in the eyes of the state due to the antifire received wisdom.

Current Malagasy local government institutions have a somewhat ambiguous role with little accountability. Popularly elected communal mayors must both carry out state mandates at the local level and communicate local needs upwards; more often than not they apply a watered-down version of the state mandate and rarely communicate local needs upwards. In addition, the delineation of responsibilities between the mayor and the state-appointed *délégué* is often vague.[14] As such, the legitimacy of the communal government is shaky.

Decision making in rural communes lacks capacity and finances, is chronically inefficient, and is hardly democratic or accountable (Madon 1996; Marcus 2000; Pfund 2000, 231). Communal *dina* often reflect top-down priorities and carry little weight among the local populace, as shown above in the example of the 1996 fire *dina*. In addition, local leadership roles are troubled by corruption (Robbins 1998, 2000; Klooster 1999; Razafintsalama and Gautschi 1999). As Marcus (2000, 217) notes, "Leaders often see it as a perk of the job that they are free from the laws they are supposed to uphold." Several mayors have allegedly interpreted the new GELOSE legislation as giving the mayor's office control of local resources such as forests, and then used this control for personal profit.[15] The conclusion is that CBNRM cannot just empower extant local authorities, for this "does not automatically resolve issues of equity, representation, and accountability, nor does it constitute community participation" (Ribot 1999, 14).

Legitimacy can be threatened as institutions evolve in response to social and political changes. In particular, when CBNRM projects seek to harness extant local institutions, results vary. The *dina* is a prime example. The state has attempted to incorporate the *dina* rule-setting institution of "traditional" law into modern law, just as conservation projects have sought to use *dina* to involve local people in resource management (Henkels 1999). As a result, the character and legitimacy of *dina* changes. In Afotsara, residents distinguish between *dina fototra*, basic *dina* which grow out of local, *fokontany*-level consensus and have legitimacy, and *dina kaominaly*, which are those created at the level of the commune and often carry little legitimacy.[16] The antifire *dina* of 1996, mentioned earlier, were top-down appropriations of the institution of *dina* that failed, and which may have tarnished its legitimacy.

The most problematic institution harnessed by CBNRM projects is the "community." The operationalization of this concept while still maintaining

14. R. Marcus, personal communication, 2001.

15. Interview, Forest Service officer, 2 July 1999.

16. Interview, 24 Nov. 1998. *Dina fototra* were associated with the stronger power of the *fokontany* during the Second Republic; *dina kaominaly* with the Third Republic when the *commune rurale* became the more important level of local governance.

legitimacy is challenging. Social scientists have identified four problems with the idea of community as used in CBNRM. First, communities are not harmonious and homogeneous as often depicted; instead they can be "contentious, unstable social groupings" (Tsing 1999, 172; see also Belsky 1999; Klooster 1999; Leach et al. 1999; Sundar 2000). Across the globe, all societies are differentiated—by wealth, gender, ethnicity, caste, and so on—and the access, use, and control of natural resources varies and is contested among these groups. For example, women's access to and use of resources often differs sharply from men's (Jarosz 1991; Leach 1995; Rocheleau and Ross 1995; Schroeder 1997), and the wealthy and the poor struggle over the control of resource access (Hahn 1982; Sheridan 1988; Peluso 1992; Suryanata 1994). Even among peasant communities that outwardly seem homogeneous, large variations in wealth and very heterogeneous patterns of resource use exist; these differences have important influences on the character of land use and thus the politics of CBNRM (Coomes and Barham 1997; Coomes and Burt 1997).

This is certainly the case in Madagascar. As noted in chapter 1, rural Malagasy society is highly stratified. In Afotsara, for example, wealth is concentrated in the hands of certain families, and village leaders are all from the ex-noble caste. Within such communities, diverse interests and group rivalries may preclude harmonious management, as mentioned in chapter 6. For example, across much of highland Madagascar, families who plant pine trees on state land complain of others deliberately burning their seedlings. The fire starters defend their actions, saying that the pines are a degradation of communal pastures (which need to be burned for maintenance), also implying that they resent the land-grabbing intentions of tree planters on state lands (planting trees is a first step to acquiring formal land title).

Second, defining the bounds of community can be troublesome. The most frequent definition used in CBNRM projects is geographical, that is the people living in a certain territory (Gauld 2000). In contrast, Leach et al. (1999, 230) view community as a "more or less temporary unity of situation, interest, or purpose." Geographic proximity may be crucial to resource use, but it hardly guarantees unity of interest. The GELOSE statutes never commit to a precise definition, calling for community institutions to represent people unified by interest but also open to membership by all territorial residents.

Third, communities are far from being "traditional" or timeless. Those things considered timeless in "traditional" Malagasy communities, such as *dina* village pacts, the *ray-aman-dreny* (elders), and mutual agricultural aid, are actually institutions that have undergone much change. *Dina* were already formalized into the official legal structures of the country in 1960 and have come to represent top-down decisions by communal leaders as much as traditional, culture-based community pacts. The *ray-aman-dreny* have likewise seen their authority change over the years, due to the economic and political changes since colonialism and the power of the churches. Mutual agricultural aid, based on local institutions for labor recruitment and organization, has diminished in the face of increased state and market influences (Kottak 1980; Vérin 1992).

Fourth, images or discourses of community can be mobilized by different actors to achieve certain goals, such as defending access to resources (Li 1996). For example, development organizations have harnessed romanticized images of traditional, sustainable local communities to create the political space in which to defend the communities from state incursions. In rural communities, individuals likewise claim and defend access to resources by using images and discourses of community, such as by appealing to ties of kinship. Neumann (1995) calls these actions strategic essentialism, where people assert ethnic, indigenous, or community identities in order to expand claims on resources (see also Pulido 1996).

In sum, "community" is a term used to refer on the one hand to a heterogeneous grouping of people involved in multiple, overlapping, historically and regionally produced institutions, and on the other hand to strategically refer to these groupings as organic and relatively homogeneous. The danger occurs when the naïve, strategic representation of community is used to operationalize decentralized resource management (Li 1996). Then, planners risk developing CBNRM that ignores intravillage divisions and favors hardly extant institutions (see Tsing 1999 on the search for romanticized tribal elders). The result is CBNRM that is not legitimate to the part of the village that is excluded (perhaps women, ex-slaves, or silk harvesters).

Giving resource management rights and responsibilities to communities guarantees neither legitimacy nor fairness. The development of CBNRM will lead to some failures or standoffs, to some successes, and in many cases to transformations in local social relations. In the Imamo area, west of Antananarivo, a lack of social cohesion—due, to a large extent, to caste differences and corrupt leadership—blocked the application of one GELOSE pilot project (Razafintsalama and Gautschi 1999). Individual interests and lack of cohesion also blocked participatory community land use planning efforts around Montagne d'Ambre National Park in northern Madagascar (Mainguet 1997). "Elite capture" (Klooster 1999; cf. Belsky 1999; Maldidier 2001), where elites use their power to control the benefits flowing from CBNRM, is a common result. As Gezon (1999) notes, changing environmental management practices directly reflect the dynamics of local social relations.

It remains to be seen whether GELOSE can facilitate legitimate institutions. GELOSE is predicated upon a tripartite contract among the state, the *commune rurale,* and the CLB. For one thing, as discussed earlier, the communal government itself faces questions of legitimacy. Furthermore, the creation of a new institution, the CLB, is potentially troubling. According to the legislation,[17] each CLB is to be an officially registered nongovernmental organization, "a voluntary association of individuals unified by the same interests and following the rules of communal life." Membership is to be open to all residents of the territory of the CLB, ignoring the potential for fractious intracommunity politics. In practice, CLBs would take shape as, for example, an association of villagers interested in exploiting a certain forest resource (e.g., wild silk or timber), or an association of cattle owners for the management of specific common pastures. The attraction to locals is the CLB's anticipated legal access to and management of the resource. The CLB's legitimacy depends upon the case, including the personalities, the politics of its formation, and its accountability to all affected locals. In Afotsara, a CLB association (*fikambana*) was founded in mid-1999 for woodland wild silk collectors. The villagers saw money flowing to the Projet Landibe in Ilaka—where government funds supported a silkworm-breeding house, a tree nursery, and a modern loom—and wanted the same. However, more than half the village refused to join, citing mistrust in the leaders (members of the village elite) and missing funds.[18] Establishing the legitimacy of the CLB can become a difficult task for GELOSE, for without legitimacy, the door is open for resistance, discord, and arson or sabotage.

SUCCESSFUL COMMUNITY-BASED FIRE MANAGEMENT

The above analysis suggests that successful CBNRM requires attention to empowerment and legitimacy. It also demonstrates how persistent ideas about resource management, as well as the sociopolitical complexities of communities, can transform outcomes. These conclusions are further demonstrated by a survey of successful community-based fire management in

17. Projet de décret relatif aux communautés de base, July 1998.

18. Interviews, Afotsara, 26–28 July 1999. By 2001, however, the CLB had succeeded in securing funds for and building a silkworm-breeding house and tree nursery.

Madagascar. Five main categories of functioning CBNRM can be identified with respect to fire management. Each has its particularities, but in all of them, locals are empowered to make and act upon decisions—sometimes because they correspond with state goals, sometimes because they are beyond the state's reach. Their decisions are legitimated because the people involved agree upon the parameters of the problem (they are not, for example, constrained by the antifire received wisdom), and because the community institutions are more or less accepted and accountable.

Hamlet-Level Fire Management

Fires burn frequently in Afotsara. However, one local hamlet is conspicuous in its fire control. In contrast to the oft-burned pastures and woodlands typical of the area, this hamlet is enveloped by overgrown fields, bushy thickets of mimosa saplings, *rambiazina* bushes, and several shady pine woodlots with raspberries in the understory. When the hamlet was founded in the mid-1880s, the area was completely covered by grassy pastures, except for a small riparian forest. Why fires no longer burn in the hamlet's little valley was clear to residents. As one told me, "We control the fires, limit them. When a fire approaches from the east or west, we fight it, stop it. Everyone is family here, we all know each other, so we control the fires. There are few strangers who pass this way; besides, everybody knows where not to burn."[19] They limit fires in the immediate vicinity of the hamlet to improve soils for fallow fields, to grow wood fuel and lumber, and to protect their houses from wildfires.

Another hamlet near Afotsara, perched in the middle of a vast pasture zone, had protected a nearby slope from fire for two years. The villagers there did their best to keep uncontrolled pasture fires from burning that slope, for they intended to harvest the thicker grass for roofing material and to burn for compost fertilizer. A third hamlet protected a grassy slope sandwiched between rice fields and dryland crop fields. It had not burned in seven years, while other nearby open slopes burned almost annually. The slope was being kept as a reserve for roofing grass, and "everybody knows that it shouldn't be burned."[20] Such hamlet-level fire control was echoed in the Tsimay case study. Here a woman told me, "We are all family in this area, therefore we talk to each other and limit the fires"; a man down the path stated that "there are lots of people watching" for unwanted burning.[21] Kepe and Scoones (1999)

19. Interview, Afotsara, 14 Apr. 1999.

20. Interview, Afotsara, 6 Dec. 1998. In Malagasy, "*samy mahalala fa tsy tokony may.*"

21. Interviews, Tsimay, 11 May 1999. In Malagasy, "*betsaka ny manara maso.*"

describe a similar case in South Africa, where families protect patches of land from fire to grow thatch and firewood. Clearly, in these situations the close-knit situation of rural life, with families living near one another, with overlapping material interests (in roofing material, wood fuel, etc.), allows them to set unwritten rules about fire control. Created by related hamlet members and with clear purpose, these rules carry plenty of legitimacy.

Fire Management in Intensive Zones

Land-use intensification, whether due to population pressure or market opportunities, leads to tighter management of fires. In most cases, uncontrolled wide-ranging fires are squeezed out of intensive zones (chapter 3). The case studies in highland Madagascar demonstrate this pattern. Behazo simply has no land to burn, for in this densely populated zone crop fields and orchards cover almost the entire area. Tsimay stopped burning because a hillside of mimosa wood fuel became more profitable than a hillside pasture. This pattern is repeated in the many areas of charcoal and construction wood production across the highlands (Bertrand 1999). Intensification even occurs in the ranching-oriented Middle West. The moderately populated region east of Tsiroanomandidy, around Ankadinondry-Sakay, is filling in due to government-sponsored and spontaneous agricultural colonization (see Raison 1984). Crop fields and eucalyptus woodlots are beginning to challenge the dominance of pastures, and burning is more tightly watched.[22] In such situations, where population pressure or economically driven intensification change land use, fire becomes unacceptable to the majority of the population. Then, controls on burning have popular legitimacy, and community-level fire rules and traditions change (cf. Gezon 1999). In these cases, locals seek to have national level fire legislation enforced, not ignored.[23] Due to changed community perspectives on fire, repressive national fire legislation becomes relevant and is enforced.

Free-Burning Extensive Pasture Management

Another fire management system that carries legitimacy and that can realistically be implemented and enforced is opportunistic free-burning pasture management, following the strategies outlined in chapter 3. Most common

22. Interview, Forest Service officer, 10 July 1999.

23. An analysis of recent criminal fire cases in the files of the Antsirabe Forest Service office demonstrates this: eleven of thirteen files consulted involve burned property, almost all in densely settled intensive agricultural zones (e.g., Betafo) or in market-oriented tree-growing zones (e.g., Andranomanelatra and Ambohibary).

in peripheral zones far from government influence—e.g., in the vast grasslands of the Middle West—this management system also exists in the open hilltop pastures of more central locations. In terms of the case studies, free-burning pasture management characterizes the vast grasslands of Kilabé, the large mountaintop pastures around Afotsara and Behazo, the derived grasslands east of Beforona, and the Fivanonana plateau north of Andringitra. While the threat of enforcement has perhaps changed the timing and frequency of fires, fires here still burn freely most years. These pastures are "*zatra may isan-taona*"—accustomed to being burned each year.[24]

Free-burning pasture management is characterized by a village consensus that a certain area be managed as pasture. Herders light pasture renewal fires opportunistically when conditions permit,[25] letting the fires run their own course. The goal is more or less a complete burn-over, in order to stimulate the growth of fresh grass shoots and to fight bush encroachment. The default condition is to burn; individuals or hamlets wishing to reserve certain areas from fire must make the effort themselves—spreading the word not to burn a certain spot, directly fighting fires that threaten houses or crops, or creating firebreaks. In these frequently burned zones, however, fires rarely present much danger due to the lack of fuel buildup. Fires are less intense, and often stop by themselves at pathways, streams, crop field edges, or previous fire scars. When fires do damage private property, the owner has little moral ground to complain, for the onus of protection falls on him or her. This was the case in Afotsara, when an August 1998 pasture fire torched a small shelter built near some mountaintop cassava fields (figure 2.3). Instead of protesting, the owner fatalistically accepted the loss.[26]

De Facto Slash-and-Burn Management

The ways in which villages in eastern zones regulated *tavy* before colonial controls, and the ways in which they continue to do so today in remote places where control is weak, are another example of community management. As described in detail in chapter 5, villagers decide upon which plots to cut and burn based upon ecological signals, cultivation needs, and traditional land rights. Access is shaped by traditions of historical precedent and lineage-based control, or in some areas near tombs, by decisions of the village elders. Plots are cut, burned, and cultivated for a few years, then left fallow.

24. Interview, young cattle herders, Afotsara, 3 Nov. 1998.

25. Opportunistically in several senses—when it is sufficiently dry, when herders happen to be in a particular place, and when enforcement is unlikely (chapter 6).

26. Interview, Afotsara, 6 Nov. 1998.

This is the de facto management system except in the most strictly enforced areas, such as around protected areas.[27] The result of village-based *tavy* management, as we saw in chapter 5, is a loss of forest cover and the slow transition of ridges and hillsides from forest to fire-maintained, nutrient-poor grassland. However, over time, fire becomes more strictly controlled in irrigated bottomland, hillside gardens, and orchards as land use is intensified (Brand 1998; Pfund 2000). While this system goes against conservation goals, it is consistent with farmer goals of crafting a livelihood with available land resources and thus is successful from their perspective as a management strategy.

Fire Management with Outside Pressure

The fifth example of CBNRM fire management is more closely related to the current GELOSE process. At Andringitra National Park, the WWF project is working with the Forest Service and local communities to create pasture and fire management zones (see chapter 3). The project and villagers divided the Namoly basin, north of the park, into eight pasture management zones (*kijanan'omby*). The cattle raisers designated leaders for each *kijanan'omby.* Each zone has a pasture-burning plan, typically encompassing late dry season fires in the lower pastures near rice fields and late rainy season fires in higher zones. The dry season fires require approval from the Forest Service; the late rainy season fires are free and also serve as firebreaks for the park.

Two factors clearly contribute to the potential success of this system. One, the WWF project team accepts fire as a legitimate management tool, and thus the project and the pasture management system established a level of trust and legitimacy with the people. Two, the externally funded project carries great political and financial weight. Not only can the project print out GIS-based maps of pasture management zones, but more importantly it has the power to ensure enforcement and implementation of the management plans. In 1996, it pushed through a regionwide collective fine for an illegal fire within the park, showing it was serious. The actual success and long-term sustainability of this system, once project funds dry up, remains to be seen.

27. People sometimes *increase* the area of forests they cut in areas where protected areas are proposed, in order to claim those resources before the boundary of the protected area is set (e.g., Ghimire 1994). Similarly, since foresters are more lenient in giving *tavy* permissions in areas that are not primary forest, state control increases the incentives to surreptitiously cut or accidentally burn primary forest—for later legal use (e.g., Coulaud 1973).

CONCLUSION

The GELOSE case demonstrates several ways in which the key variables of empowerment and legitimacy can block or transform CBNRM projects. Empowerment of locals is restricted, in part, by the continued weight of a powerful antifire received wisdom, and legitimacy is threatened by CBNRM's entry into the complex arena of community institutions and power. In contrast, the above cases of community-level fire management work because they address precisely these obstacles. They ignore the antifire received wisdom and accept fire as an appropriate resource management tool in specific agropastoral contexts, thus addressing material interests and gaining popular support. In addition, the institutions that make the decisions have sufficient legitimacy and power to do so—whether they are family-based hamlets, broader villages with more or less common interests (and, in the case of intensification, with a state that will back them up in case of conflict), or a powerful foreign-funded project that, by accepting fire, may have gained some legitimacy. In each case, the fire management system was specific to both the local context and the local institutional and social situation.

The challenges presented by CBNRM are many, yet so are the opportunities. If locals are empowered to manage specific natural resources through legitimate and accountable institutions, then CBNRM programs stand a chance of achieving equitable and sustainable resource use. If, however, CBNRM programs lack empowerment, if only responsibilities and not rights are transferred to the local level in a top-down attempt to expand control, the result may be resistance, corruption, and unsustainability. If CBNRM programs ignore the dynamics of community institutions, the result, such as noncompliance, elite capture, or reshuffled local power relations, may be inequitable or may work against the goals. Finally, if CBNRM programs intersect with powerful and contradictory environmental received wisdoms, the result may be a changed focus if not outright blockage.

The GELOSE process may not meet the goals originally conceived by Bertrand (1994) and Rajaonson et al. (1995): to end a century-long impasse between the peasants and the state over landscape burning and instead to seek to manage fires through a sustainable, flexible, and regionally contextualized decentralization of resource management. In addition to issues of empowerment and legitimacy, factors that threaten the successful implementation of GELOSE to fire management include its financial costs, its cumbersome complexity, and lingering issues over land tenure security (Pfund 2000; Maldidier 2001).

GELOSE is most likely to succeed in cases where resource management does not conflict with extant laws or the antifire received wisdom, where community issues are addressed carefully, and where the potential financial benefit makes it worthwhile. Pilot efforts indicate that the more successful GELOSE contracts may come in the forestry sector, where community institutions are given exploitation rights to certain forests (Maldidier 2001). This is due to donor interest in forest management, its lack of conflict with the antifire received wisdom or extant legislation, and the palpable financial interests to the local community of forest resources. Establishing GELOSE contracts is a lengthy twenty-two-step process, with intrusive expert studies into the local resource, in which significant costs accrue to the villagers. Where profit is likely (e.g., timber producing forests) or where donors may expedite or underwrite the process (e.g., near parks, reserves, and forests), the process will be more successful. As a result, while GELOSE may address forestry and possibly *tavy,* it may not adequately address the management of pasture and woodland fires. Some hope comes from the current third phase of the Environmental Action Plan (the PE3), which aims among other things to transfer 500,000 ha of pastures to community management. Yet the PE3 goals simultaneously include the continued "fight against bush fires" (RDM 2002), and President Ravalomanana's recent antifire pronouncements (see next chapter) lead one to doubt that communities will be given real freedom to burn.

This is an example of how development interventions can evolve or change from their originally stated intentions (Ferguson 1994; Scott 1998). The outcomes of CBNRM programs include some utter failures and some successes; the result depends on how questions of empowerment and legitimacy are addressed in widely varying regional historical and environmental contexts. This variation is well illustrated both by the above cases of non-GELOSE successful fire management and by the experience of GELOSE to date. Each GELOSE contract is different, reflecting the dynamics of local actors, resources, and politics.

The momentum of GELOSE has led to the creation of hundreds of CBNRM institutions across the island.[28] They vary in characteristics due to the actors involved (e.g., the different agencies and foreign donors), the resource in question, and local institutional and environmental histories. Some are full-fledged GELOSE contracts, such as at Sakaraha (near Zombitse forest in the southwest), at Brickaville (between the Beforona case study

28. The momentum of GELOSE also explains why a forester would tell me "the politics of repression are out of date" (interview, 7 Aug. 2001).

and the east coast), at the Andapa case study in the northeast (Garreau et al. 2001), and in the *tapia* woodlands of Imamo. Others include "pre-GELOSE" agreements, park periphery management pacts like at Andringitra, as well as alternative or competing CBNRM structures such as GPF (participatory forest management) and GCF (contractual forest management).[29] In Ambatolampy, for example, a local nongovernmental organization, the Union Forestière d'Ambatolampy, now guides the village-led management and exploitation of nearby state forest plantations, while near Fort Dauphin, a GPF has been set up for the community management of the Fanjahira state-owned forest plantations.[30]

The key change in GELOSE since its inception is its hesitancy to approach fire management. This was one of the original goals (Rajaonson et al. 1995). GELOSE is more successful when applied to forest or water resource management. To succeed in addressing the Malagasy fire problem as originally intended, the state must accord local communities sufficient freedom in fire management, and the onerous bureaucratic process for GELOSE should be simplified. More critically, however, the government must accept fire as a legitimate resource management tool.

29. The politics of aid come into play here, as different donors put their support behind competing approaches. In 1999–2000 there was an important falling out between the French, supporters of GELOSE, and the Americans and Swiss, who preferred the simpler, forest-specific GCF (included in the 1997 Forest Policy). GCF applies only to lands under the purview of the Forest Service (e.g., classified forests); the area is opened to local use (logging, grazing, plant collection) based on a *dina* between the Forest Service and the *commune rurale* or *fokontany*. However, the area remains Forest Service domain. GELOSE, in contrast, can be applied to any form of state domain, involves a mediator, and aims to give the local community tenure rights to the resource.

30. Interviews, UFA Ambatolampy, 29 Apr. 1999; J. L. Pfund, 7 Aug. 2001.

9 [CONCLUSION

> Fellow Malagasy who love our ancestral land, let us protect Madagascar, so it becomes a Green Island, an Island of Development. The work of development is proceeding. But there are still some obstacles to remove. Among the most important is forest fires and hill fires. How can we develop when all the nation's efforts turn to ashes and smoke? . . .
>
> PRESIDENT MARC RAVALOMANANA,
> SPEECH ON 28 SEPTEMBER 2002

> But the grass of Madagascar is in its glory on the great hills. Burnt year after year by long sweeping fires, it springs up again with a profusion and a fullness which clasps huge rocks within its soft embrace.
>
> LONDON MISSIONARY SOCIETY MEMBER
> JOSEPH MULLENS (1875, 200)

THE MALAGASY "fire problem" is a telling case of environmental conflict, a long drawn-out struggle over the character and use of natural resources. The analysis of this conflict can help us understand other conflicts over the use and management of natural resources. To illustrate, let me take the next few pages to compare it to four examples from across the globe. In the United States, a conflict over private grazing rights on federal lands has simmered for years. In Kenya and Tanzania, the pastoralist Maasai have struggled for most of a century against the appropriation of their lands for national parks. In Indonesia and other Southeast Asian countries, villages have competed with state-run forestry agencies for control of forest resources. In maritime Canada, Micmac natives recently clashed violently with New Brunswick fishermen over access to lobster fisheries.[1]

Madagascar's fire problem is helpful in understanding such natural resource conflicts. Simply chronicling different interest groups and their points of view is insufficient, because the path a conflict takes is fundamentally shaped by crucial contextual factors. Knowing that the Malagasy state has long sought to snuff out the island's fire habit and that the farmers depend upon fire for their agropastoral livelihoods is important, but it does not explain the persistent impasse over fire. In the same way, just knowing the interests of the Micmac and the New Brunswick fishermen, or of the Maasai and the Kenyan Wildlife Service, or of the Javanese and the Indonesian State Forestry Corporation, is not sufficient to explain the character of the conflicts, nor to develop solutions.

Crucial to understanding and resolving these conflicts is an appreciation of their broader context. There are at least five aspects to this context. First, the ecological or natural context of a conflict shapes the paths and directions it may take. In Madagascar, fire's ambiguity, anonymity, and self-propagation distinguish it from other contested management techniques such as hunting, timber harvesting, grazing, or fishing. Such ecological characteristics give advantages to certain interests over others. For example, a successful invasive species favors those who benefit from that species; people dependent on a plant intolerant of shade suffer if others grow trees nearby; and the complex yet inevitable process of fire is a trump card in the hands of those who benefit from it. Furthermore, the ecosystem within which a practice occurs bounds the possibilities for conflict. For example, Madagascar's fire ecology, with one long dry season, differs from that in Kenya, which has two shorter dry seasons. Different soils and plant communities lead to different grazing impacts around the American West. The ecology of lobster along the coast of New Brunswick—found near the shore (where the Micmac live) in the fall and then offshore in the spring—shapes the parameters of that conflict. Each case of conflict has a basic ecological context that can both limit possibilities and provide opportunities for resolution.

Second, the social and economic context is a critical determinant of people's interests and possibilities in resource conflicts. Peoples' access to forest, range, or fisheries resources, whether in New Brunswick, East Africa, Indone-

1. These examples, to which I refer several times over the following paragraphs, are respectively based upon McCarthy (1999), Neumann (1998), Peluso (1992), and Sarah Curry, McGill University, personal communication, November 2001.

sia, the American West, or Madagascar, is crucial to their socioeconomic livelihoods. To peasants involved in extensive subsistence farming and animal husbandry, certain resource states (such as oft-burned rangelands or long-fallow *tavy* hillsides) better facilitate livelihood production than others, given their restricted access to capital and labor. Peasants practicing intensive agriculture in densely settled areas near market centers prefer different resource states, including fire conservancy. Other interest groups' social and economic contexts demand completely different resource uses. For example, colonial capitalists had access to the capital and technology necessary to exploit natural resources for quick profit (so they preferred hardwood-rich forests or large-scale plantations), while modern conservationists in an era of prosperity and globalization focus on preserving habitats and natural processes (thus preferring vegetation with little human influence).

Third, the nature of state power and institutions shapes the parameters of natural resource conflicts that involve the state (as most do). Madagascar's nonmonolithic state (with multiple viewpoints hidden behind an official antifire policy) is relatively weak and provides some of the opportunities exploited by the peasants. Elsewhere, the specific character of institutions and national politics determines the shape of the playing field. In the western United States, the mandates of many federal land management agencies have metamorphosed from emphasizing economic resources (mining, logging, and ranching) to broader concerns with ecological and aesthetic interests. Canada's attempts to improve its legacy of relations with its native people, for example, led its Supreme Court to side with the natives in treaty disputes over resource rights. In Indonesia, the turbulent years of occupation and revolution during and after World War II threw the old forms and institutions of colonial forest control into question, allowing forest villagers to assert additional resource claims.

Fourth, ideas about environmental change and about appropriate resource use are important components in resource conflicts. Such ideas, like the antifire received wisdom in Madagascar, emerge out of material interests, social struggle, and ideologies through a long historical process. Environmental received wisdoms trace their roots to concerns over human impacts on the environment and to the self-serving interests of colonial resource exploiters. The ideas that persist are precisely those that form a simple, coherent story and at the same time serve the ideological and material needs of the elite (Dove 1983; Bergeret 1993; Leach and Mearns 1996). They persist as they are institutionalized into bureaucracies and because the peasants rarely engage with the ideas or discourses directly; instead they engage with their material effects in specific historical and regional contexts (Leach and Fairhead 2000). People harness such ideas in resource conflicts, seeking to justify approaches and mobilize funds and opinions. Typically, locals are labeled irresponsible resource managers who are degrading resources that could be put to better use or oversight by state forestry operations (Indonesia), game parks (Kenya), government fisheries experts (New Brunswick), or conservationists (western U.S.). Continuing conflict serves to further the persistence of received wisdoms, for in conflicts they are both hardened and constantly tested (and thus adapted and improved). In Madagascar, I demonstrated how the persistent antifire received wisdom continues to shape the discursive space of environmental debate and bounds the scope of new, innovative approaches like GELOSE.

Finally, all of the above observations highlight how historical precedents are critically important to the unfolding and structuring of events. Ecological legacies mean that certain transitions between ecological states are irreversible—long-time *tavy* hilltops of fern and grass cannot revert quickly to rain forest; a depleted fishery does not necessarily rebound if exploitation stops. Economic developments constantly create new needs and new situations on top of the social systems and endowments in labor, capital, land, and technology inherited from the past. Ideas originating in the biases and material needs of colonial administrators become entrenched in the imagination and structures of current-day institutions, shaping current events. Historical struggles over power in resource use create a legacy of mistrust and repression (or alternatively of cooperation), which may block or shape change.

Environmental conflicts, or struggles over resource access and character, are rooted in history, in ideas, in the realm of power, in the socioeconomic context, and in the very biological and physical forms and processes of the resources themselves. They are struggles between different elements of society, different interests and narratives, utilizing different strategies. It is state versus peasants, timber companies versus farmers, forest ideologies versus pastoralist dreams, and so on. But, as the case of Madagascar exemplifies, the solutions to many of the world's natural resource conflicts lie not only in understanding the interest groups but also in understanding the contextual factors that shape or limit the way in which a conflict unfolds.

Resource conflicts are much more than a contest between two actors with differing interests, or a clear-cut decision between environmental salvation and economic interests. The heuristic abstracts I harnessed to make sense of the struggle over fire in Madagascar—"good fire" and "bad fire," "state" and "peasant," "criminalization" and "resistance," "community"—are crucial analytical categories, yet they are not sufficient. One must also consider the entangled, complex middle ground between these abstract concepts, such as the ecological complexity of fire, the intertwined nature of peasants and state, or the very political nature of community empowerment. Neither the distinct categories nor their slippery middles are sufficient in themselves to explain resource conflicts. In addition, as I argued above, it is crucial to understand the broader regional biological, physical, and social context.

I have argued that the central state's antifire received wisdom, which confronted the peasants' livelihood-based attachment to burning, led to a conflict over fire. I argued further that it is because of fire's ambiguity and the state's de facto ambivalence that fire-setting peasants have succeeded in defending their desired resource character and access (*tavy* or free-burning pasture management, when agroeconomically suitable). Shaped by the character of Malagasy ecosystems and the particular history of social and political development on the island, this conflict has persisted for over one hundred years.

More specifically, the development and persistence of an antifire received wisdom—based in Western, urban culture, rooted in the ideas and institutions of scientific forestry and biodiversity conservation, and fueled by the economic interests of the colony and later by the conservation interests of powerful international donors—led the central forces within the Malagasy state to seek to ban traditional burning practices. The central state criminalized burning, both ideologically and legally. The peasants, in general, resisted this frontal attack on one of their most useful, well-adapted, efficient, and inexpensive resource management tools. Except in areas of intensive land use, fire was the obvious tool to maintain pastures, prepare crop fields, control insects, and fight wildfire, ideal to the conditions of labor, capital, and ecological resources at the hands of the peasants. To resist the criminalization of fire, the peasants took advantage of all opportunities presented to them—especially of fire's inevitability, its easy anonymity, its self-propagation, as well as of the state's hesitations and multivalence—in order to burn illegally and avoid enforcement. As a result, despite state repression and antifire rhetoric, Madagascar is still aflame.

This kind of struggle over resource access and character typifies environmental conflicts around the globe. Behind most land use conflicts lie different groups of people, different institutions, with differing goals, needs and ideologies, and who use the tools at hand—e.g., creating powerful discursive arguments or striking matches—to defend their interests. In order to analyze these conflicts, it is not sufficient to just understand the scientific evidence about the character and dynamics of the resource in question. One must also comprehend the historical situation, the power relations, the economic context, the livelihood and material interests, and the ideological baggage of the groups in question, as well as how they meld and blend at their contact points.

MOVING FORWARD

This stalemate, this impasse that is the Malagasy "fire problem," will persist without end unless the conditions producing it are addressed. Two avenues seem possible. One is to strengthen the resolve of the state to enforce its restrictions, leaving no room for exceptions, inconsistencies, or corruption. This, however, would be both politically untenable and environmentally dangerous. No government could successfully implement such an unpopular ban, not without threatening its credibility or without resistance, as peasants could still take advantage of fire's characteristics to accomplish their burning. Even if fire *were* successfully stopped in highland and western zones, the resulting unburned, fuel-rich zones would be dangerously prone to devastating wildfires. As Minnich (1987) noted so elegantly, the size and intensity of fire are inversely proportional to its frequency.

The second, more reasonable approach is to address the criminalization of fire and the antifire received wisdom that produces it. Instead of arguing against burning, the most realistic stance is that there has always been fire in Madagascar, there always will be, and there always should be. Fire is a useful management technique that fits closely with the current needs and capabilities of Malagasy land managers, the farmers and herders. In challenging the antifire received wisdom, such "good" fires can be "decriminalized," removing the need for resistance and thus the complications producing the long-term stalemate over fire.

State elites, the Forest Service, and international environmental actors have largely been too reluctant to accept that some fires are good, useful, and even desirable. They should accept fire as not just a necessary evil, but as a complex, critical tool for resource management. This view has already infiltrated parts of these institutions, but the overall discourse that frames discussions continues to emphasize fire as a negative agent of degradation. A report from a workshop on *tavy* and bushfires held by the Multi-Donor Secretariat in April 2001 recognized *tavy* and pasture fires as "economically efficient" and "totally understandable strategies for rural societies colonizing their space,"[2] yet the workshop's overall goal was to seek strategies to reduce these two damaging practices. The recent *document stratégique* outlining the goals of the PE3, the third phase of the Environmental Action Plan, laments that agricultural needs prevent the eradication of bushfires, and pushes a continued program of "fight against bushfires" (RDM 2002, 5, 32). J. Andriamampianina (1998, 9–10), the recent director general of the Office National de l'Environnement, is much less compromising: "People still talk of pasture fires as a necessary evil and of *tavy* slash-and-burn fires as a subsistence method . . . I, however, think all fires should be completely banned. Fires are a destructive force, and there will be no agricultural improvement while fires destroy the vegetation." Andriamampianina's concern with peasant pyromania reflects a widespread view in government and urban society that all fires are bad and lead only to degradation. As chapter 8 argues, this ideology even straitjackets forward-looking policy efforts like GELOSE.

In his first year in office, President Marc Ravalomanana has reinvigorated the fight against fire with strong rhetoric and policy. In August 2002, his government signed a new decree (see appendix 2) that links the funding of communal development projects, such as roads, schools, and health centers, to the performance of the *commune rurale* with respect to fire. Each district is to rank communes by their success and effort in stopping burning, and a certificate of meritorious behavior is to be required before funds are disbursed for development projects. Ravalomanana introduced this policy in a speech dedicated to the fight against fires on 28 September 2002. His speech does not recognize any useful fires; he just states "if we want to develop quickly, we must stop the burning of forests and hills." Likewise, on 16 May 2003 he called for an antifire awareness campaign to educate the farmers about the negative consequences of fire. Rural communities are said to be afraid, for Ravalomanana is known to follow words with action: already twenty people were jailed for illegal *tavy* in the east.[3] Ravalomanana's policies echo a hundred years of failures and set the basis for a continued impasse.

Were the antifire received wisdom to be replaced with a pragmatic appreciation of fire's useful, if complex, role in land management, then GELOSE could be a great step forward in solving the fire problem. However,

2. Note de Réflexion of the multidonor group (Groupe des Bailleurs de Fonds) entitled "Problématique, constats et questions concernant la pratique du tavy et des feux de brousse à Madagascar," June 2001.

3. *L'Express de Madagascar,* 17 May 2003; interview, Forest Service agent, Antananarivo, 6 June 2003.

a number of pieces of legislation would need to be adapted to reflect this ideological reorientation. In particular, the fire regulations dating from 1960 would warrant updating. Despite a strongly antifire preamble, Ordonnance 60-127 is in part already a relatively enlightened piece of legislation, in that it distinguishes between bad fires (wildfires) and useful fires (crop field and cleaning fires, pasture fires, preventive fires). However, its cumbersome legal requirements for "exceptional" fire permits are followed by neither the peasants nor the Forest Service and deserve to be scrapped.

Community-based natural resource management strategies like GELOSE are a useful step forward in reshaping how Madagascar manages its fires. While GELOSE is bogged down by its cumbersome and expensive procedures, and currently hindered in its approach to fire by the antifire received wisdom, its ideological orientation towards community management is positive. While some may have seen GELOSE as a chance to have local communities enforce national antifire legislation, instead it should be an opportunity for local communities to best regulate and oversee *legitimate* forms of burning. Unencumbered by the antifire received wisdom and the national bureaucracy, and empowered with legitimate authority, local communities may be best placed to handle the locally specific characteristics of fire use, including opportunistic burning, the complex rotation of pasture or *tavy* fires, and the timing of burns to best benefit cattle or crops.

What would such management look like? In highland and western zones, where fire is both a natural part of landscape processes as well as an integral cultural landscape management tool, people use fires as an efficient and inexpensive means of pasture, woodland, and crop field management. If a community were allowed to manage its own fires, with the blessing of the central state and without the restrictions of top-down rules or the antifire rhetoric, the result might be as follows. In less-densely populated areas the social consensus would be to allow pastures to burn. Since enforcement would not be feared, villagers might openly plan their burning activities, and fire starters might stick around to monitor the flames. Communities would probably require few formalities—at most to inform village leaders ahead of time; burning would likely take place in an opportunistic, casual style. More strict management would occur only when problems arise: a roof catches on fire, a pine plantation is torched. In these cases, the *ray-aman-dreny* (elders) and the disputants would convene to determine a resolution. If the problem were serious or irresolvable, they would call upon higher levels of government to assist. Several such cases might lead to changes in the community consensus. For example, if enough people began to desire to claim portions of the once-common pasture to grow trees or pineapples for the urban market, the consensus would change (not without contestation and conflicts, of course) and acceptable burning would likely be more limited.

In the humid forest zones of the east, villages might use similar strategies to highland or western villages for the management of pasture and crop field fires in grassland zones. In areas of forest or secondary forest, villages would manage their cutting and *tavy* fires with the same goals as elsewhere: seeking a livelihood. They would allocate *tavy* lands based upon traditions of resource access and decide where to cut and burn based on ecological signals, cultivation needs, and labor availability. Perhaps, if villagers were assured of control over all village lands, there would be less speculative clearing than has occasionally occurred in the past.[4] However, this kind of community control (in Madagascar in the current political and economic situation) would nearly certainly lead to a progressive deforestation of most village lands.

Clearly, there are environmental consequences to what would happen if communities gained control over their burning activities. Most dramatically, deforestation in forest zones would nearly certainly continue apace. However, very little would change elsewhere. Grasslands and woodlands would be maintained in their ecological state by a cycle of fires, and small fires would serve a number of useful tasks as they do today, from preparing a field for plowing or cleaning out canals. If anything, fires may become somewhat better controlled, as people no longer need to burn clandestinely. The consequences of fire for erosion, soil nutrients, and air quality would also persist, but as argued earlier in this book, some of these concerns over fire-caused environmental degradation are either exaggerated or matters of perspective. For example, soils clearly erode faster in areas that have been burned, and this has been demonstrated several times in Madagascar (see chapter 2). However, no researcher has asked why herders would insist on continuing burning despite increased erosion, investigating whether there are biological and/or biogeochemical processes that serve to maintain productivity despite soil loss (e.g., Ojima et al. 1994).

Instead of wasting energy on trying to achieve a blanket reduction or eradication of "damaging" pasture fires and *tavy*, policy makers should instead focus on a few key issues. The first is the place-specific control of the

4. Some farmers have been known to clear more primary forest than they really need, in order to create more *savoka* (secondary forest) and thereby fight state attempts to grab the same land as a forest reserve or protected area. Once land is considered *savoka*, it is much easier for the farmers to legally claim the land and get authorization to *tavy* within it (Coulaud 1973; Ghimire 1994).

environmental consequences of fire: protecting forests with firebreaks, rehabilitating particularly degraded hillsides, or regulating some burning around cities for air quality reasons. This may also include some level of fuel management, by prescribing burning in areas where fuel buildup could lead to dangerous wildfires.

Second, efforts should focus on the control of criminal and accidental fires. As chapter 2 notes, people do burn out of jealousy, revenge, or protest, and some fires do become wildfires due to poor precautionary measures. If state efforts towards fire control were focused around supporting village-based institutions in addressing these fires, instead of working against all fires including those crucial to rural livelihoods, then the state may gain credibility among rural residents.

Third, the policy question of deforestation due to *tavy* remains critical. This issue is not over fire per se but over the political control of remaining forests. Over the past century, Malagasy farmers, the colonial and independent governments, and the global environmental community have been—in effect—fighting over the control of forest resources. Each group has used the means available to it, such as anti-*tavy* legislation, protected-areas legislation, project funding, or physical acts of cutting and burning (possession being nine-tenths of the law). Region by region, these groups are still struggling to delimit which forests are to be conserved and which zones may be used for agriculture. The crucial thing will be to find some compromise that is tenable, which (hopefully) keeps at least a small part of Madagascar under relatively natural forest cover, which permits the continuation of *tavy* in other zones, and which might eventually lead to the reversal of the deforestation trend.

In the end, the "fire problem" in Madagascar comes down to a struggle over the access to and character of natural resources. It is a struggle fought in the complex interwoven planes of ideology, material interests, science, politics, history, and resource ecology. In simplistic terms, Malagasy peasants want continued access to a landscape that meets their livelihood needs in a way that is appropriate to their human, technical, and financial resources, while important parts of the state and their allies want to change the character of the landscape—by removing fire—due to material interests (beef and timber export) or concerns with environmental degradation. Hopefully, as the twenty-first century unfolds, a more enlightened, tolerant view among leaders will be able to encourage a better, less conflictual management of fire in Madagascar. While Madagascar would still be aflame, it would no longer have a fire problem.

Appendix 1
Official Statistics of Fire Extent

THE TABLE BELOW presents official statistics on the areal extent of fires in Madagascar, as graphed in figure 1.2. The data are very unreliable, and should be regarded with extreme caution. Data quality and sources are discussed on the following pages.

Province	Antsiranana (Diego Suarez)	Fianarantsoa	Mahajanga (Majunga)	Toamasina (Tamatave)	Antananarivo (Tananarive)	Toliary (Tuléar)	TOTAL
1968	na	na	na	na	na	na	301,300
1969	5,950	152,187	155,188	5,646	1,041,698	320,952	1,681,621
1970	na	na	na	na	na	na	na
1971	18,681	504,948	201,276	20,871	1,728,100	397,460	2,871,336
1972	42,251	895,717	114,381	19,709	190,000	337,110	1,599,168
1973	22,625	628,912	13,281	36,716	2,525,761	244,556	3,471,851
1974	94,783	1,024,679	45,863	237,377	66,164	214,718	1,683,584
1975	48,332	411,056	60,569	52,015	2,651,445	123,110	3,346,527
1976	17,007	234,923	269,302	318,586	1,431,403	271,719	2,542,940
1977	16,092	119,673	238,055	23,913	1,808,019	177,486	2,383,238
1978	na	na	na	na	na	na	1,384,151
1979	na	na	na	na	na	na	1,547,415
1980	na	71,452	69,971	1,040	148,036	2,788	> 293,287
1981	971	556,117	252	na	475,800	6,160	> 1,039,300
1982	na	496,870	296,769	19,190	101,386	7,383	> 921,598
1983	26,317	1,125,000	881,500	5,100	1,137,000	244,615	3,419,532
1984	157,067	528,979	42,271	31,568	191,429	159,800	1,111,114
1985	210,006	21,325	278,628	34,810	168,193	21,676	734,638
1986	1,656	176,125	212,841	4,815	502,121	61,082	958,640
1987	4,702	50,416	90,487	59,879	537,733	42,133	785,350

Province	Antsiranana (Diego Suarez)	Fianarantsoa	Mahajanga (Majunga)	Toamasina (Tamatave)	Antananarivo (Tananarive)	Toliary (Tuléar)	TOTAL
1988	126,663	35,121	54,449	19,975	79,285	59,173	374,666
1989	123,643	27,648	278,628	34,810	49,147	67,233	581,109
1990	6,412	202,613	155,647	29,421	107,732	32,483	534,308
1991	na	85,992	47,770	262	48,580	1,343	> 183,947
1992	11,631	108,803	41,526	5,045	57,392	14,963	239,360
1993	3,083	208,630	96,086	3,707	64,577	25,666	401,749
1994	2,371	54,245	162,295	5,934	314,793	12,037	551,675
1995	3,544	44,709	201,847	1,873	816,952	169,168	1,238,093
1996	9,118	160,848	271,827	3,851	639,902	111,768	1,197,314
1997	1,510	57,725	138,452	133	385,371	71,010	654,201
1998	542	56,793	346,441	17,886	311,828	87,620	821,110
1999	6,830	167,401	202,400	4,830	740,328	160,400	1,282,189
2000	255	25,774	368,374	1,096	549,420	18,347	963,266
2001	1,526	38,520	251,634	563	0	524,759	817,002
2002	576	21,627	105,661	471	210,402	555,178	893,915

NOTES AND SOURCES

na = not available.

All years: Unless otherwise noted, source is Rapports Annuels of the DEF (1969, 1971–1977, 1982–1984, 1988–1990, 1992–1996) and a spreadsheet attributed to "SCB/DGEF/Tananarive" (1980–2002). Consulted at the DEF offices, Nanisana, Antananarivo. In many cases, my calculation of the sum of provincial figures did not match the total figure given; I gave preference to the former.

1968: Prairie fires only. From Forest Service speech at the opening ceremony for the annual reforestation campaign (ANM Vice-Pres. 850).

1969: From document "Feux de brousse d'après les renseignements communiqués pendant l'année 1969" by DEF (ANM Vice-Pres. 850). Different statistic given in DEF annual report 1969: 1,740,000 ha.

1976–1977: Two other documents give different total figures. RDM (1980), a government report on fires, gives 1,545,000 ha for 1976 and 1,500,000 ha for 1977 (figures interpolated from graphs). Andriamampionona (1992), citing DEF, gives 2,130,213 ha for 1976 and 2,243,797 ha for 1977.

1978–1979: Source: Andriamampionona (1992), who cites DEF and the 1988 Réunion Interministerielle sur les feux de brousse.

1982: The Rapport Annuel and the SCB/DGEF spreadsheet differ by 50,000 ha. I use the latter source, which is the higher estimate. The SCB/DGEF spreadsheet total does not reflect provincial totals.

1984: From SCB/DGEF spreadsheet; numbers differ slightly from *Rapport Annuel* due to rounding.

1987: Rakatonindrina (1989) reports a different total figure: 1,274,852 ha.

1988–1989: From SCB/DGEF spreadsheet, except figure for Mahajanga province in 1988, which is deduced from total in ONE/Instat (1994).

1990–1991: From Rapport Annuel. Note that the SCB/DGEF spreadsheet reports smaller totals: 242,079 ha for 1990 and 107,762 for 1991. Other documents, citing DEF, give similar numbers. ONE/Instat (1994) reports 242,079 and 109,946, respectively, while the Ministère de l'Agriculture (1991) cites figures of 231,001 and 107,455. I use the larger estimates.

1993: From SCB/DGEF spreadsheet, which reports higher figures than the Rapport Annuel.

1995: While most sources agree, Andrianavosoa (1996) reports the much larger figure of 3,034,060 ha, citing as his source the DEF.

The data in the above table should be read with extreme caution, due to problems of definition, consistency, and reporting. The actual extent of fire is likely grossly underreported.

The agency responsible for assembling statistics on fire extent is the Service des Eaux et Forêts (DEF). Trimestrial reports from each *cantonnement forestier* are grouped into Annual Reports from each *circonscription forestière*, then summarized at provincial and national levels. It is unclear exactly how the figures are arrived at in each forest agent's territory. Some are simply the sum of areas burned as reported in citations (*procès-verbaux*) for illegal fires together with authorized burn areas. Others include estimates of fires from field tours by foresters. Considering that one agent is typically responsible for thousands of square kilometers and has only limited time, motivation, and transportation available, many fires go unreported.[1]

In addition, it is unclear which fires are to be reported and which not. Pasture fires and wildfires in natural forests and tree plantations are clearly reported. *Tavy* slash-and-burn fires in the rain forest may be reported—as fires destructive of forests—or not, for they are better categorized as agricultural clearance. Not reported, most likely, are the thousands of crop field fires that cover only a few hundred square meters at a time. In most years, prairie fires (e.g., grassland and pasture fires) make up 95 to 99 percent of official fire extent.

An additional problem may be poor training at estimating areas. In an interview,[2] a retired Forest Service agent noted that his agents consistently underestimated fire areas in the field, guessing 4 to 8 ha for a scar that covered 20 ha (I noted a similar trend while visiting a fire scar in Andringitra National Park).

Political activities directly influence fire reporting. For example, in 1991, a six-month general strike paralyzed the country as the people protested the authoritarian regime of Didier Ratsiraka. Fires that year were significant across the island (e.g., Andriamampionona 1992), yet Forest Service statistics report one of the lowest fire years ever. As forest agents were not working for at least half that year, only few fires were reported.

Finally, the statistics reported are often inconsistent or error prone. Frequently, the provincial totals do not add up to the total given. Also, figures re-

1. One Forest Service agent stated that if an agent spent only one day in the field during a particular month, and saw only one fire of one hectare extent on that day, that would be the total reported (interview, 15 June 1998).

2. Interview, 20 May 1999.

ported in different government documents frequently do not match. Most years, figures differ from one source to another by only thousands of hectares, perhaps due to rounding errors. But a few years show surprising differences. For example, for 1990, the Ministère de l'Agriculture (1991), citing DEF, reports 231,001 ha burned. DEF itself, in its annual report, cites 534,308 ha, but in a spreadsheet from the DEF's Service de la Conservation de la Biodiversité, the total reported is 242,079 ha (this figure is also reported by ONE/Instat 1994).

The Forest Service itself, as well as others (e.g., Schnyder and Serretti 1997) point out that the statistics are absolutely unreliable. According to the Annual Report in 1996, for example, "the [statistics] of vegetation fires . . . in no case reflect on-the-ground realities. These statistics are incomplete and clearly lower than reality." However, the statistics are the only ones in existence, and they continue to be widely and uncritically used in documents and reports. With the above reservations in mind, please regard the statistics with caution.

Appendix 2
Summary of Colonial and Malagasy Legislation, Government Acts, and Important Circulars Relevant to Vegetation Fires, Madagascar, 1896–2002

Legislation	Date	Source	Comments
Circulaire 286 from Governor Gallieni	28 Dec. 1897	JOM 15jan1898	Encourages local authorities to use their powers to stop pasture fires until more complete regulations can be established.
Décret établissant le régime forestier applicable à la colonie de Madagascar	10 Feb. 1900	JOM 7apr1900	Bans fires within 200 m of forests, except by owner and with governor's permission. However, several authors state that all pasture fires were banned in this year (Grandidier 1918; Bertrand and Sourdat 1998).
Circulaire from Governor Gallieni	22 Mar. 1904	JOM 30mar1904	Reauthorizes all pasture fires not closer than 2 km from forests, due to economic importance of cattle and to fight locusts.
Arrêté réglementant les feux de brousse	26 Feb. 1907	JOM 9mar1907	Tries again to limit fires by banning all fires not necessary for pasture renewal or insect control. Fires must be announced ahead of time to local authorities.
Circulaire au sujet des incendies de forêt	26 May 1909	JOM 12jun1909	Asks officers to ban all fires near forests, especially *tavy* (except where forest already gone)
Décisions locales réglementant les feux de brousse	1907 to 1919	JOM (various)	Province-level rules about authorized pasture fires, based on a template created by the central government.
Arrêté interdisant l'emploi du feu . . .	12 Apr. 1913	JOM 10may1913	Bans all *tavy* and cultivation fires.

Legislation	Date	Source	Comments
Circulaire au sujet de l'interdiction d'employer le feu . . .	29 Apr. 1913	JOM 10may1913	Explains above *arrêté,* stating that the time to phase out *tavy* fires has come.
Décret établissant le régime forestier applicable à Madagascar	28 Aug. 1913	Dez 1968	Due to desire for soil conservation, bans all tavy and pasture fires, except pasture fires lit more than 2 km from forest, in well demarcated zones, and these will be phased out.
Circulaire au sujet de l'incendie des pâturages et des feux de brousse	9 Nov. 1914	JOM 14nov1914	Asks officers not to confuse pasture fires and *tavy.*
Circulaire au sujet des feux de brousse	18 Jan. 1915	JOM 23jan1915	On the difficulty of getting farmers to apply for authorization; thus decentralizes the process
Circulaire au sujet des *tavy* et feux de brousse	8 Oct. 1915	JOM 9oct1915	States that pasture fires should be replaced by improving pastures; asks officers to tolerate *tavy* only in the most extreme cases.
Circulaire au sujet des *tavy*	15 Nov. 1920	JOM 20nov1920	Governor repeats call to replace fires with alternative techniques and calls for officers to forbid *tavy.*
Arrêté créant deux brigades forestières mobiles en vue de la répression des *tavy,* incendies . . .	24 Mar. 1926	JOM 3apr1926	Creates two enforcement brigades with the specific purpose of repressing forest infractions like *tavy.*
Circulaire au sujet des questions agricoles et forestières	4 Feb. 1929	JOM 9feb1929	States that the goal of the colony's agrarian policies is to sedentarize people and stop fires. Encourages officers to enforce ban on *tavy* (a destructive practice that has always been forbidden) and a tight control on other fires (governor's approval required to burn).
Décret réorganisant le régime forestier en Madagascar et Dépendances	15 Jan. 1930	JOM 22nov1930	Article 36: bush and pasture fires are forbidden within 500 m of forest; a 15 m firebreak required around all fires. Articles 58, 59: Penalty for pasture fires is fine of 100–2,000 frs and/or 1–6 months prison. In case the fire setter is

Legislation	Date	Source	Comments
			unknown, the local community is legally responsible.
Arrêté soumettant les mises à feu au régime de la déclaration préalable dans les districts de . . .	6 Nov. 1936	JOM 21nov1936	In specified zones, all prairie fires must be annouced 15 days in advance by written notice to the *chef de district* (as stipulated by 1930 decree).
Décret portant organisation du régime forestier applicable à Madagascar	25 Sep. 1937	JOM 13nov1937	Modifies the 1930 forest decree. Article 36: states that all fires (forest clearings, bushfires, pasture fires) are banned, but then allows *chefs de district* to authorize fires further than 2 km from forest. Decree strengthens repressive measures for all forest crimes. Forest Service agents authorized to write *procès-verbaux* (citations) without consulting the police.
Circulaire relative à l'application du décret du 25 sep. 1937 . . .	27 June 1938	JOM 2jul1938	Explains that the 1937 decree totally bans *tavy* (but allows cultivation in *savoka*). Pasture fires are very damaging, but may be authorized temporarily.
Circulaire relative aux incendies, destruction et défrichement de forêts et aux feux de brousse	26 Nov. 1941	JOM 6dec1941	Encourages officers to take a much harder line against burning; *tavy* is absolutely prohibited; pasture fires temporarily necessary.
Décret modifiant les articles 58 et 59 du décret du 25 jan. 1930	9 Dec. 1941	JOM 20dec1941	Modifies Art. 58 and 59 of the 1930 decree such that local communities (*fokonolona*), represented by their chiefs, are legally responsible for *all* fires within their neighborhood.
Décret validant le décret du 9 déc. 1941	2 Feb. 1944	JOM 29apr1944	Decree by Free French government revalidating previous decree (from Vichy government).
Arrêté 1884-SE/SF réglementant l'application des articles 36, 58, et 59 du décret forestier de 1930	22 Aug. 1950	JOM 2sep1950	Prohibits all grazing in pastures burned without authorization for 6 months and prohibits cultivation in cut or burned state forests.

Legislation	Date	Source	Comments
Décret 55-582 relatif à la protection des forêts dans les territoires d'Afrique relevant du ministre de la France d'outre-mer	20 May 1955	JOM 23feb1957	Fires allowed only for cultivation or pasture renewal. Where there are dangerous wildfires, preemptive controlled burning of all savannas or in and around forests at the beginning of the dry season may be authorized.
Arrêté 25-SE/FOR/CG relatif à la protection des forêts à Madagascar	14 Jan. 1957	JOM 23feb1957	Specifies procedures to satisfy (and strengthen) Décret 55-582. Crop fires require prior authorization and are allowed only on delimited crop lands or on nonsloping lands. *Chefs de district* will decide on pasture zones and period of fires; *chefs de province* can ban all fires if not necessary for pasture renewal, if fires are particularly damaging, or if there is an exceptional drought. *Chefs de province* can regulate the circulation of people or cattle in certain zones and times to reduce the danger of fire. Regarding Art. 58 and 59 of the 1930 decree, collectivities can only be held responsible if the crime is committed within their territory.
Ordonnance 60-127 fixant le régime des défrichements et des feux de végétation	3 Oct. 1960	JOM 15oct1960	Bans *tavy.* Defines three categories of "vegetation fires": • Crop and field edge fires. No permission needed. • Pasture renewal fires. Free during the rainy season and may exceptionally be authorized during other periods. • Wildfires. Uncontrolled fires without economic benefits; banned at all times. Forest Service may use preventive fires and backfires. Infractions to be punished by 15,000 to 300,000 MGF fine and/or prison for 6 mos.–3 years. Allowing cattle to

Legislation	Date	Source	Comments
			pasture on burned land without permission to be fined 100 MGF per head. Collectivities are responsible for fires with unknown author and may be fined in money or labor.
Ordonnance 60-128 fixant la procédure applicable à la répression des infractions à la législation forestière . . .	3 Oct. 1960	JOM 15oct1960	Based on 1930 forest decree and modifications, presents in one text the procedures for forest agents to deal with all infractions. Seeks to speed up the process. Maintains the possibility of out-of-court settlement and idea of collective responsibility.
Arrêté 058 portant autorisation des feux de pâturage	7 Jan. 1961	RDM 1980	Gives dates when pasture fires are legal without authorization, based on a principle of counter-season fires (wet season), e.g.: • Tananarive Province, western areas: 15 Dec.–15 Feb. • Tananarive Province, all other areas: 1 Jan.–1 Mar. • Fianarantsoa Province, central areas: 15 Dec.–15 Apr. • Fianarantsoa Province, western areas: 15 Dec.–1 Apr.
Décret 61-078 fixant les modalités d'application de l'ordonnance 60-128	8 Feb. 1961	JOM 18feb1961	Applies the above law, fixes the procedures for repression, including general templates for *procès-verbaux.*
Décret 61-079 fixant les modalités d'application de l'ordonnance 60-127	8 Feb. 1961	JOM 18feb1961	Fixes the procedure for getting exceptional pasture fire authorizations. Stipulates use of *dina* to create fire-defense committees in each collectivity.
Décret 61-261 instituant une prime de rendement . . .	26 May 1961	JOM 2jun1961	Creates financial bonuses for forest agents for enforcement.
Ordonnance 72-023 tendant a renforcer la répression de certaines infractions . . .	18 Sep. 1972	JOM 30sep1972	Strengthens measures to enforce fire legislation (among others) by no longer accepting attenuating circumstances.

Legislation	Date	Source	Comments
Ordonnance 72-039 abrogeant l'article 25 de l'ordonnance 60-127	30 Oct. 1972	JOM 2dec1972	Speeds up the enforcement process for fire crimes in order to assure "rapid and exemplary repression."
Ordonnance 75-028 modifiant certaines dispositions de l'ordonnance 60-127	22 Oct. 1975	JOM 1nov1975	Makes the fines stipulated by Ord. 60-127 harsher, raising the prison sentence to 5–10 years.
Ordonnance 76-019 portant création d'un tribunal spécial économique	21 May 1976	Andriamampionona 1992	Creates the Tribunal Economique Spécial in order to quickly judge economic crimes, including illegal fires.
Ordonnance 76-030 édictant des mesures exceptionnelles pour la poursuite des auteurs de feux sauvages	21 Aug. 1976	JOM 27aug1976	Establishes fines for withholding information, requires *fokonolona* to report wildfires to nearest legal officer and authorizes them to hold suspects and witnesses for up to 24 hours. Prescribes sentence of forced labor for life for lighting wildfires with at least three of the following circumstances (nighttime, in a group of 2 or more people, while carrying arms, with violence, by using a technique of delayed fire starting, with the aid of a vehicle).
Ordonnance 77-068	1977	Andriamampionona 1992	Moves wildfire crimes to the Tribunal Criminel Spécial, which could theoretically give the death sentence.
Décret 82-313 instituant la tenue de cahier des charges des pâturages	19 July 1982	JOM 7aug1982	Under the rubric of the "exceptional fire authorizations" created by Ord. 60-127, each cattle-raising province will create a plan of pasture burning rotations. Includes provisions to cultivate fodder grass.
Décret 87-143	20 Apr. 1987	Rakatonindrina 1989	Reinforces anti-*tavy* measures and implicates villages in forest protection.
Loi 90-033, la chartre de l'environnement	20 Dec. 1990	JOM no. 2035; RDM 1990	The environmental charter puts into law the national environmental action plan, stating that

Legislation	Date	Source	Comments
			the environment is a primary responsibility of the state.
Loi 96-025 relative à la gestion locale des ressources naturelles renouvelables	30 Sep. 1996	JOM 14oct1996	The GELOSE legislation. Legislates the contractual transfer of the management of renewable natural resources to "basic local communities."
Loi 97-017 portant révision de la législation forestière	8 Aug. 1997	JOM 25aug1997; Henkels 1999	Revised forest legislation. Among other things, aligns itself with national decentralization policy and revalidates Ordonnance 60-127.
Décret 97-1200 portant adoption de la politique forestière malagasy	1997	JOM 1dec1997	A policy document accompanying Loi 97-017. States that the new legislation is based on the need to change from a 1980s focus on food sufficiency and balance of payments to liberalization and decentralization. The principles include making local actors more reponsible, and seeking better management of fires with local participation.
Décret 2002-793 définissant les mesures incitatives à la prévention et à l'éradication des feux de brousse	7 Aug. 2002	DEF	Establishes a committee in each *fivondronana* that is to rank *communes* by their effort and success in fighting fires; the funding of communal development projects is linked to good performance.

References

Adams, J. S., and T. O. McShane. 1992. *The Myth of Wild Africa.* New York: Norton.

AGM (Association des Géographes de Madagascar). 1969. *Atlas de Madagascar.* Tananarive: Bureau pour le Developpement de la Production Agricole and Association des Géographes de Madagascar.

Agrawal, A., and J. Ribot. 1999. Accountability in decentralization: a framework with South Asian and West African cases. *Journal of Developing Areas* 33: 473–502.

Albignac, R. 1989. Impact de l'homme sur l'environnement à Madagascar: raisons sociales et économiques. 223–226 in *Environnement et Gestion des Ressources Naturelles dans la Zone Africaine de l'Océan Indien,* ed. M. Maldague, K. Matuka, and R. Albignac. Paris: UNESCO.

Allen, P. M. 1995. *Madagascar: Conflicts of Authority in the Great Island.* Boulder: Westview.

Althabe, G. 1969. *Oppression et Libération dans l'Imaginaire.* Paris: François Maspero.

Amanor, K. S. 2002. Bushfire management, culture, and ecological modernisation in Ghana. *IDS Bulletin* 33:65–74.

Andriamampianina, J. 1998. Introduction générale. 9–10 in *Feux et déforestation à Madagascar,* ed. A. Bertrand and M. Sourdat. Antananarivo: CIRAD/ORSTOM/CITE.

Andriamampionona, A. 1992. Les feux de brousse dans la région de l'Itasy. Report, Programme FAO/PNUD/MAG/88.032, Jan.–Mar. 1992.

Andrianarivo, J. A. 1990. Analysis of Forest Cover Changes and Estimation of Lemur Population in Northwestern Madagascar Using Satellite Digital Data. Ph.D. dissertation, Duke University.

Andrianavosoa, C. 1996. Madagascar brûle-t-il? *Revue de l'Océan Indien Madagascar* 162 (Dec. 1996): 4–5.

anonymous. 1904. Les feux de brousse. *Bulletin Economique, Madagascar et Dépendences* 4(1): 25–29.

anonymous. 1955. Quelques problèmes locaux: les travaux de «fokonolona», les feux de pâturage. *Bulletin de Madagascar* 111: 741–743.

Apt, J. 1996. The astronaut's view of home. *National Geographic* 190(5): 1–30.

Aquaterre. 1997. *L'atelier de travail sur l'élevage et l'environnement à Madagascar.* Antananarivo: Coopération Française et Aquaterre.

Aubréville, A. 1952. Prospections en chambre (XXVI). *Bois et Forêts des Tropiques* 21: 42–47.
———. 1954. Il n'y aura pas de guerre de l'eucalyptus à Madagascar. *Bulletin de Madagascar* 92: 75–78.
———. 1971. La destruction des forêts et des sols en pays tropical. *Adansonia* 11: 5–39.
Bailly, C., J. de Vergnette, and G. Benoit de Coignac. 1967. Essai de mise en valeur d'une zone des Hautes Plateaux par l'aménagement rationnel: Effet de cet aménagement sur les pertes en terre et le ruissellement. Report, Centre Technique Forestier National, IRAM.
Baron, R. 1890. A Malagasy forest. *Antananarivo Annual* 4(14): 196–211.
———. 1891. The flora of Madagascar. *Antananarivo Annual* 4(15): 322–357.
Barrett, C. B. 1997. Food marketing liberalization and trader entry: evidence from Madagascar. *World Development* 25(5): 763–777.
Bartlett, H. H. 1955. *Fire in Relation to Primitive Agriculture and Grazing in the Tropics: Annotated Bibliography.* Ann Arbor: University of Michigan Department of Botany.
———. 1956. Fire, primitive agriculture, and grazing in the tropics. 692–720 in *Man's Role in Changing the Face of the Earth,* ed. W. L. J. Thomas. Chicago: University of Chicago Press.
Basse, É. 1934. Les groupements végétaux du Sud-ouest de Madagascar. *Annales des Sciences Naturelles, Dixième Série, Botanique* 16(2): 94–229.
Bassett, T. J., and Z. Koli Bi. 2000. Environmental discourses and the Ivorian savanna. *Annals of the Association of American Geographers* 90(1): 67–95.
Batchelder, R. B., and H. F. Hirt. 1966. *Fire in Tropical Forests and Grasslands.* United States Army Natick Laboratories, Earth Sciences Division 67-41-ES.
Batterbury, S., and A. Bebbington. 1999. Environmental histories, access to resources, and landscape change. *Land Degradation and Development* 10(4): 279–289.
Batterbury, S., T. Forsyth, and K. Thomson. 1997. Environmental transformations in developing countries: hybrid research and democratic policy. *Geographical Journal* 163(2): 126–132.
Battistini, R., and P. Vérin. 1972. Man and the environment in Madagascar. 311–338 in *Biogeography and Ecology in Madagascar, Monographiae Biologicae 21,* ed. R. Battistini and G. Richard-Vindard. The Hague: Junk.
Beaujard, P. 1995. Les rituels en riziculture chez les Tañala de l'Ikonogo (Sud-Est de Madagascar). 249–273 in *Civilisations de Madagascar,* ed. S. Evers and M. Spindler. Leiden: IIAS.
Bebbington, A. 1999. Capitals and capabilities: a framework for analyzing peasant viability, rural livelihoods, and poverty. *World Development* 27: 2021–2044.
———. 2000. Reencountering development: livelihood transitions and place transformations in the Andes. *Annals of the Association of American Geographers* 90: 495–520.
Bebbington, A. J., and S. P. J. Batterbury. 2001. Transnational livelihoods and landscapes: political ecologies of globalization. *Ecumene* 8: 369–380.
Behnke, R. H., and I. Scoones. 1992. *Rethinking Range Ecology: Implications for Rangeland Management in Africa.* International Institute for Environment and Development Paper no. 33. London: Overseas Development Institute.
Beinart, W. 1996. Soil erosion, animals, and pasture over the longer term: environmental destruction in Southern Africa. 54–72 in *The Lie of the Land,* ed. M. Leach and R. Mearns. Oxford: James Currey.
Bekker, F. 1878. Pastor Bekker paa Ilaka skriver. *Norsk Misjions-Tidene* 33(19): 379–381.

Belsky, J. M. 1999. Misrepresenting communities: the politics of community-based rural ecotourism in Gales Point Manatee, Belize. *Rural Sociology* 64: 641–666.

Berg, G. M. 1981. Riziculture and the founding of monarchy in Imerina. *Journal of African History* 22: 289–308.

Bergeret, A. 1993. Discours et politiques forestières coloniales en Afrique et à Madagascar. *Revue Français d'Histoire d'Outre-Mer* 79(298): 23–47.

Berry, S. 1989. Social institutions and access to resources. *Africa* 59(1): 41–55.

Bertrand, A. 1994. Revue documentaire préalable à l'élaboration d'une politique et d'une stratégie de gestion des feux de végétation à Madagascar. CIRAD-Forêt, Nov. 1994.

———. 1995. La sécurisation foncière, condition de la gestion viable de ressources naturelles renouvelables? 313–327 in *Sustainable land management in African semi-arid and sub-humid regions,* ed. F. Ganry and B. Campbell. Montpellier: CIRAD.

———. 1997. Les relations entre la sylviculture paysanne et l'élevage laitier en situation d'intensification agricoles sur les Hautes Terres malgaches: les exemples de Manjakandriana et du Vakinankaratra. *L'atelier de travail sur l'élevage et l'environnement à Madagascar.* Antananarivo: Aquaterre.

———. 1999. Le boisement, le bail, et la législation environnementale à Madagascar: trois articles courts. *African Studies Quarterly* 3(2): <web.africa.ufl.edu/asq>.

Bertrand, A., and M. Sourdat. 1998. *Feux et déforestation à Madagascar, revues bibliographiques.* Antananarivo: CIRAD/ORSTOM/CITE.

Bissakonou, J. 1995. Analyse Economique des Ménages dans la zone d'intervention du projet WWF/Andringitra, Madagascar. Thèse pour Diplome d'Ingénieur d'Agronomie Tropicale, CNEARC/WWF Madagascar/ESAT.

Biswell, H. H. 1989. *Prescribed Burning in California Wildlands Vegetation Management.* Berkeley: University of California Press.

Blaikie, P. 1989. Environment and access to resources in Africa. *Africa* 59: 18–40.

———. 1999. A review of political ecology: issues, epistemology, and analytical narratives. *Zeitschrift für Wirtschaftsgeographie* 43(3–4): 131–147.

Blaikie, P., and H. Brookfield. 1987. *Land Degradation and Society.* New York: Methuen.

Blanc-Pamard, C. 1989. Au voleur! Economie de crise et tactiques paysannes: le cas du manioc sur les Hautes Terres malgaches. 198–208 in *Tropiques: Lieux et Liens,* ed. B. Antheaume, C. Blanc-Pamard, J.-L. Chaleard, et al. Paris: Editions de l'ORSTOM.

———. 1998. «La moitié du quart»: Une ethnographie de la crise à Tananarive et dans les campagnes de l'Imerina (Madagascar). *Natures, Sciences, Sociétés* 6(4): 20–32.

Blanc-Pamard, C., J. Bonnemaison, and H. Rakoto Ramiarantsoa. 1997. Tsarahonenana 25 ans après: un terroir «où il fait toujours bon vivre». 25–61 in *Dynamique des Systèmes Agraires,* ed. C. Blanc-Pamard and J. Boutrais. Paris: Editions de l'ORSTOM.

Blanc-Pamard, C., and H. Rakoto Ramiarantsoa. 2000. *Le Terroir et son Double: Tsarahonenana 1966–1992, Madagascar.* Paris: IRD Editions.

Bloch, M. 1995. People into places: Zafimaniry concepts of clarity. 63–77 in *The Anthropology of Landscape,* ed. E. Hirsch and M. O'Hanlan. New York: Oxford University Press.

Bloesch, U. 1997. La problématique du feu dans les aires protégées d'Ankarafantsika et leurs zones périphériques. Rapport de Mission, Conservation International, 24 Nov.

———. 1999. Fire as a tool in the management of a savanna/dry forest reserve in Madagascar. *Applied Vegetation Science* 2: 117–124.

Blumler, M. A. 1998. Biogeography of land-use impacts in the Near East, 215–236 in *Nature's*

Geography: New Lessons for Conservation in Developing Countries, ed. K. S. Zimmerer and K. R. Young. Madison: University of Wisconsin Press.

Bonnemaison, J. 1976. *Tsarahonenana: Des Riziculteurs de Montagne dans l'Ankaratra.* Atlas des Structures Agraires à Madagascar. ORSTOM. Paris: Mouton.

Borie, J. M. 1989. Place et intégration de l'arbre dans l'exploitation agricole des petits périmétres irrigués de l'O.D.R. (Madagascar). Mémoire de fin d'études, CIRAD/CNEARC.

Boserup, E. 1965. *The Conditions of Agricultural Growth.* Chicago: Aldine.

Bosser, J. 1954. Les paturages naturels de Madagascar. *Mémoires de l'Institut Scientifique de Madagascar* Série B, 5: 65–77.

———. 1969. *Graminées des Pâturages et des Cultures à Madagascar.* Mémoire ORSTOM no. 35. Paris: ORSTOM.

Bosshard, A., and F. Klötzli. 1997. Plateau d'Andohariana, RNI Andringitra, Madagascar: Rapport de mission, WWF Madagascar/PCDI Andringitra, 9–16 février 1997.

Bosshard, A., and T. Mermod. 1996. Réserve Nationale Intégrale 5 Andringitra (Madagascar), plateau de l'Andohariana: la Végétation et stratégies pour sa conservation. Rapport de mission pour WWF/PCDI Andringitra du 27.9–27.10.1995, ETH, Janvier 1996.

Bourdariat, M. A. J. 1911. Les Forêts de Madagascar: leur mise en valeur. Congrès de l'Afrique Orientale, Paris, 9–14 Octobre 1911.

Bourgeat, F., and G. Aubert. 1972. Les sols ferralitiques à Madagascar. *Madagascar Revue de Géographie* 20: 1–23.

Bradt, H., D. Schuurman, and N. Garbutt. 1996. *Madagascar Wildlife: A Visitor's Guide.* Bucks, UK: Bradt.

Brady, N. C. 1990. *The Nature and Property of Soils,* 10th ed. New York: Macmillan.

Braithwaite, R. W. 1991. Aboriginal fire regimes of monsoonal Australia in the 19th century. *Search* 22(7): 247–249.

Brand, J. 1998. Das Agro-Ökologische System am Ostabhang Madagaskars: Ressourcen- und Nutzungsdynamik unter Brandrodung. Inauguraldissertation der Philosophischen-naturwissenschaftlichen Fakultät, Universität Bern.

———. 1999. La dégradation des ressources naturelles sous culture sur brûlis: étude de cas de la région de Beforona, falaise est de Madagascar, 141–158 in *African Mountain Development in a Changing World,* ed. H. Hurni and J. Ramamonjisoa. Antananarivo: African Mountains Association, UNU, African Highlands Initiative.

Brand, J., and J. L. Pfund. 1998. Site-and watershed-level assessment of nutrient dynamics under shifting cultivation in eastern Madagascar. *Agriculture, Ecosystems, and Environment* 71: 169–183.

Brookfield, H., and C. Padoch. 1994. Appreciating agrodiversity: a look at the dynamism and diversity of indigenous farming practices. *Environment* 36: 271–289.

Brown, M. 1995. *A History of Madagascar.* London: Damien Tunnacliffe.

Bruzon, V. 1994. Les pratiques du feu en Afrique subhumide: examples des milieux savanicoles de la Centrafrique et de la Côte d'Ivoire. 147–162 in *A la Croisée des Parcours: Pasteurs, Eleveurs, Cultivateurs,* ed. C. Blanc-Pamard and J. Boutrais. Paris: Editions ORSTOM.

Bryant, R. 1997. *The Political Ecology of Forestry in Burma: 1824–1994.* Honolulu: University of Hawai'i Press.

———. 1998. Resource politics in colonial south-east Asia, 29–51 in *Environmental Challenges in South-East Asia,* ed. V. T. King. Richmond, Surrey: Curzon.

Bryant, R. L., and S. Bailey. 1997. *Third World Political Ecology.* London: Routledge.

Burney, D. A. 1996. Climate change and fire ecology as factors in the Quaternary biogeography of Madagascar. 49–58 in *Biogéographie de Madagascar,* ed. W. R. Lourenço. Paris: Editions de l'ORSTOM.

———. 1997. Theories and facts regarding Holocene environmental change before and after human colonization. 75–89 in *Natural and Human-induced Change in Madagascar,* ed. B. D. Patterson and S. M. Goodman. Washington: Smithsonian Institution Press.

Butzer, K. W. 1989. Cultural ecology. 192–208 in *Geography in America,* ed. G. L. Gaile and C. J. Willmott. Columbus: Merrill.

———. 1990. The realm of cultural-human ecology: adaptation and change in historical perspective. 685–701 in *The Earth as Transformed by Human Action,* ed. B. L. Turner II, W. C. Clark, R. W. Kates, J. F. Richards, J. T. Mathews, and W. B. Meyer. Cambridge: Cambridge University Press.

Cahoon, D. R. J., B. J. Stocks, J. S. Levine, W. R. I. Cofer, and K. P. O'Neill. 1992. Seasonal distribution of African savanna fires. *Nature* 359: 812–815.

Callet, R. P. 1908. *Tantaran'ny Andriana eto Madagascar.* Tananarive: Imprimerie Officielle.

———. 1953–1958. *Histoire des Rois. Tantaran'ny Andriana.* Collection des Documents Concernant Madagascar et les Pays Voisins, IV. Tananarive: Académie Malgache.

Campbell, B. (ed.). 1996. *The Miombo in Transition.* Bogor, Indonesia: Center for International Forestry Research (CIFOR).

Canaby, M. 1933. Les forêts de Madagascar. *Revue de Géographie Commerciale* 57: 62–80.

Carney, J. A. 1996. Converting the wetlands, engendering the environment: the intersection of gender with agrarian change in Gambia. 165–187 in *Liberation Ecologies,* ed. R. Peet and M. Watts. London: Routledge.

Catat, L. [ca. 1895]. *Voyage à Madagascar (1889–1890).* Paris: l'Univers Illustré.

Chapotte. 1898. Les forêts de Masoala. *Colonie de Madagascar. Notes, Reconnaissances et Explorations* 4:870–889.

Charon, A. 1897. Etude sur les prairies et l'élevage du boeuf dans le pays Sihanaka et le Haut-Bouéni. *Colonie de Madagascar: Notes, Reconnaissances et Explorations* 2(12): 561–591.

Chauvet, B. 1972. The forest of Madagascar, 191–200 in *Biogeography and Ecology in Madagascar, Monographiae Biologicae 21,* ed. R. Battistini and G. Richard-Vindard. The Hague: Junk.

Clements, F. E. 1916. *Plant Succession: An Analysis of the Development of Vegetation.* Carnegie Institute of Washington Publications no. 242.

Cline-Cole, R. 1996. Dryland forestry: manufacturing forests and farming trees in Nigeria. 122–139 in *The Lie of the Land,* ed. M. Leach and R. Mearns. Oxford: James Currey.

Conklin, H. C. 1954. The ethnoecological approach to shifting agriculture. *Transactions of the New York Academy of Science* 17: 133–142.

Connell, J. H., and R. O. Slatyer. 1977. Mechanisms of succession in natural communities and their role in community stability and organization. *American Naturalist* 111: 1119–1144.

Conservation International, DEF, CNRE, and FTM. [ca. 1995]. Formations végétales et domaine forestier national de Madagascar (carte 1:1,000,000). Antananarivo: CI/DEF/CNRE/FTM.

Coomes, O. T., and B. L. Barham. 1997. Rain forest extraction and conservation in Amazonia. *Geographical Journal* 163(2): 180–188.

Coomes, O. T., and G. J. Burt. 1997. Indigenous market-oriented agroforestry: dissecting local diversity in western Amazonia. *Agroforestry Systems* 37: 27–44.

Coomes, O. T., F. Grimard, and G. J. Burt. 2000. Tropical forests and shifting cultivation: secondary forest fallow dynamics among traditional farmers of the Peruvian Amazon. *Ecological Economics* 32: 109–124.

Cori, G., and P. Trama. 1979. *Types d'élevage et de vie rurale à Madagascar.* Travaux et Documents de Géographie Tropicale no. 37. Bordeaux: Centre d'Etudes de Géographie Tropicale, Centre National de la Recherche Scientifique.

Coudreau, J. 1937. La forêt malgache: Son role dans l'économie générale du pays; sa conservation; son amélioration. *Bulletin Economique, Madagascar* 1937(1): 75–96.

Coulaud, D. 1973. *Les Zafimaniry: un Groupe Ethnique de Madagascar à la Poursuite de la Forêt.* Antananarivo: Fanontam-Boky Malagasy.

Cruikshank, B. 1999. *The Will to Empower: Democratic Citizens and Other Subjects.* Ithaca: Cornell University Press.

Crummey, D. (ed.). 1986. *Banditry, Rebellion, and Social Protest in Africa.* Portsmouth, N.H.: Heinemann.

Crutzen, P. J., and M. O. Andreae. 1990. Biomass burning in the tropics: impact on atmospheric chemistry and biogeochemical cycles. *Science* 250: 1669–1678.

Dahl, Ø. 1998. *Merkverdige Madagaskar: øya mellom øst og vest.* Oslo: Spartacus Forlag.

Dahle, L. 1884. Geographical fictions with regard to Madagascar. *Antananarivo Annual* 2:403–407.

Dandoy, G. 1973. *Terroirs et Economies Villageoises de la Région de Vavatenina (Côte Orientale Malgache).* Vol. 1. *Atlas des Structures Agraires à Madagascar.* ORSTOM. Paris: Mouton.

Dark, P. [ca. 1989]. Uncompleted dissertation manuscript on ecological and economic changes among the Betsimisaraka agriculturalists of northeastern Madagascar. Yale University Department of Anthropology.

Davis, M. 1998. *Ecology of Fear.* New York: Vintage.

DEF (Direction des Eaux et Forêts). 1982. La politique du ministère en matiere forestière: Direction des Eaux et Forêts et de la Conservation des Sols, Ministère de la Production Agricole et de la Réforme Agraire, Repoblika Demokratika Malagasy, 29 Nov.

———. 1996. *Inventaire Ecologique Forestier National.* Report, République de Madagascar, Ministère de l'Environnement, Plan d'Actions Environnementales PE1, Direction des Eaux et Forêts, DFS Deutsche Forstservice GmbH, Entreprise d'Etudes de Développement Rural Mamokatra, Foiben-Taosarintanin'i Madagasikara, Nov.

de Montgolfier, J. 1991. Les feux de forêts: idées fausses et idées justes. *Sécheresse* 2:183–188.

Denevan, W. M. 1992. The pristine myth: the landscape of the Americas in 1492. *Annals of the Association of American Geographers* 82(3): 369–385.

Deschamps, H. 1965. *Histoire de Madagascar,* 3rd ed. Paris: Editions Berger-Levrault.

Desjeux, D. 1979. *La Question Agraire à Madagascar.* Paris: Editions l'Harmattan.

Dewar, R. E. 1984. Extinctions in Madagascar: the loss of the subfossil fauna. 574–593 in *Quaternary Extinctions: a Prehistoric Revolution,* ed. P. S. Martin and R. G. Klein. Tucson: University of Arizona Press.

Dez, J. 1966. Les feux de végétation: aperçus psycho-sociologiques. *Bulletin de Madagascar* 247: 1211–1229.

———. 1968. La limitation des feux de végétation. *Tany Malagasy/Terre Malgache* 4: 97–124.

———. 1970. Eléments pour une étude de l'économie agro-sylvie-pastorale de l'Imerina ancienne. *Tany Malagasy/Terre Malgache* 8: 9–60.

Dommergues, Y. 1954. Action du feu sur la microflore des sols de prairie. *Mémoire de l'Institut Scientifique de Madagascar, Serie D* D-6: 149–158.

Donque, G. 1979. Bilan de dix-sept années de recherches du Laboratoire de Géographie de l'Université de Madagascar. *Bulletin de l'Academie Malgache* 55(1–2): 309–316.

Dougill, A. J., D. S. G. Thomas, and A. L. Heathwaite. 1999. Environmental change in the Kalahari: integrated land degradation studies for nonequilibrium dryland environments. *Annals of the Association of American Geographers* 89(3): 420–442.

Dove, M. R. 1983. Swidden agriculture and the political economy of ignorance. *Agroforestry Systems* 1(1): 85–99.

Durbin, J. 1994. Integrating Conservation and Development: the Role of Local People in the Maintenance of Protected Areas in Madagascar. Doctor of Philosophy Thesis in Ecology, University of Kent, Canterbury.

Durbin, J. C., and J. A. Ralambo. 1994. The role of local people in the successful maintenance of protected areas in Madagascar. *Environmental Conservation* 21(2): 115–120.

Dwyer, E., J. M. C. Pereira, J.-M. Grégoire, and C. C. DaCamara. 1999. Characterization of the spatio-temporal patterns of global fire activity using satellite imagery for the period April 1992 to March 1993. *Journal of Biogeography* 27: 57–69.

Edmond, R. 1996. Analyses floristiques et écologique de la 'savane subalpine' d'Andohariana (secteur Nord Andringitra). Rapport de mission pour PCDI Andringitra, Laboratoire d'Ecologie Végétale, Faculté des Sciences, Université d'Antananarivo, 11–28 Feb. 1996.

Edwards, M., S. Armbruster, P. Blaikie, I. Kalland, J. Klein, H. Lein, G. Olsson, I. Pareliussen, J. Ratsirarson, and B. Réau. 2002. I hver sin verden: skogforvaltning og samfunnsutvikling på Madagaskars høyland. 81–97 in *Forskning på tvers: Tverrfaglige forskningsprosjekter ved NTNU.* Trondheim: Tapir Trykk.

EESR Lettres. 1992. Synthèse de Conférence. 25–30 in *Colloque pour chercheurs en sciences sociales et opérateurs économiques sur changements sociaux dans la région du Vakinankaratra.* Antananarivo: Tranompirinty FLM.

Einrem, J. [1912] 1936. *Gjennem Grønne Briller.* Samlede Verker. Bergen: A. S. Lunde.

EIU. 1998; 2001. *Madagascar: Country Profile.* London: Economist Intelligence Unit.

Elliot, G. F. S. 1892. Notes on a botanical trip in Madagascar. *Antananarivo Annual* 4(16): 394–398.

Ellis, R. W. 1838. *History of Madagascar.* Vol. 1. London: Fisher, Son, and Co.

———. 1859. *Three Visits to Madagascar during the Years 1853–1854–1856.* New York: Harper and Brothers.

Escobar, A. 1995. *Encountering Development.* Princeton: Princeton University Press.

Esoavelomandroso, M. 1986. La forêt dans le Mahafale aux XIXe et XXe siècle. Unpublished proceedings of *Séminaire de l'U.E.R. d'Histoire "Arbres et Plantes à Madagascar."*

Espeland, W. N. 1998. *The Struggle for Water: Politics, Rationality, and Identity in the American Southwest.* Chicago: University of Chicago Press.

Fairhead, J., and M. Leach. 1996. *Misreading the African Landscape.* Cambridge: Cambridge University Press.

———. 1998. *Reframing Deforestation.* London: Routledge.

Falloux, F., and L. M. Talbot. 1993. *Crisis and Opportunity.* London: Earthscan.

Fanony, F. 1989. Un modèle de stratégie de conservation de la forêt à Madagascar: L'example d'Andrianampoinimerina. 349–352 in *Environnement et Gestion des Ressources Naturelles dans la Zone Africaine de l'Océan Indien*, ed. M. Maldague, K. Matuka, and R. Albignac. Paris: UNESCO.

Favre, J.-C. 1996. Traditional utilization of the forest. 33–40 in *Ecology and Economy of a Tropical Dry Forest in Madagascar*, ed. J. U. Ganzhorn and J.-P. Sorg. *Primate Report* 46-1, pp. 33–40.

Feeley-Harnik, G. 2001. *Ravenala madagascariensis* Sonnerat: the historical ecology of a "flagship species" in Madagascar. *Ethnohistory* 48: 31–86.

Ferguson, J. 1994. *The Anti-politics Machine*. Minneapolis: University of Minnesota Press.

Flacourt, E. d. [1658] 1995. *Histoire de la Grande Isle Madagascar*, edited and annotated by Claude Allibert. Paris: Karthala/INALCO.

Fortmann, L. 1995. Talking claims: discursive strategies in contesting property. *World Development* 23(6): 1053–1063.

Frappa, C. 1947. L'apiculture et la sériculture. 345–346 in *Madagascar*, ed. M. d. Coppet. Paris: Encyclopédie de l'Empire Français.

Frost, P. 1996. The ecology of miombo woodlands. 11–57 in *The Miombo in Transition: Woodlands and Welfare in Africa*, ed. B. Campbell. Bogor, Indonesia: CIFOR.

Fujisaka, S., G. Escobar, and E. Veneklaas. 1998. Plant community diversity relative to human land uses in an Amazon forest colony. *Biodiversity and Conservation* 7: 41–57.

Gade, D. W. 1985. Savanna woodland, fire, protein and silk in highland Madagascar. *Journal of Ethnobiology* 5(2): 109–122.

———. 1996. Deforestation and its effects in highland Madagascar. *Mountain Research and Development* 16(2): 101–116.

Gade, D. W., and A. N. Perkins-Belgram. 1986. Woodfuels, reforestation, and ecodevelopment in highland Madagascar. *GeoJournal* 12(4): 365–374.

Gallegos, C. M. 1997. Unrealized potential: Madagascar. *Journal of Forestry* 95(2): 10–15.

Garreau, J.-M., L. Alexandris, and M.-R. Manantsara. 2001. La GELOSE: condition préalable pour l'innovation et le développement dans le contexte de la culture sur brûlis? 49–53 in *Culture sur Brûlis: Vers l'Application des Résultats de Recherche*, ed. P. Kistler, P. Messerli, and S. Wohlhauser. Antananarivo: Projet BEMA/FOFIFA, Projet EPB/ESSA-Forêt.

Gasse, F., and 17 others. 1994. A 36 ka environmental record in the southern tropics: Lake Tritrivakely (Madagascar). *Comptes Rendus de l'Académie Scientifique de Paris* 318 série II: 1513–1519.

Gauld, R. 2000. Maintaining centralized control in community-based forestry: policy construction in the Philippines. *Development and Change* 31: 229–254.

Gauthier, M., and T. Ravoavy. 1997. Appui à l'établissement des contrats locaux dans le cadre de GELOSE. USAID/Tropical Research and Development, Jan. 1997.

Gautier, E. F. 1902. *Madagascar: Essai de Géographie Physique*. Paris: Augustin Challamel.

Gautier, L., C. Chatelain, and R. Spichiger. 1999. Déforestation, altitude, pente et aires protégées: une analyse diachronique des défrichments sur le pourtour de la Réserve Spéciale de Manongarivo (NW de Madagascar). 255–279 in *African Mountain Development in a Changing World*, ed. H. Hurni and J. Ramamonjisoa. African Mountain Association/UNU/African Highlands Initiative.

Geist, H. J., and E. F. Lambin. 2002. Proximate causes and underlying driving forces of tropical deforestation. *BioScience* 52: 143–150.

Genini, M. 1996. Deforestation. *Primate Report* 46-1: 49–56.

Geoffroy, M. 1923. L'élevage à Madagascar. In *Première Foire Commerciale Officielle*. Tananarive: Imprimerie de l'Imerina.

———. 1931. Les feux de brousse (traduction). La pratique en est-elle bonne ou mauvaise: nouvelles lumières tirées d'expériences faites à l'Ecole Agriculture de Cedara (Natal). *Bulletin Economique, Madagascar* 52(March 1931): 79–82.

Gezon, L. L. 1997. Political ecology and conflict in Ankarana, Madagascar. *Ethnology* 36(2): 85–100.

———. 1999. From adversary to son: political and ecological process in northern Madagascar. *Journal of Anthropological Research* 55(1): 71–97.

Gezon, L. L., and B. Z. Freed. 1999. Agroforestry and conservation in northern Madagascar: hopes and hindrances. *African Studies Quarterly* 3(2):<web.africa.ufl.edu/asq>.

Ghai, D., and J. M. Vivian (eds.). 1992. *Grassroots Environmental Action: People's Participation in Sustainable Development*. London: Routledge.

Ghimire, K. B. 1994. Parks and people: livelihood issues in national parks management in Thailand and Madagascar. *Development and Change* 25: 195–229.

Gilibert, J., P. Dubois, P. Granier, P. Grenier, and H. Serres. 1974. Lutte contre les feux sauvages dans une exploitation de ranching. *Bulletin de Madagascar* 24: 465–480.

Gill, A. M., R. H. Groves, and I. R. Noble (eds.). 1981. *Fire and the Australian Biota*. Canberra: Australian Academy of Science.

Gillon, D. 1983. The fire problem in tropical savannas. 617–641 in *Tropical Savannas, Ecosystems of the World 13*, ed. F. Bourlière. Amsterdam: Elsevier.

Girod-Genet, L. 1898. Renseignements forestiers, tournée du Chef du Service des Forêts de Tananarive à Antanifotsy à Ambositra à Fianarantsoa. *Journal Officiel de Madagascar* 294: 2349–2350; 295: 2356.

———. 1899. Les forêts à Madagascar. *Colonie de Madagascar. Notes, Reconnaissances et Explorations* 5(25): 51–85.

Goldammer, J. G. (ed.). 1990. *Fire in the Tropical Biota*. Berlin: Springer Verlag.

Goldammer, J. G., J. L. Pfund, M. R. Helfert, K. P. Lulla, and STS-61 Mission Crew. 1996. Use of the earth observation system in the space shuttle program for research and documentation of global vegetation fires: a case study from Madagascar. 236–240 in *Biomass Burning and Global Change*, vol. 1, ed. J. S. Levine. Cambridge, MA: MIT Press.

Goody, J. 1980. Rice-burning and the Green Revolution in northern Ghana. *Journal of Development Studies* 16:136–155.

Gottesfeld, L. M. J. 1994. Aboriginal burning for vegetation management in northwest British Columbia. *Human Ecology* 22: 171–188.

Goujon, Bailly, de Vergnette, Benoit de Cognac, and Roche. 1968. Conservation des sols en Afrique et à Madagascar. *Bois et Forêts des Tropiques* 119: 3–12; 120: 3–14.

Grandidier, M. G. 1918. L'Elevage à Madagascar. In *Congrès d'Agriculture Coloniale, 1 Mar. 1918*. Paris: Augustin Challamel.

Grangeon. 1906. Etude sur le landibe. *Bulletin Economique de Madagascar* 2ème trim.: 121–127.

———. 1910. Les bois de tapia. *Bulletin Economique de Madagascar* 10(2ème sem.): 181–185.

Granier, P. 1965. Le rôle de l'élevage extensif dans la modification de la végétation à Madagascar. *Bulletin de Madagascar* 235: 1047–1055.

Granier, P., and H. Serres. 1966. A propos de l'incidence des feux de brousse sur l'évolution des pâturages. *Bulletin de Madagascar* 16(243): 789–791.

Gray, L. 1999. Is land being degraded? A multi-scale investigation of landscape change in southwestern Burkina Faso. *Land Degradation and Development* 10: 329–343.

Green, G. M., and R. W. Sussman. 1990. Deforestation history of the eastern rain forests of Madagascar from satellite images. *Science* 248: 212–215.

Grégoire, J.-M. 1993. Description quantitative des régimes de feu en zone soudanienne d'Afrique de l'Ouest. *Sécheresse* 4: 37–45.

Grossman, L. 1998. *The Political Ecology of Bananas.* Chapel Hill: University of North Carolina Press.

Grove, R. H. 1995. *Green Imperialism.* Cambridge: Cambridge University Press.

Guillermin, A. 1947. Les forêts. 23–42 in *Madagascar,* ed. M. d. Coppet. Paris: Encyclopédie de l'Empire Français.

Gupta, A. 1995. Blurred boundaries: the discourse of corruption, the culture of politics, and the imagined state. *American Ethnologist* 22(2): 374–402.

Hackel, J. D. 1999. Community conservation and the future of Africa's wildlife. *Conservation Biology* 13: 726–734.

Hahn, S. 1982. Hunting, fishing, and foraging: common rights and class relations in the postbellum South. *Radical History Review* 26: 37–64.

Hardenbergh, S. H. B., G. Green, and D. Peters. 1995. The relationship of human resource use, socioeconomic status, health, and nutrition near Madagascar's protected areas: an assessment of assumptions and solutions. 57–58 in *Environmental Change in Madagascar,* ed. B. D. Patterson, S. M. Goodman, and J. L. Sedlock. Chicago: Field Museum.

Harper, J. 2002. *Endangered Species: Health, Illness, and Death among Madagascar's People of the Forest.* Durham: Carolina Academic Press.

Harrison, P., and F. Pearce. 2000. *AAAS Atlas of Population and Environment.* Berkeley: University of California Press.

Hay, D., P. Linebaugh, J. G. Rule, E. P. Thompson, and C. Winslow (eds.). 1975. *Albion's Fatal Tree: Crime and Society in Eighteenth-Century England.* New York: Pantheon.

Head, L. 2000. *Second Nature.* Syracuse: Syracuse University Press.

Hecht, S. 1985. Environment, development, and politics: capital accumulation and the livestock sector in eastern Amazonia. *World Development* 13(6): 663–684.

Heim, R. 1935. L'état actuel des dévastations forestières à Madagascar. *Revue de Botanique Appliquée et d'Agriculture Tropicale* 15: 418–426.

Helfert, M. R., and C. A. Wood. 1986. Shuttle photos show Madagascar erosion. *Geotimes* 31(3): 4–5.

Henkels, D. M. 1999. Une vue d'ensemble du droit environnemental Malgache. *African Studies Quarterly* 3(2): <web.africa.ufl.edu/asq>.

Herinivo, A. 1994. Inventaire floristique et approche sur l'écologie des savanes de la région de Tsiandro (partie orientale du plateau de Bemaraha). Mémoire DEA, Université d'Antananarivo.

Hill, R., P. Griggs, and Bamanga Bubu Ngadimunku. 2000. Rainforests, agriculture, and aboriginal fire-regimes in wet tropical Queensland, Australia. *Australian Geographical Studies* 38:138–157.

Ho, P. 2000. The myth of desertification on China's northwestern frontier. *Modern China* 26: 348–395.

Hoeltgen, D. 1994. Where the hills catch fire. *Ceres* 145: 42–45.

Hoerner, J. M. 1982. Les vols de boeufs dans le sud malgache. *Madagascar Revue de Géographie* 41: 85–105.

Holland, P., and S. Olson. 1989. Introduced versus native plants in austral forests. *Progress in Physical Geography* 13(2): 260–293.

Holmes, H. 1997. Mudhopping in Madagascar. *Sierra*, January–February, 22–23.

Homewood, K. M., and W. A. Rodgers. 1991. *Maasailand Ecology.* Cambridge: Cambridge University Press.

Horn, S. P. 1998. Fire management and natural landscapes in Chirripo Paramo, Chirripo National Park, Costa Rica. 125–146 in *Nature's Geography* ed. K. S. Zimmerer and K. R. Young. Madison: University of Wisconsin Press.

Humbert, H. 1927. Principaux aspects de la végétation à Madagascar: La destruction d'une flore insulaire par le feu. *Mémoires de l'Académie Malgache* Fascicule V.

———. 1938. Les aspects biologiques du problème des feux de brousse et la protection de la nature dans les zones intertropicales. *Bulletin des Séances—Institut Royal Colonial Belge* 9: 811–835.

———. 1947. La végétation. 47–62 in *Madagascar,* ed. M. d. Coppet. Paris: Encyclopédie de l'Empire Français.

———. 1949. La dégradation des sols à Madagascar. *Mémoires de l'Institut de Recherche Scientifique de Madagascar* D1(1): 33–52.

———. 1953. Le probléme du recourse au feux courants. *Revue Internationale de Botanique Appliqué et d'Agronomie Tropicale* 363: 19–28.

———. 1955. Les territoires phytogéographiques de Madagascar: Leur cartographie. *Année Biologique* 31(5–6): 439–448.

Huntsinger, L., and J. W. Bartolome. 1992. Ecological dynamics of *Quercus* dominated woodlands in California and southern Spain: a state-transition model. *Vegetatio* 99–100: 299–305.

Huntsinger, L., and S. McCaffrey. 1995. A forest for the trees: forest management and the Yurok environment, 1850 to 1994. *American Indian Culture and Research Journal* 19(4): 155–192.

Hurni, H. 1983. Soil erosion and soil formation in agricultural ecosystems, Ethiopia and northern Thailand. *Mountain Research and Development* 3(2): 131–142.

Isaacman, A. 1982. The Mozambique cotton cooperative: the creation of a grassroots alternative to forced commodity production. *African Studies Review* 25: 5–25.

IUCN. 1972. *Comptes rendus de la Conférence internationale sur la Conservation de la Nature et de ses Ressources à Madagascar, Tananarive 7–11 Octobre,* 1970. Publications UICN Nouvelle Série, Document Supplémentaire no. 36. Morges, Switzerland: IUCN.

IUCN, UNEP, and WWF. 1980. *World Conservation Strategy.* Gland, Switzerland: IUCN.

Ives, J. D., and B. Messerli. 1989. *The Himalayan Dilemma.* London: Routledge.

Jacoby, K. 2001. *Crimes against Nature.* Berkeley: University of California Press.

Jarosz, L. 1991. Women as rice sharecroppers in Madagascar. *Society and Natural Resources* 4: 53–63.

———. 1996. Defining deforestation in Madagascar. 148–164 in *Liberation Ecologies,* ed. R. Peet and M. Watts. London: Routledge.

Jenkins, M. D. 1987. *Madagascar: An Environmental Profile.* Gland, Switzerland, and Cambridge, U.K.: IUCN.

Jolly, A. 1989. The Madagascar challenge: human needs and fragile ecosystems. 189–215 in *Environment and the Poor: Development Strategies for a Common Agenda,* ed. H. J. Leonard. New Brunswick: Transaction Books.

———. 1990. On the edge of survival. 110–121 in *Madagascar: A World Out of Time,* ed. F. Lanting. New York: Aperture.

Jolly, A., and R. Jolly. 1984. Malagasy economics and conservation: a tragedy without villains. 211–217 in *Key Environments: Madagascar,* ed. A. Jolly, P. Oberlé, and R. Albignac. Oxford: IUCN/Pergamon.

Julien, M. G. 1900. *Code des 305 Articles Promulgué le 29 Mars 1881.* Tananarive: Imprimerie Officielle.

Kalaora, B., and A. Savoye. 1986. *La Forêt Pacifiée: Sylviculture et Sociologie au XIXe Siècle.* Collection Alternatives Paysannes. Paris: L'Harmattan.

Kaufmann, J. C. 2001. La Question des Raketa: colonial struggles with prickly pear cactus in southern Madagascar, 1900–1923. *Ethnohistory* 48: 87–121.

Kayll, A. J. 1974. Use of fire in land management. 483–511 in *Fire and Ecosystems,* ed. T. T. Kozlowski and C. E. Ahlgren. New York: Academic Press.

Kepe, T., B. Cousins, and S. Turner. 2001. Resource tenure and power relations in community wildlife: the case of Mkambati area, South Africa. *Society and Natural Resources* 14: 911–925.

Kepe, T., and I. Scoones. 1999. Creating grasslands: social institutions and environmental change in Mkambati Area, South Africa. *Human Ecology* 27(1): 29–53.

Kistler, P., P. Messerli, and S. Wohlhauser (eds.). 2001. *Culture sur Brûlis: Vers l'Application des Résultats de Recherche.* Antananarivo: Projet BEMA/FOFIFA et Projet EPB/ESSA Forêt.

Klooster, D. 1999. Community-based forestry in Mexico: can it reverse processes of degradation? *Land Degradation and Development* 10: 365–381.

———. 2000. Community forestry and tree theft in Mexico: resistance or complicity in conservation? *Development and Change* 31: 281–305.

Koechlin, J. 1972. Flora and vegetation of Madagascar. 145–190 in *Biogeography and Ecology in Madagascar,* ed. R. Battistini and G. Richard-Vindard. The Hague: Junk.

———. 1993. Grasslands of Madagascar. 291–301 in *Natural Grasslands,* ed. R. T. Coupland. Amsterdam: Elsevier.

Koechlin, J., J.-L. Guillaumet, and P. Morat. 1974. *Flore et Végétation de Madagascar.* Vaduz: J. Cramer.

Kottak, C. P. 1980. *The Past in the Present.* Ann Arbor: University of Michigan Press.

Kozlowski, T. T., and C. E. Ahlgren (eds.). 1974. *Fire and Ecosystems.* New York: Academic Press.

Kramer, R. A., D. D. Richter, S. Pattanayak, and N. P. Sharma. 1997. Ecological and economic analysis of watershed protection in Eastern Madagascar. *Journal of Environmental Management* 49: 277–295.

Kremen, C., V. Razafimahatratra, R. P. Guillery, J. Rakotomalala, A. Weiss, and J.-S. Ratsisompatrarivo. 1999. Designing the Masoala National Park in Madagascar based on biological and socioeconomic data. *Conservation Biology* 13: 1055–1068.

Kuhlken, R. 1999. Settin' the woods on fire: rural incendiarism as protest. *Geographical Review* 89: 343–363.

Kuhnholtz-Lordat, G. 1938. *La Terre Incendiée.* Nimes: Editions de la Maison Carrée.

Kull, C. A. 1995. Land use in the highlands of Madagascar: the explanation of land use change and implications on development policy. M.A. thesis, University of Colorado.

———. 1996. The evolution of conservation efforts in Madagascar. *International Environmental Affairs* 8(1): 50–86.

———. 1998. Leimavo revisited: agrarian land-use change in the highlands of Madagascar. *Professional Geographer* 50(2): 163–176.

———. 1999. Observations on repressive environmental policies and landscape burning strategies in Madagascar. *African Studies Quarterly* 3(2): <web.africa.ufl.edu/asq>.

———. 2000a. Deforestation, erosion, and fire: degradation myths in the environmental history of Madagascar. *Environment and History* 6(4): 21–50.

———. 2000b. Isle of Fire: the Political Ecology of Grassland and Woodland Burning in Highland Madagascar. Ph.D. dissertation, University of California, Berkeley.

Kunckel d'Herculais, J. c. 1890. *Invasion des Acridiens vulgo Sauterelles en Algérie, Documents Annexes* (in collection of FOFIFA-MRA, Ambatobe, Antananarivo).

Lafort, C. 1897. De Mananjary à Fianarantsoa. *Colonie de Madagascar: Notes, Reconnaissances et Explorations* 1: 223–234.

Lambin, E. F., and 25 others. 2001. The causes of land-use and land-cover change: moving beyond the myths. *Global Environmental Change* 11:261–269.

Laney, R. M. 1999. Agricultural Change and Landscape Transformations in the Andapa Region of Madagascar. Ph.D. dissertation, Clark University.

———. 2002. Disaggregating induced intensification for land-change analysis: a case study from Madagascar. *Annals of the Association of American Geographers* 92:702–726.

Laris, P. 2002. Burning the seasonal mosaic: preventative burning strategies in the wooded savanna of southern Mali. *Human Ecology* 30:155–186.

Larson, P. M. 1995. Multiple narratives, gendered voices: remembering the past in highland central Madagascar. *International Journal of African Historical Studies* 28(2): 295–325.

Lavauden, L. 1931. Le déboisement et la végétation de Madagascar. *Revue de botanique appliquée et d'agriculture coloniale* 122: 817–824.

———. 1934. Histoire de la législation et de l'administration forestière à Madagascar. *Revue des Eaux et Forêts* 72: 949–960.

Leach, M. 1995. *Rainforest Relations.* Washington: Smithsonian Institution Press.

Leach, M., and J. Fairhead. 2000. Fashioned forest pasts, occluded histories? International environmental analysis in West African locales. *Development and Change* 31: 35–59.

Leach, M., J. Fairhead, and K. S. Amanor (eds.). 2002. *Science and the Policy Process: Perspectives from the Forest. IDS Bulletin* 33(1).

Leach, M., and R. Mearns (eds.) 1996. *The Lie of the Land.* Portsmouth, N.H.: Heinemann.

Leach, M., R. Mearns, and I. Scoones. 1999. Environmental entitlements: dynamics and institutions in community-based natural resource management. *World Development* 27(2): 225–247.

Le Bourdiec, F. 1974. *Hommes et Paysages du Riz à Madagascar.* Antananarivo: FTM.

Le Bourdiec, P. 1972. Accelerated erosion and soil degradation. 227–259 in *Biogeography and Ecology in Madagascar,* ed. R. Battistini and G. Richard-Vindard. The Hague: Junk.

Leigh, J. H., and J. C. Noble. 1981. The role of fire in the management of rangelands in Australia. 471–496 in *Fire and the Australian Biota,* ed. A. M. Gill, R. H. Groves, and I. R. Noble. Canberra: Australian Academy of Science.

Leisz, S., A. Robles, and J. Gage. 1994. *Land and Natural Resource Tenure Security in Madagascar.* Prepared for the Land Tenure Center, University of Wisconsin-Madison.

Lewis, H. T. 1989. Ecological and technological knowledge of fire: Aborigines versus park rangers in northern Australia. *American Anthropologist* 91(4): 940–961.

Li, T. M. 1996. Images of communities: discourse and strategy in property relations. *Development and Change* 27: 501–527.

——— (ed.). 1999. *Transforming the Indonesian Uplands.* Amsterdam: Harwood.

Lowry, P. P. I., G. E. Schatz, and P. B. Phillipson. 1997. The classification of natural and anthropogenic vegetation in Madagascar. 93–123 in *Natural Change and Human Impact in Madagascar,* ed. S. M. Goodman and B. D. Patterson. Washington: Smithsonian Institution Press.

Madon, G. 1996. Gestion locale sécurisée des ressources renouvelables et du foncier (GELOSE), étude de faisabilité. MARGE/ONE.

Mainguet, M. 1997. Elevage et environnement au sein du PCDI/complexe Montagne d'Ambre. *L'Atelier de travail sur l'élevage et l'environnement à Madagascar.* Antananarivo: Aquaterre.

Maldidier, C. 2001. *La décentralisation de la gestion des ressources renouvelables à Madagascar: les premiers enseignements sur les processus en cours et les méthods d'intervention.* Report. Antananarivo: ONE.

Maltby, E., C. J. Legg, and M. C. F. Proctor. 1990. The ecology of severe moorland fire on the North York Moors: effects of the 1976 fires, and subsequent surface and vegetation development. *Journal of Ecology* 78: 490–518.

MARA. 1996. Rural-Urban Dynamics in the Fianarantsoa and Mahajanga High Potential Zones (Madagascar Regional Analysis Project). Report, SARSA—Clark University and IMaTeP, 17 July 1996.

Marchal, J.-Y. 1968. Antanety-Ambohidava: terroir du Moyen-Ouest Malgache. *Madagascar Revue de Géographie* 13: 91–160.

———. 1974. *La Colonisation Agricole au Moyen-Ouest Malgache: la Petite Région d'Ambohimanambola (Sous-Préfecture de Betafo).* Atlas des Structures Agraires à Madagascar. ORSTOM. Paris: Mouton.

Marcus, R. R. 2000. Cultivating Democracy on Fragile Grounds: Environmental Institutions and Non-elite Perceptions of Democracy in Madagascar and Uganda. Ph. D. dissertation, University of Florida.

———. 2001. Seeing the forest for the trees: integrated conservation and development projects and local perceptions of conservation in Madagascar. *Human Ecology* 29: 381–397.

Marcus, R. R., and C. A. Kull. 1999. Setting the stage: the politics of Madagascar's environmental efforts. *African Studies Quarterly* 3(2): 1–7 <web.africa.ufl.edu/asq>.

Martin, R. E., and D. B. Sapsis. 1991. Fire as agents of biodiversity: pyrodiversity promotes biodiversity. Proceedings of the Symposium on Biodiversity of Northwestern California, Santa Rosa, California, 28–30 Oct. 1991.

Matzke, N. J. 2003. Remote Sensing and Geostatistical Analysis of Anthropogenic Biomass Burning in Madagascar with the DMSP-OLS Fire Product. Master's thesis final draft, University of California at Santa Barbara.

Mayaux, P., V. Gond, and E. Bartholomé. 2000. A near–real time forest cover map of Madagascar derived from SPOT-4 VEGETATION data. *International Journal of Remote Sensing* 21:3139–3144.

Mbow, C., T. T. Nielsen, and K. Rasmussen. 2000. Savanna fires in east-central Senegal: distribution patterns, resource management, and perceptions. *Human Ecology* 28: 561–583.

McCarthy, J. 1999. The Political and Moral Economies of Wise Use. Ph.D. dissertation, University of California at Berkeley.

McConnell, W. J. 2001. How and why people and institutions matter beyond economy: people and trees in Madagascar. *Global Change Newsletter (IGBP)* 47:20–22.

———. 2002a. Madagascar: emerald isle or paradise lost? *Environment* 44:10–22.

———. 2002b. Misconstrued land use and tenure niches in Vohibazaha. *Land Use Policy* 19: 217–230.

Messerli, P. 2000. Use of sensitivity analysis to evaluate key factors for improving slash-and-burn cultivation systems on the eastern escarpment of Madagascar. *Mountain Research and Development* 20: 32–41.

———. [ca. 1999]. *Projet BEMA: Réconcilier la culture sur brûlis avec la conservation.* IC/FOFIFA/ESSA/FNRS/EPFZ/CDE.

Messerli, P., and J.-L. Pfund. 1999. Improvements of slash-and-burn cultivation systems, an experience of systemic analysis in the Beforona region, Madagascar. 83–101 in *African Mountain Development in a Changing World,* ed. H. Hurni and J. Ramamonjisoa. African Mountain Association, UNU, African Highlands Initiative.

Metzger, G. 1951. Les feux de prairie et l'élevage extensif à Madagascar. *Bulletin de Madagascar* 28: 17–21.

Michel, L. 1897. Excursion dans la province de Andévorante. *Colonie de Madagascar: Notes, Reconnaissances et Explorations* 2: 113–117, 467–471.

Middleton, K. 1999. Who killed the Malagasy cactus? Science, environment, and colonialism in southern Madagascar (1924–1930). *Journal of Southern African Studies* 25: 215–248.

Ministère de l'Agriculture. 1974–1991. *Annuaire des Statistiques Agricoles.* Antananarivo: (Agricultural Ministry, under various names) Repoblikan'i Madagasikara.

Minnich, R. A. 1987. Fire behavior in southern California chaparral before fire control: the Mount Wilson burns at the turn of the century. *Annals of the Association of American Geographers* 77(4): 599–618.

———. 1989. Chaparral fire history in San Diego County and adjacent northern Baja California: an evaluation of natural fire regimes and the effects of suppression management. 37–47 in *The California Chaparral: Paradigms Reexamined,* ed. S. C. Keeley. Los Angeles: Natural History Museum of Los Angeles County.

Mistry, J. 1998. Decision-making for fire use among farmers in savannas: an exploratory study in the Distrito Federal, central Brazil. *Journal of Environmental Management* 54: 321–334.

Mitchell, A. 2000. Could this child's struggle to survive help to spark a wave of extinction that threatens humanity? A special report from the vanishing forests of Madagascar. (Toronto) *Globe and Mail.* 15 Apr. 2000, sec. A, 13, 16–17, 20.

Mohr, C. 1971. "Fires of anger" scar Madagascar. *New York Times.* 3 Jan. 1971, sec. 1, p. 7.

Moore, D. S. 1996. Marxism, culture, and political ecology: environmental struggles in Zimbabwe's Eastern Highlands. 125–147 in *Liberation Ecologies,* ed. R. Peet and M. Watts. London: Routledge.

———. 1998. Subaltern struggles and the politics of place: remapping resistance in Zimbabwe's eastern highlands. *Cultural Anthropology* 13(3): 344–381.

Moran, E. F. 1982. *Human Adaptability.* Boulder: Westview.

Morell, V. 1999. Restoring Madagascar. *National Geographic,* February, 60–69.

Moss, C. F. 1876. Over swamp, moor, and mountain: being the journal of a visit to Antongodrahoja, and home by Ambatondrazaka. *Antananarivo Annual* 1(2): 131–149.

Mottet, G. 1988. Quelques exemples d'érosion anthropique dans une montagne tropicale: l'Ankaratra (Madagascar). 143–146 in *L'Homme et la Montagne Tropicale.* Bordeaux: SEPANRIT.

Muldavin, J. S. S. 1997. Environmental degradation in Heilongjiang: policy reform and agrarian dynamics in China's new hybrid economy. *Annals of the Association of American Geographers* 87(4): 579–613.

Mullens, J. 1875. *Twelve Months in Madagascar.* London: James Nisbet.

Murphy, D. 1985. *Muddling Through in Madagascar.* Woodstock, N.Y.: Overlook.

Myers, N. 1999. Pushed to the edge. *Biodiversity* 3: 20–22.

Myers, R. L., and P. A. Peroni. 1983. Approaches to determining aboriginal fire use and its impact on vegetation. *Bulletin ESA* 64: 217–218.

NAS. 1980. *Firewood Crops.* Washington: National Academy of Sciences.

Nelson, R., and N. Horning. 1993. AVHRR-LAC estimates of forest area in Madagascar, 1990. *International Journal of Remote Sensing* 14:1463–1475.

Netting, R. M. 1968. *Hill Farmers of Nigeria.* Seattle: University of Washington Press.

———. 1993. *Smallholders, Householders.* Stanford: Stanford University Press.

Neumann, R. P. 1995. Local challenges to global agendas: conservation, economic liberalization, and the pastoralists' rights movement in Tanzania. *Antipode* 27(4): 363–382.

———. 1998. *Imposing Wilderness.* Berkeley: University of California Press.

———. 2000. Wilderness with a history: social control and landscape change in south-central Tanzania. Annual Meeting of the Association of American Geographers, Pittsburgh, 4–8 April.

———. 2001. Disciplining peasants in Tanzania: from state violence to self-surveillance in wildlife conservation. 305–327 in *Violent Environments,* ed. N. L. Peluso and M. Watts. Ithaca: Cornell University Press.

Neuvy, G. 1979. Aménagement régional à Madagascar: La cuvette d'Andapa. *Madagascar Revue de Géographie* 35:55–139.

———. 1986. Facteurs inhibiteurs de la production rizicole à Madagascar. 81–97 in *Crise Agricole et Crise Alimentaire dans les Pays Tropicaux,* ed. P. Vennetier. Bordeaux: Centre National de la Recherche Scientifique.

———. 1989. Le développement agricole du bassin supérieur de la Lokoho, à Madagascar. *Cahiers d'Outre Mer* 42: 135–154.

Nichols, J. 1974. *The Milagro Beanfield War.* New York: Ballantine.

Nielsen, T. T., and K. Rasmussen. 2001. Utilization of NOAA AVHRR for assessing the determinants of savanna fire distribution in Burkina Faso. *International Journal of Wildland Fire* 10: 129–135.

Nietschmann, B. Q. 1973. *Between Land and Water.* New York: Seminar Press.

Nye, P. H., and D. J. Greenland. 1964. Changes in the soil after clearing tropical forest. *Plant and Soil* 21:101–112.

Odum, E. 1969. The strategy of ecosystem development. *Science* 164: 260–272.

Ojima, D. S., D. S. Schimel, W. J. Parton, and C. E. Owensby. 1994. Long- and short-term effects of fire on nitrogen cycling in tallgrass prairie. *Biogeochemistry* 24: 67–84.

Oliver, S. P. [ca. 1863]. *Madagascar and the Malagasy: With Sketches in the Provinces of Tamatave, Betanimena, and Ankova.* London: Day and Son.

Olson, S. 1984. The robe of the ancestors: forests in the history of Madagascar. *Journal of Forest History* 28: 174–186.

———. 1987. Red destinies: the landscape of environmental risk in Madagascar. *Human Ecology* 51(1): 67–89.

———. 1988. Environments as shock absorbers, examples from Madagascar. *Environmental Review* 12(4): 61–80.

ONE/Instat. 1994. *Rapport sur l'Etat de l'Environnement à Madagascar.* Antananarivo: PNUD/Banque Mondiale.

Otto, J. S., and N. E. Anderson. 1982. Slash-and-burn cultivation in the highlands South: a problem in comparative agricultural history. *Comparative Studies in Society and History* 24: 131–147.

Padoch, C., and N. L. Peluso (eds.). 1996. *Borneo in Transition.* Kuala Lumpur: Oxford University Press.

Panda, S. M. 1999. Towards a sustainable natural resource management of tribal communities: findings from a study of swidden and wetland cultivation in remote hill regions of eastern India. *Environmental Management* 23: 205–216.

Parfit, M. 1996. The essential element of fire. *National Geographic*, September, 116–139.

Parrot, A. 1924. Déboisement et reboisement à Madagascar. *Bulletin Economique* 4ème trimestre: 192–195.

Paulian, R. 1953. Observations sur les Boroceras de Madagascar, papillons séricigènes. *Le Naturaliste Malgache* 5(1): 69–86.

———. 1984. Madagascar: a microcontinent between Africa and Asia. 1–26 in *Key Environments: Madagascar*, ed. A. Jolly, P. Oberlé, and R. Albignac. Oxford: Pergamon.

Pavageau, J. 1981. *Jeunes Paysans Sans Terre.* Paris: l'Harmattan.

Peet, R., and M. Watts (eds.). 1996. *Liberation Ecologies.* London: Routledge.

Peluso, N. L. 1992. *Rich Forests, Poor People.* Berkeley: University of California Press.

———. 1996. Fruit trees and family trees in an anthropogenic forest: ethics of access, property zones, and environmental change in Indonesia. *Comparative Studies in Society and History* 38(3): 510–548.

Peluso, N. L., and M. Watts (eds.). 2001. *Violent Environments.* Ithaca: Cornell University Press.

Perrier de la Bâthie, H. 1921. La végétation Malgache. *Annales du Musée Colonial de Marseille* Sér. 3, v. 9: 1–266.

———. 1927. Le Tsaratanana, l'Ankaratra, et l'Andringitra. *Mémoires de l'Académie Malgache* 3:15–71.

———. 1936. *Biogéographie des Plantes de Madagascar.* Paris: Société d'Editions Géographiques, Maritimes et Coloniales.

Peters, W. J. 1994. Attempting to Integrate Conservation and Development among Resident Peoples of the Ranomafana National Park, Madagascar. Ph.D. dissertation, North Carolina State University.

———. 1998. Transforming the integrated conservation and development (ICDP) approach: observations from the Ranomafana National Park Project, Madagascar. *Journal of Agricultural and Environmental Ethics* 11: 17–47.

———. 1999. Understanding conflicts between people and parks at Ranomafana, Madagascar. *Agriculture and Human Values* 16: 65–74.

Peters, W. J., and L. F. Neuenschwander. 1988. *Slash and Burn.* Moscow: University of Idaho Press.

Petit, G. 1934. Cultures extensives à Madagascar et disparition de la forêt orientale. *Bulletin de l'Association de Géographie Française,* March, 35–39.

Pfund, J. L. 2000. Culture sur Brûlis et Gestion des Ressources Naturelles: Evolution et Perspectives de Terroirs Ruraux du Versant Est de Madagascar. Thèse EPFZ no 13966, Docteur dès Sciences Naturelles, Ecole Polytechnique Fédérale de Zurich.

Phillips, W. S. 1962. Fire and vegetation of arid lands. Tall Timbers Fire Ecology Conference, Tallahassee, Fla., 1–2 Mar.

Pile, S., and M. Keith (eds.) 1997. *Geographies of Resistance.* London: Routledge.

Platon, P. 1953. Y-a-t-il un avenir pour la sériciculture? *Bulletin de Madagascar* 77: 22–26.

Price, C. T. 1989. *Missionary to the Malagasy: The Madagascar Diary of the Rev. Charles T. Price, 1875–1877.* New York: Peter Lang.

Prochaska, D. 1986. Fire on the mountain: resisting colonialism in Algeria. 229–252 in *Banditry, Rebellion, and Social Protest in Africa,* ed. D. Crummey. London: James Currey.

Projet Andringitra. 1996. Monographie du Sous-secteur Namoly. WWF: PCDI Andringitra/ Pic d'Ivohibe, Cellule suivi-évaluation, June 1996.

Projet MADIO. 1997. *Aperçu de l'Etat des Campagnes Malgaches en 1997.* Antananarivo: Institut National de la Statistique.

Pulido, L. 1996. Ecological legitimacy and cultural essentialism: Hispano grazing in the Southwest. *Capitalism, Nature, Socialism* 7: 37–59.

Pyne, S. J. 1990. Fire conservancy: the origins of wildland fire protection in British India, America, and Australia. 319–336 in *Fire in the Tropical Biota,* ed. J. G. Goldammer. Berlin: Springer Verlag.

———. 1995. *World Fire.* New York: Henry Holt.

———. 1997. *Vestal Fire.* Seattle: University of Washington Press.

Pyne, S. J., P. L. Andrews, and R. D. Laven. 1996. *Introduction to Wildland Fire.* New York: Wiley.

Rabearimanana, G. 1994. Le Boina. 1–153 in *Paysanneries Malgaches dans la Crise,* ed. J.-P. Raison. Paris: Karthala.

Rabetaliana, H., M. Randriambololona, and P. Schachenmann. 1999. The Andringitra National Park in Madagascar. *Unasylva* 50(196): 25–30.

Rabetaliana, H. 1997. La réserve d'Andringitra et le site de Zombitse. *L'atelier de travail sur l'élevage et l'environnement à Madagascar, Antananarivo.* Antananarivo: Aquaterre.

———. 2001. Le *tavy* dans la région de Fianarantsoa: l'expérience de la préservation du corridor forestier entre Ranomafana et Andringitra. 81–84 in *Culture sur Brûlis: Vers l'Application des Résultats de Recherche,* ed. P. Kistler, P. Messerli, and S. Wohlhauser. Antananarivo: Projet BEMA/FOFIFA, Projet EPB/ESSA-Forêt.

Rabetaliana, H., and P. Schachenmann. 1999. Coordinating traditional values, scientific research and practical management to enhance conservation and development objectives in the Andringitra Mountains: lessons learned! *African Studies Quarterly* 3(2): 61–67 <web.africa.ufl.edu/asq>.

Rahamalivony, R. 1989. Les rôles et formes du reboisement à Madagascar. 123–132 in *Environnement et Gestion des Ressources Naturelles dans la Zone Africaine de l'Océan Indien,* ed. M. Maldague, K. Matuka, and R. Albignac. Paris: UNESCO.

Raharijaona, V. 1990. Etude du peuplement de l'espace d'une vallée des Hautes Terres centrales de Madagascar: archéologie de la Manandona (XVème–XIXème siècle). Thèse de Doctorat, Université d'Antananarivo.

———. 1994. Répartition des villages anciens dans une vallée des Hautes-Terres centrales: archéologie de la Manandona. *Taloha* 12: 51–76.

Raharijaona, V., and S. Kus. 1989. Forêt et foyer, feu et eau: sources de la vie parmi les Tanala de Kelialina. Paper given at Séminaire National de Réflexion sur la Politique en Sciences Sociales, CIDST, Antananarivo.

Raharivelo, V. M. N. 1995. Evolution de l'état des bassins versants et protection contre les crues de la ville et des plaines d'Antananarivo. *Akon'ny Ala* 16: 24–26.

Raison, J.-P. 1968. Mouvements et commerce des bovins dans la region de Mandoto (Moyen-Ouest de Madagascar). *Madagascar Revue de Géographie* 12: 7–58.

———. 1970. Paysage rural et démographie, Leimavo (nord du Betsileo, Madagascar). *Etudes Rurales* 37–38–39: 345–377.

———. 1984. *Les Hautes Terres de Madagascar et leurs Confins Occidentaux.* Paris: Editions Karthala.

Rajaonson, B., L. P. Randriamarolaza, D. Randrianaivo, E. Ratsimbazafy, V. Rejo Tsiresy, and A. Bertrand. 1995. Elaboration d'une politique et d'une stratégie de gestion des feux de végétation à Madagascar. Office National de l'Environnement et OSIPD, Janvier 1995.

Rajosefasolo, V. 1992. Ny olona, ny tany, ary ny tontolo iainana ao Vakinankaratra. 47–60 in *Colloque pour chercheurs en sciences sociales et opérateurs économiques sur changements sociaux dans la région du Vakinankaratra, Antsirabe.* Antananarivo: Tranompirinty FLM.

Rakatonindrina, R. 1989. Les grands problèmes actuels de la conservation des écosystèmes: défrichement et feux de brousse. 111–122 in *Environnement et Gestion des Ressources Naturelles dans la Zone Africaine de l'Océan Indien,* ed. M. Maldague, K. Matuka, and R. Albignac. Paris: UNESCO.

Rakoto, I. (ed.). 1997. *L'Esclavage à Madagascar: Actes du Colloque International sur l'Esclavage Antananarivo (24–28 Sept. 1996).* UNESCO/Projet Route de l'Esclave. Antananarivo: Institut de Civilisations-Musée d'Art et d'Archéologie.

Rakotoarisetra, F. N. 1997. Monographie de l'Uapaca densifolia dans la forêt d'Ambohitantely. Mémoire de fin d'études, Université d'Antananarivo, Ecole Supérieure des Sciences Agronomiques.

Rakotoarivelo, L. A. 1993. Analyse sylvicole d'une forêt sclérophylle de moyenne altitude à Uapaca bojeri (tapia) de la région d'Arivonimamo. Mémoire de fin d'études, Université d'Antananarivo, Ecole Supérieure des Sciences Agronomiques.

Rakotomanana, J. L. 1993. Le transfert de fertilité dans les écosystèmes des hautes terres de Madagascar. 385–391 in *Bas-Fonds et Riziculture,* ed. M. Raunet. Montpellier: CIRAD.

Rakoto Ramiarantsoa, H. 1985. Développement à contre-sens: un aménagement hydro-agricole qui n'a pas donné les résultats escomptés—localisation et caractères du secteur étudié (Réseau hydro-agricole du PC 23 BIRD, cuvette du Lac Alaotra, Madagascar). 409–420 in *Les Politiques de l'Eau en Afrique,* ed. G. Conac, C. Savonnet-Guyot and F. Conac. Paris: Economica.

———. 1993. Ligneux et terroir d'altitude dans le Vakinankaratra: l'importance des formations de mimosas et de pins dans la gestion de l'occupations de l'espace. Exemple du Fokontany de Faravohitra, Fivondronana de Faratsiho. FOFIFA and CIRAD, Oct. 1993, ATP 41/90.

———. 1995a. *Chair de la Terre, Oeil de l'Eau.* Paris: ORSTOM.

———. 1995b. Les boisements d'eucalyptus dans l'est de l'Imerina (Madagascar). 83–103 in *Terre, Terroir, Territoir,* ed. C. Blanc-Pamard and L. Cambrézy. Paris: ORSTOM.

Rakotovao, L. H., V. Barre, and J. Sayer. 1988. *L'Equilibre des Ecosystèmes Forestiers à Madagascar.* Gland and Cambridge: IUCN.

Rakotovao Andriankova, S., M. Razafindrabe, and A. Bertrand. 1997. Vers la gestion communautaire locale des feux de végétation à Madagascar. *Akon'ny Ala* 20: 8–22.

Ramamonjisoa, B. S. 2001. Le rôle de la filière pour le développement durable des zones de culture sur brûlis. 77–80 in *Culture sur Brûlis: Vers l'Application des Résultats de Recherche,* ed. P. Kistler, P. Messerli, and S. Wohlhauser. Antananarivo: Projet BEMA/ FOFIFA, Projet EPB/ESSA-Forêt.

Ramamonjisoa, J. 1994. Le Vakinankaratra. 153–234 in *Paysans Malgaches dans la crise,* ed. J.-P. Raison. Paris: Editions Karthala.

———. 1995. Le Processus de Développement dans le Vakinankaratra, Hautes Terres Malgaches. Thèse de Doctorat d'Etat, Université de Paris I—Panthéon/Sorbonne.

Ramanantsoa, G. 1968. Effort de reboisement et exploitation de la forêt a Madagascar. *Tany Malagasy/Terre Malgache* 4: 195–202.

Rambeloarisoa, G. 1995. Etude des feux de brousse dans trois Firaisampokontany d'Antananarivo-Atsimondrano concernés par le reboisement villageois. *Akon'ny Ala* 17: 4–13.

Rambeloarisoa, R. 1999. Gestion et aménagement du peuplement de "Tapia" en vue de la production de "Landibe"—Mission à Ilaka-Centre du 18 au 24 Mars 1999. Rapport, Projet Landibe (Présidence).

Ramos-Neto, M. B., and V. R. Pivello. 2000. Lightning fires in a Brazilian savanna national park: rethinking management strategies. *Environmental Management* 26:675–684.

Randriambelo, T., S. Baldy, M. Bessafi, M. Petit, and M. Despinoy. 1998. An improved detection and characterization of active fires and smoke plumes in south-eastern Africa and Madagascar. *International Journal of Remote Sensing* 19:2623–2638.

Randriamboavonjy, G. 2000. La dynamique de la forêt de tapia (Uapaca bojeri) pour le développement de la sériciculture dans la région d'Ambohimanjaka. Mémoire de fin d'études, Université d'Antananarivo, Ecole Supérieure des Sciences Agronomiques.

Randrianarijaona, P. 1983. The erosion of Madagascar. *Ambio* 12: 308–311.

Randrianarison, J. 1976. Le boeuf dans l'économie rurale de Madagascar (1ère partie). *Madagascar Revue de Géographie* 28: 9–122.

Randrianatoandro, A. L. N. V. 1990. La migration organisée dans le Moyen-Ouest est-elle une mesure de la mauvaise répartition de la population? Mémoire de Maîtrise, Université d'Antananarivo.

Rangan, H. 1996. From Chipko to Uttaranchal: development, environment, and social protest in the Garhwal Himalayas, India. 205–226 in *Liberation Ecologies,* ed. R. Peet and M. Watts. London: Routledge.

Raonintsoa, P. N. 1996. The role of the forest in the regional economy, 41–47 in *Ecology and Economy of a Tropical Dry Forest in Madagascar,* ed. J. U. Ganzhorn and J.-P. Sorg. Primate Report 46-1.

Rasoavarimanana, M.-A. 1997. Le marais d'altitude de Torotorofotsy et son environnement socio-économique. *Agriculture et Développement,* no. 14: 3–10.

Ratovoson, C. 1979. Les problèmes du tavy sur la côte est Malgache. *Madagascar Revue de Géographie* 35:141–163.

Ratsimamanga, A. R. 1968. *Dictionnaire des Noms Malgaches des Végétaux.* Antananarivo: Institut Malgache de Recherches Appliqués.

Ratsirarson, J. 1997. Etude comparative de la situation de la Réserve Naturelle No. 5 de l'Andringitra et ses zones périphériques de la période 1960/1975 a la période 1990/1996. Unpublished consultant's report, WWF, April 1997.

Ratsirarson, J., and S. M. Goodman (eds.). 2000. *Monographie de la Forêt d'Ambohitantely. Recherches pour le Développement, Série Sciences biologiques no 16.* Antananarivo: CIDST.

Rauh, W. 1995. *Succulent and Xerophytic Plants of Madagascar.* Mill Valley, Calif.: Strawberry.

Ravoavy, L. 2001. Culture sans brûlis: une alternative au tavy? 13–18 in *Culture sur Brûlis: Vers l'Application des Résultats de Recherche,* ed. P. Kistler, P. Messerli, and S. Wohlhauser. Antananarivo: Projet BEMA/FOFIFA et Projet EPB/ESSA Forêts.

Razafiarivony, M. 1995. Le riz, un aspect de l'identité culturelle Malgache en question, 237–247 in *Cultures of Madagascar,* ed. S. Evers and M. Spindler. Leiden: IIAS.

Razafindrabe, M., and P. Rakotondrainibe. 1998. Bilan et perspectives d'après les revues bibliographiques de Michel Sourdat et Alain Bertrand. 147–153 in *Feux et Déforestation à Madagascar: Revues Bibliographiques,* ed. A. Bertrand and M. Sourdat. Antananarivo: CIRAD/ORSTOM/CITE.

Razafindrabe, M., S. Rakotovao, and A. Bertrand. 1996. Promoting fire management by local communities in Madagascar. Voices from the Commons, International Association for the Study of Common Property, Berkeley, 5–8 June.

Razafindrabe, M., and J. T. Thomson. 1994. Rapport sur les recherches relatives à la gouvernance locale à Madagascar. Project KEPEM, USAID/ARD, August 1994.

Razafintsalama, A. S., and M. Gautschi. 1999. Etude des structures socio-organisationelles des villages pour l'identification et la formalisation d'un organe de gestion, dans le processus de transfert de gestion de la forêt de tapia (Uapaca bojeri) dans la région d'Arivonimamo. Rapport final de stage, Intercoopération, Projet FDP, ESSA-Forêts, et EPF Zürich, juillet 1999.

Razakaboana, F. 1967. Les possibilités d'amélioration des pâturages malgaches. *Bulletin de Madagascar* 253: 494–504.

Razanadravao, S. 1990. Etude rurale du sillon de Manandona, Vakinankaratra. Mémoire de Maîtrise, Université d'Antananarivo.

Razanajoelina, L. 1993. Etude de l'action des feux tardifs/précoces sur quelques espèces savanicoles (Réserve d'Ankarafantsika); Etudes et amélioration des paturages. Nov 1991–Juin 1993. Report, Projet UNESCO/PNUD MAG/88/007.

Razoharinoro-Randriamboavonjy. 1971. Note sur Madagascar pendant la deuxième guerre mondiale: la question du riz. *Bulletin de Madagascar* 305–306: 820–868.

RDM (République Démocratique de Madagascar). 1980. Rapport du Groupe Interministériel d'Etudes sur les Feux de Brousse. Rapport, Repoblika Demokratika Malagasy, Novembre 1980.

———. 1988. Stratégie Nouvelle sur la Lutte contre les Feux de Brousse. Report, Eaux et Forêts.

———. 1990. *Charte de l'Environnement.* Antananarivo: Foi et Justice.

———. 2002. *Programme Environnement III: Document Stratégique.* Ministère de l'Environnement/PNAE/Ministère des Eaux et Forêts.

Réau, B. 1996. Dégradation de l'environnement forestier et réaction paysannes: les migrants Tandroy sur la côte Ouest de Madagascar. Thèse de Doctorat, Université Bordeaux III.

Reiners, W. A., and G. E. Lang. 1979. Vegetal patterns and processes in the balsam fir, White Mountains, New Hamshire. *Ecology* 60: 403–417.

Ribot, J. C. 1998. Theorizing access: forest profits along Senegal's charcoal commodity chain. *Development and Change* 29: 307–341.

———. 1999. Decentralisation, participation, and accountability in Sahelian forestry: legal instruments of political-administrative control. *Africa* 69(1):23–65.

Richard, A. F., and R. E. Dewar. 2001. Politics, negotiation, and conservation: a view from Madagascar. 535–544 in *African Rain Forest Ecology and Conservation,* ed. W. Weber, L. J. T. White, A. Vedder, and L. Naughton-Treves. New Haven: Yale University Press.

Richard, A. F., and S. O'Connor. 1997. Degradation, transformation, and conservation: the past, present, and possible future of Madagascar's environment. 406–418 in *Natural Change and Human Impacts in Madagascar,* ed. S. M. Goodman and B. D. Patterson. Washington: Smithsonian Institution Press.

Richards, A. I. 1939. *Land, Labour, and Diet in Northern Rhodesia: An Economic Study of the Bemba Tribe.* London: Oxford University Press.

Richards, P. 1985. *Indigenous Agricultural Revolution.* Boulder: Westview.

Robbins, P. 1998. Authority and environment: institutional landscapes in Rajasthan, India. *Annals of the Association of American Geographers* 88: 410–435.

———. 2000. The rotten institution: corruption in natural resource management. *Political Geography* 19: 423–443.

Rocheleau, D., and L. Ross. 1995. Trees as tools, trees as text: struggles over resources in Zambrana-Chacuey, Dominican Republic. *Antipode* 27(4): 407–428.

Roe, E. 1991. Development narratives, or making the best of blueprint development. *World Development* 19(4): 287–300.

Roffet, C. 1995. Madagascar, entre conservation et développement. *Habbanae* 36: 7–9.

Rollin, D. 1997. Quelles améliorations pour les systèmes de culture de sud-ouest malagache? *Agriculture et Développement,* no. 16: 57–72.

Rossi, G. 1979. L'érosion à Madagascar: l'importance des facteurs humains. *Cahiers d'Outre Mer* 32(128): 355–370.

Rudel, T. K., D. Bates, and R. Machinguiashi. 2002. A tropical forest transition? Agricultural change, out-migration, and secondary forests in the Ecuadorian Amazon. *Annals of the Association of American Geographers* 92:87–102.

Saboureau, P. 1950. La place de la forêt dans l'économie malgache. *Marchés Coloniaux:* 1707–1709.

Sahlins, P. 1994. *The War of the Demoiselles in Nineteenth-Century France.* Cambridge: Harvard University Press.

Sakaël, K. 1994. Tavy et riziculture de bas-fonds (nord-est de Madagsascar). Mémoire pour le Diplôme d'Agronomie Tropicale, Centre National d'Etudes Agronomiques des Régions Chaudes, France.

Salomon, J. N. 1978. Fourrés et forêts sèches du Sud-Ouest Malgache. *Madagascar Revue de Géographie* 32: 19–39.

———. 1981. Réalité et conséquences de la déforestation dans l'Ouest malgache. *Omaly sy Anio* 13: 329–336.

Sandford, S. 1983. *Management of Pastoral Development in the Third World.* Chichester: Wiley.

Sauer, C. O. 1956. The agency of man on earth. 49–69 in *Man's Role in Changing the Face of the Earth,* ed. W. L. J. Thomas. Chicago: University of Chicago Press.

———. 1975. Man's dominance by use of fire. *Geoscience and Man* 10: 1–13.

Sayer, J. A., C. S. Harcourt, and N. M. Collins. 1992. *The Conservation Atlas of Tropical Forests: Africa.* London: IUCN and Macmillan.

Schatz, G. E. 2001. *Generic Tree Flora of Madagascar.* Kew: Royal Botanical Gardens; St. Louis: Missouri Botanical Garden.

Schmitz, A. 1996. *Contrôle et utilisation du feu en zones arides et subhumides africaines.* Rome: FAO.

Schnyder, A., and S. Serretti. 1997. Les Feux de Brousse et la Réserve Spéciale d'Ambohitantely (Ankazobe, Madagascar). Travaux de diplôme, Ecole Polytechnique Fédérale de Zurich (ETH) and ESSA-Forêts, Université d'Antananarivo.

Schroeder, R. A. 1997. "Re-claiming" land in the Gambia: gendered property rights and environmental intervention. *Annals of the Association of American Geographers* 87(3): 487–508.

———. 1999. *Shady Practices.* Berkeley: University of California Press.

Schüle, W. 1990. Landscapes and climate in prehistory: interactions of wildlife, man, and fire. 273–318 in *Fire in the Tropical Biota,* ed. J. G. Goldammer. Berlin: Springer Verlag.

Scoones, I. 1996. Range management science and policy: politics, polemics and pasture in Southern Africa. 34–53 in *The Lie of the Land,* ed. M. Leach and R. Mearns. Oxford: James Currey.

———. 1999. New ecology and the social sciences: what prospects for a fruitful engagement? *Annual Review of Anthropology* 28: 479–507.

Scott, J. C. 1976. *The Moral Economy of the Peasant.* New Haven: Yale University Press.

———. 1985. *Weapons of the Weak.* New Haven: Yale University Press.

———. 1998. *Seeing Like a State.* New Haven: Yale University Press.

Seddon, N., J. Tobias, J. W. Yount, J. R. Ramanampamonjy, S. Butchart, and H. Randrianizahana. 2000. Conservation issues and priorities in the Mikea Forest of south-west Madagascar. *Oryx* 34:287–304.

SEM (Service de l'Elevage de Madagascar). 1949. Les feux de brousse à Madagascar. *Bulletin Agriculturel du Congo Belge:* 1945–1950.

Sharp, J. P., P. Routledge, C. Philo, and R. Paddison (eds.). 2000. *Entanglements of Power.* London: Routledge.

Sheridan, T. E. 1988. *Where the Dove Calls.* Tuscon: University of Arizona Press.

Sibree, J. 1870. *Madagascar and Its People.* London: Religious Tract Society.

Sigaut, F. 1975. *L'Agriculture et le Feu.* Paris: Mouton.

Sirois, M.-C., H. A. Margolis, and C. Camiré. 1998. Influence of remnant trees on nutrients and fallow biomass in slash-and-burn agroecosystems in Guinea. *Agroforestry Systems* 40:227–246.

Sivaramakrishnan, K. 1995. Colonialism and forestry in India: imagining the past in present politics. *Comparative Studies in Environment and History* 37(1): 3–40.

———. 1996. The politics of fire and forest regeneration in colonial Bengal. *Environment and History* 2(2): 145–194.

———. 1997. A limited forest conservancy in southwest Bengal, 1864–1912. *Journal of Asian Studies* 56(1): 75–112.

Smith, A. P. 1997. Deforestation, fragmentation, and reserve design in Western Madagascar. 415–441 in *Tropical Forest Remnants,* ed. W. Lawrence and O. W. Bierregard. Chicago: University of Chicago Press.

Société Fanalamanga. 1992. Renforcement de l'organisation de la lutte contre les feux dans les plantations de la FANALAMANGA. Rapport, Fanalamanga, Avril 1992.

Sorrensen, C. L. 2000. Linking smallholder land use and fire activity: examining biomass burning in the Brazilian lower Amazon. *Forest Ecology and Management* 128: 11–25.

Spack, S. 2001. Identification d'un certain nombre de stratégies endogènes communautaires dans la région périphérique de la Réserve Spéciale de Manongarivo. 46–48 in *Culture sur Brûlis: Vers l'Application des Résultats de Recherche,* ed. P. Kistler, P. Messerli, and S. Wohlhauser. Antananarivo: Projet BEMA/FOFIFA, Projet EPB/ESSA-Forêts.

Spence, M. D. 1999. *Dispossessing the Wilderness.* New York: Oxford University Press.

SSG (Service de Statistique Générale). 1953. *Annuaire Statistique de Madagascar, Vol. 1 1938–1951.* Tananarive: Imprimerie Officielle.

Stewart, O. C. 1956. Fire as the first great force employed by man. 115–133 in *Man's Role in Changing the Face of the Earth,* ed. W. L. J. Thomas. Chicago: University of Chicago Press.

———. 2002. *Forgotten Fires.* Norman: University of Oklahoma Press.

Stocking, M. 1987. Measuring land degradation. 49–63 in *Land Degradation and Society,* ed. P. Blaikie and H. Brookfield. London: Methuen.

———. 1996. Soil erosion: breaking new ground. 140–154 in *The Lie of the Land,* ed. M. Leach and R. Mearns. Portsmouth, N.H.: Heinemann.

Straka, H. 1996. Histoire de la végétation de Madagascar oriental dans les derniers 100 millénaires. 37–47 in *Biogéographie de Madagascar,* ed. W. R. Lourenço. Paris: ORSTOM.

Sundar, N. 2000. Unpacking the "joint" in Joint Forest Management. *Development and Change* 31: 255–279.

Sundar, N., A. Mishra, and N. Peter. 1996. Defending the Dalki forest: "joint" forest management in Lapanga. *Economic and Political Weekly* 31(45–46): 3021–3025.

Suryanata, K. 1994. Fruit trees under contract: tenure and land use change in upland Java, Indonesia. *World Development* 22(10): 1567–1578.

Sussman, R. W., G. M. Green, and L. K. Sussman. 1994. Satellite imagery, human ecology, anthropology, and deforestation in Madagascar. *Human Ecology* 22(3): 333–354.

Suyanto, S., G. Applegate, and L. Tacconi. 2001. Community-based fire management, land tenure, and conflict: insights from Sumatra. Report, CIFOR. Accessed at www.cifor.org/fire-project/fireweb/content3c2.htm.

Swaney, D., and R. Wilcox. 1994. *Madagascar and Comoros: A Travel Survival Kit.* Hawthorn, Australia: Lonely Planet.

Swift, J. 1996. Desertification: narratives, winners, and losers. 73–90 in *The Lie of the Land,* ed. M. Leach and R. Mearns. Oxford: James Currey.

Tachez, C. 1996. Opportunités et méthodes d'animation pour la gestion locale communautaire des feux de brousse. Rapport de mission d'appui, réalisée du 16 août au 11 oct 1996 sur les projets Bemaraha et Montagne d'Ambre, Madagascar. Vétérinaires Sans Frontières, Oct. 1996.

Thomas, W. L. J. (ed.). 1956. *Man's Role in Changing the Face of the Earth.* Chicago: University of Chicago Press.

Thompson, E. P. 1975. *Whigs and Hunters.* London: Allen Lane.

Thuy, G. 1898. Six semaines dans le Sud-Ouest. *Colonie de Madagascar. Notes, Reconnaissances et Explorations* 3: 26–64.

Tiffen, M., M. Mortimore, and F. Gichuki. 1994. *More People, Less Erosion.* Chichester: Wiley.

Tricart, J. 1953. Erosion naturelle et érosion anthropogène à Madagascar. *Revue de Géomorphologie Dynamique* 4 suppl.: 225–230.

Tsing, A. L. 1993. *In the Realm of the Diamond Queen.* Princeton: Princeton University Press.

———. 1999. Becoming a tribal elder, and other green development fantasies. 159–202 in *Transforming the Indonesian Uplands,* ed. T. M. Li. Amsterdam: Harwood Academic Publishers.

Turner, B. L., II, and S. B. Brush. 1987. *Comparative Farming Systems.* New York: Guilford.

Turner, B. L., II., G. Hyden, and R. W. Kates (eds.). 1993. *Population Growth and Agricultural Change in Africa.* Gainesville: University Press of Florida.

Turner, M. D. 1998. The interaction of grazing history with rainfall and its influence on annual rangeland dynamics in the Sahel. 237–261 in *Nature's Geography,* ed. K. S. Zimmerer and K. R. Young. Madison: University of Wisconsin Press.

Uhart, E. 1962. Les reboisements et le développement de Madagascar. *Bois et Forêts des Tropiques* 83: 15–29.

Uphoff, N., and J. Langholz. 1998. Incentives for avoiding the tragedy of the commons. *Environmental Conservation* 25: 251–261.

USAID. 1997a. Congressional Presentation FY 1997 on Madagascar. www.usaid.gov/pubs/cp97/countries/mg.htm.

———. 1997b. Strategic objectives agreement, USAID and Government of Madagascar. www.info.usaid.gov/countries/mg/sntzsoag.htm.

Vale, T. R. (ed.). 2002. *Fire, Native Peoples, and the Natural Landscape.* Washington: Island.

van Rensburg, H. J. 1952. Grass burning experiments on the Msima River stock farm, southern highlands, Tanganyika. *East African Agricultural Journal* 17(3): 119–129.

van Wilgen, B. W., C. S. Everson, and W. S. W. Trollope. 1990. Fire management in southern Africa: some examples of current objectives, practices, and problems. 179–215 in *Fire in the Tropical Biota,* ed. J. G. Goldammer. Berlin: Springer Verlag.

Vea, J. 1992. *Le développement de la production laitière sur les Hautes Terres Malgaches du Vakinankaratra.* Antananarivo: Tranompirinty FLM.

Vérin, P. 1992. Socio economic factors in economic development in Madagascar. Report, USAID Doc. Id. PN-ABM-995.

———. 1994. *Madagascar,* 3rd ed. Paris: Karthala.

Vignal, P. 1956a. La disparition de la forêt malgache des Hauts Plateaux. *Bois et Forêts des Tropiques* 49: 3–8.

———. 1956b. Les reboisements en Pinus patula de la Haute-Matsiatra. *Bois et Forêts des Tropiques* 45(Jan-Fev): 15–25.

———. 1963. Les phénomènes de météorologie dynamique et la disparition des formations forestières malgaches d'altitude. *Bois et Forêts des Tropiques* 89: 31–35.

Vogl, R. J. 1974. Effects of fire on grasslands. 139–194 in *Fire and Ecosystems,* ed. T. T. Kozlowski and C. E. Ahlgren. New York: Academic.

Watts, M. J. 1985. Social theory and environmental degradation. 14–32 in *Desert Development,* ed. Y. Gradus. Dordrecht: D. Reidel.

Wells, M. P., and K. Brandon. 1993. The principles and practise of buffer zones and local participation in biodiversity conservation. *Ambio* 22(2–3): 157–162.

Wells, N. A., and B. Andriamihaja. 1993. The initiation and growth of gullies in Madagascar: are humans to blame? *Geomorphology* 8:1–46.

———. 1997. Extreme gully erosion in Madagascar and its natural and anthropogenic causes. 44–74 in *Natural Change and Human Impact in Madagascar,* ed. S. M. Goodman and B. D. Patterson. Washington: Smithsonian Institution Press.

Western, D., and R. M. Wright. 1994. *Natural Connections: Perspectives in Community-Based Conservation.* Washington: Island.

Westoby, M., B. Walker, and I. Noy-Meir. 1989. Opportunistic management for rangelands not at equilibrium. *Journal of Range Management* 42: 266–274.

Wollenberg, E., D. Edmunds, and L. Buck. 2000. Using scenarios to make decisions about the future: anticipatory learning for the adaptive co-management of community forests. *Landscape and Urban Planning* 47: 65–77.

World Bank. 1988. *Madagascar Environmental Action Plan.* Preliminary version. World Bank, USAID, Coop. Suisse, UNESCO, UNDP, WWF.

WWF. 1992. Madagascar, in *WWF International Conservation Programme 1992–1993.* Gland, Switzerland: WWF (unpublished document).

Young, R., and H. Fosbrooke. 1960. *Smoke in the Hills.* Evanston: Northwestern University Press.

Zahner, P. 1990. *Protection Intégrée en Rizières à Madagascar. Projet lutte integrée (PLI).* Antananarivo: MRSTD/MPAPF; Bern: Coopération Suisse.

Zerner, C. (ed.). 2000. *People, Plants, and Justice: The Politics of Nature Conservation.* New York: Columbia University Press.

Zimmerer, K. S. 1996. *Changing Fortunes.* Berkeley: University of California Press.

———. 2000. The reworking of conservation geographies: nonequilibrium landscapes and nature-society hybrids. *Annals of the Association of American Geographers* 90(2): 356–369.

Zimmerer, K. S., and K. R. Young, ed. 1998. *Nature's Geography: New Lessons for Conservation in Developing Countries.* Madison: University of Wisconsin Press.

Index

(Page numbers in bold type refer to a figure or table.)

Titles in Print

127. PETER G. GOHEEN, *Victorian Toronto, 1850 to 1900: Pattern and Process of Growth,* 1970
132. NORMAN T. MOLINE, *Mobility and the Small Town, 1900–1930,* 1971
152. MARVIN W. MIKESELL, ed., *Geographers Abroad: Essays on the Problems and Prospects of Research in Foreign Areas,* 1973
186. KARL W. BUTZER, *Recent History of an Ethiopian Delta: The Omo River and the Level of Lake Rudolf,* 1971
194. CHAUNCY D. HARRIS, *Annotated World List of Selected Current Geographical Serials,* 4th ed., 1980
206. CHAUNCY D. HARRIS, *Bibliography of Geography,* part 2, *Regional,* volume 1, *The United States of America,* 1984
207–8. PAUL WHEATLEY, *Nagara and Commandery: Origins of the Southeast Asian Urban Traditions,* 1983
209. THOMAS S. SAARINEN, DAVID SEAMON and JAMES L. SELL, eds., *Environmental Perception and Behavior: An Inventory and Prospect,* 1984
210. JAMES L. WESCOAT JR., *Integrated Water Development: Water Use and Conservation Practice in Western Colorado,* 1984
213. RICHARD LOUIS EDMONDS, *Northern Frontiers of Qing China and Tokugawa Japan: A Comparative Study of Frontier Policy,* 1985
216. NANCY J. OBERMEYER, *Bureaucrats, Clients, and Geography: The Bailly Nuclear Power Plant Battle in Northern Indiana,* 1989
217–18. MICHAEL P. CONZEN, ed., *World Patterns of Modern Urban Change: Essays in Honor of Chauncy D. Harris,* 1986
222. MARILYN APRIL DORN, *The Administrative Partitioning of Costa Rica: Politics and Planners in the 1970s,* 1989
223. JOSEPH H. ASTROTH JR., *Understanding Peasant Agriculture: An Integrated Land-Use Model for the Punjab,* 1990

224. RUTHERFORD H. PLATT, SHEILA 6. PELCZARSKI, and BARBARA K. BURBANK, eds., *Cities on the Beach: Management issues of Developed Coastal Barriers,* 1987
225. GILLATZ, *Agricultural Development in Japan: The Land Improvement District in Concept and Practice,* 1989
226. JEFFREY A. GRITZNER, *The West African Sahel: Human Agency and Environmental Change,* 1988
228–29. BARRY C. BISHOP, *Kamali under Stress: Livelihood Strategies and Seasonal Rhythms in a Changing Nepal Himalaya,* 1990
230. CHRISTOPHER MUELLER-WILLE, *Natural Landscape Amenities and Suburban Growth: Metropolitan Chicago, 1970–1980,* 1990
233. RISA PALM and MICHAEL E. HODGSON, *After a California Earthquake: Attitude and Behavior Change,* 1992
234. DAVID M. KUMMER, *Deforestation in the Postwar Philippines,* 1992
235. MICHAEL P. CONZEN, Thomas A. Rumney, and Graeme Wynn, *A Scholar's Guide to Geographical Writing on the American and Canadian Past,* 1993
236. SHAUL EPHRAIM COHEN, *The Politics of Planting: Israeli-Palestinian Competition for Control of Land in the Jerusalem Periphery,* 1993
237. CHAD F. EMMETT, *Beyond the Basilica: Christians and Muslims in Nazareth,* 1995
238. EDWARD T. PRICE, *Dividing the Land: Early American Beginnings of Our Private Property Mosaic,* 1995
239. ALEX G. PAPADOPOULOS, *Urban Regimes and Strategies: Building Europe's Central Executive District in Brussels,* 1996
240. ANNE KELLY KNOWLES, *Calvinists Incorporated: Welsh Immigrants on Ohio's Industrial Frontier,* 1996
241. HUGH PRINCE, *Wetlands of the American Midwest: A Historical Geography of Changing Attitudes,* 1997
242. KLAUS FRANTZ, *Indian Reservations in the United States: Territory, Sovereignty, and Socioeconomic Change,* 1999
243. JOHN A. AGNEW, *Place and Politics in Modern Italy,* 2002
245. CHARLES M. GOOD, *The Steamer Parish: The Rise and Fall of Missionary Medicine on an African Frontier,* 2003